THE
FINITE ELEMENT
METHOD

Its Fundamentals and Applications in Engineering

Zhangxin Chen

University of Calgary, Canada

T0324946

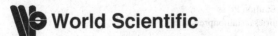 **World Scientific**

NEW JERSEY · LONDON · SINGAPORE · BEIJING · SHANGHAI · HONG KONG · TAIPEI · CHENNAI

Published by

World Scientific Publishing Co. Pte. Ltd.

5 Toh Tuck Link, Singapore 596224

USA office: 27 Warren Street, Suite 401-402, Hackensack, NJ 07601

UK office: 57 Shelton Street, Covent Garden, London WC2H 9HE

British Library Cataloguing-in-Publication Data

A catalogue record for this book is available from the British Library.

THE FINITE ELEMENT METHOD
Its Fundamentals and Applications in Engineering

ISBN-13 978-981-4350-56-3
ISBN-10 981-4350-56-7
ISBN-13 978-981-4350-57-0 (pbk)
ISBN-10 981-4350-57-5 (pbk)

Desk Editor: Tjan Kwang Wei

Typeset by Stallion Press
Email: enquiries@stallionpress.com

Printed in Singapore.

This book is dedicated to my dear late father
Furong Chen and late mentor Richard E. Ewing

Preface

A numerical method for solving a differential equation problem is to discretize this problem, which has infinitely many degrees of freedom, to produce a discrete problem, which has finitely many degrees of freedom and can be solved using a computer. Examples include the classical finite difference and finite element methods. The advantages of the finite element method over the finite difference method are that general boundary conditions, complex geometry, and variable material properties can be handled relatively easily. The clear structure and versatility of the finite element method makes it possible to develop general purpose software for applications. Furthermore, it has a solid theoretical foundation that gives added reliability, and in many situations it is possible to obtain concrete error estimates for finite element solutions. The finite element method was first introduced by Courant in 1943. From the 1950s to the 1970s, it was developed by engineers and mathematicians into a general method for the numerical solution of partial differential equations.

This book offers a fundamental and practical introduction to the finite element method, its variants, and their applications. It is based on material that I have used in a graduate course offered in the Department of Mathematics, Southern Methodist University, Dallas, Texas, USA and in the Department of Chemical and Petroleum Engineering, University of Calgary, Calgary, Alberta, Canada for many years. According to my experience, engineering graduate students (toward whom this book is mainly directed) often do not have a strong background in functional analysis, particularly in Sobolev spaces, which is needed for the finite element analysis. Therefore, these spaces are not be used in this book. Instead, an admissible function space is introduced for each underlying partial differential equation; this space is the minimum requirement that gives a meaningful variational formulation on which the finite element method is based for the equation. This admissible space also allows us to introduce every concept in the simplest possible setting and to maintain a level of treatment that is as rigorous as possible without being unnecessarily abstract.

In Chapter 1, the finite element method is introduced in one dimension with two purposes: to introduce the terminology and to summarize the basic ingredients that are required for the development of this method. The treatment of boundary conditions of different types, the formulation of system matrices, and the definition of shape (basis) functions using different approaches are also considered in this chapter. In addition, computer programming issues are addressed. All these considerations are extended to two dimensions in Chapter 2. In Chapter 3, a general variational formulation is given for the finite element method, and some concrete examples are described to illustrate this general formulation.

Various finite elements are introduced in one, two, and three space dimensions, respectively, in Chapters 4–6. These elements include triangular (tetrahedral), rectangular (hexahedral), and pentahedral elements and elements of both Lagrangian and Hermitian types. Different approaches, such as global and local coordinates and interpolation functions, are described to derive shape (basis) functions. Application of the finite element method to transient and nonlinear partial differential problems is briefly discussed in Chapter 7. Chapters 8–10 are devoted to the application of this method to solid mechanics, fluid mechanics, and fluid flow in porous media, respectively. Basic differential equations are stated for each application, and their finite element formulation is given.

The last chapter — Chapter 11 — is devoted to the introduction of a wide range of variants of the standard finite element method discussed in Chapters 1–7. These include the control volume, multipoint flux approximation, nonconforming, mixed, discontinuous, characteristic, adaptive, and multiscale finite element methods. Some of these methods are more often used for practical engineering and science problems.

This book can serve as a course that provides an introduction to numerical methods for partial differential equations for graduate students. It can be even taught at undergraduate level. This book can also be used as a reference book for engineers, mathematicians, and scientists interested in numerical solutions. The necessary prerequisites are relatively moderate: a basic course in advanced calculus and some acquaintance with partial differential equations. The problem section in each chapter plays a role in the presentation, and the student should spend time to solve the exercise problems.

The finite element method introduced in this book may have a different definition in the engineering community, such as the Galerkin, direct, or residual finite element method. Also, the concepts — basis, trial, and shape functions are interchangeably used. The development of a convergence theory for the finite element method is beyond the scope of this book. Only the convergence in terms of error estimates between an exact solution and its finite element solution is stated. The solution techniques for solving the linear systems arising from the finite element discretization of partial differential equations are not covered.

I take this opportunity to thank the people who have helped, in different ways, in the preparation of this book. In particular, many of my colleagues and students at Southern Methodist University and University of Calgary made invaluable comments and suggestions at the early stages of this book. Dr. Hongsen Chen has contributed to the C++ Finite Element Programs, and Mr. Zhong He has helped in graphing some of the figures in this book. I would also like to thank Ms. Jamie McInnis, my Project Manager for reading the entire manuscript and making invaluable suggestions. Several methods introduced in this book have been similarly covered by Chen (2005) and Chen et al. (2006). To reach more readers in Asia, thanks to the invitation of the World Scientific Publishing, these methods are reconsidered in the present book.

Zhangxin Chen, Calgary, Alberta, Canada
Ph.D. and Professor
NSERC/AERI/Foundation CMG Chair in Reservoir Simulation
iCORE Industrial Chair in Reservoir Modeling
Director, Schlumberger iCentre for Simulation & Visualization
December 2010

Contents

Preface   vii

List of Figures   xvii

List of Tables   xxi

1. **One-Dimensional Model Problems**   1
   - 1.1. Examples of one-dimensional problems 1
   - 1.2. The finite element method 4
     - 1.2.1. Basis (shape) functions 6
     - 1.2.2. Linear systems 7
   - 1.3. Boundary conditions 9
     - 1.3.1. Nonhomogeneous Dirichlet boundary conditions 10
     - 1.3.2. General boundary conditions 11
   - 1.4. Local coordinate formulation 12
     - 1.4.1. Element matrices 12
     - 1.4.2. Local coordinate transformation 13
   - 1.5. Computer programming considerations 15
   - 1.6. Equivalence and error estimates 18
   - 1.7. Exercises 21

2. **Two-Dimensional Model Problems**   27
   - 2.1. Two-dimensional differential problems 27
   - 2.2. The finite element method 30
     - 2.2.1. Green's formula 30
     - 2.2.2. Variational formulation 31
     - 2.2.3. Basis (shape) functions 33
     - 2.2.4. Linear systems 35

2.3.   Extensions to general boundary conditions   38
    2.3.1.   Nonhomogeneous Dirichlet boundary
         conditions   38
    2.3.2.   General boundary conditions   40
2.4.   Local coordinate formulations   42
    2.4.1.   Local element matrices   42
    2.4.2.   Construction of triangulations   43
    2.4.3.   Assembly of stiffness matrices   45
    2.4.4.   Local coordinate transformation   46
2.5.   Programming considerations   48
    2.5.1.   Numbering of nodes   49
    2.5.2.   Matrix storage   51
    2.5.3.   Computer program   51
2.6.   Error estimates   58
2.7.   Exercises   60

3. **General Variational Formulation**   **69**
3.1.   Continuous variational formulation   69
3.2.   The finite element method   70
3.3.   Examples   71
3.4.   Exercises   79

4. **One-Dimensional Elements and their Properties**   **81**
4.1.   Element classification   82
4.2.   Different approaches for deriving basis functions   82
    4.2.1.   Global coordinate approach   82
    4.2.2.   Local coordinate transformation approach   84
    4.2.3.   Interpolation function approach   85
4.3.   Lagrangian elements   85
4.4.   Hermitian elements   86
4.5.   Exercises   87

5. **Two-Dimensional Elements and their Properties**   **89**
5.1.   Rectangular and quadrilateral elements   89
    5.1.1.   Lagrangian rectangular elements   90
    5.1.2.   Serendipity elements   93
    5.1.3.   Hermitian rectangular elements   94
    5.1.4.   Quadrilateral elements   96
5.2.   Triangular elements   97
    5.2.1.   Natural coordinates in two dimensions   98
    5.2.2.   Lagrangian triangular elements   100

5.2.3. Hermitian triangular elements 104
5.3. Exercises 109

6. Three-Dimensional Elements and their Properties 111
6.1. Hexahedral elements 111
6.1.1. Lagrangian hexahedral elements 112
6.1.2. Serendipity elements 113
6.2. Tetrahedral elements 114
6.2.1. Natural coordinates in three dimensions 115
6.2.2. Natural coordinates in d-dimensions 117
6.2.3. Lagrangian tetrahedral elements 118
6.2.4. Hermitian tetrahedral elements 119
6.3. Pentahedral elements 120
6.4. Isoparametric elements 122
6.5. Choice of an element 124
6.6. General domains 124
6.7. Quadrature rules 128
6.7.1. One dimension 128
6.7.2. Rectangles and bricks 129
6.7.3. Triangles and tetrahedra 130
6.8. Exercises 132

7. Finite Elements for Transient and Nonlinear
Problems 135
7.1. Finite elements for transient problems 135
7.1.1. A one-dimensional model problem 136
7.1.2. A semi discrete scheme in space 138
7.1.3. Fully discrete schemes 140
7.2. Finite elements for nonlinear problems 144
7.2.1. Linearization approach 145
7.2.2. Extrapolation time approach 146
7.2.3. Implicit time approximation 148
7.2.4. Explicit time approximation 149
7.3. Exercises 150

8. Application to Solid Mechanics 151
8.1. Plane stress and plane strain 151
8.1.1. Kinematics 152
8.1.2. Equilibrium 153
8.1.3. Material laws 154
8.1.4. Boundary conditions 157
8.1.5. The finite element method 158

8.2. Three-dimensional solids 161
8.3. Axisymmetric solids 165
      8.3.1. Anisotropic material 167
      8.3.2. Isotropic material 167
8.4. Exercises 169

9. Application to Fluid Mechanics   171
9.1. Equations of fluid dynamics 172
9.2. A characteristic-based splitting method 174
      9.2.1. An explicit characteristic-based method 175
      9.2.2. Application to fluid mechanics 178
      9.2.3. Solution schemes in time 184
      9.2.4. Remarks on the splitting method 185
9.3. The finite element method 187
9.4. The nonconforming finite element method 189
9.5. The mixed finite element method 191
9.6. The Navier–Stokes equations 194
9.7. Exercises 195

10. Application to Porous Media Flow   197
10.1. Single-phase flow 197
      10.1.1. Basic differential equations 197
      10.1.2. Units 198
      10.1.3. Different forms of flow equations 199
      10.1.4. Boundary and initial conditions 203
10.2. Two-phase flow 204
      10.2.1. Basic differential equations 204
      10.2.2. Alternative differential equations 205
      10.2.3. Boundary conditions 210
10.3. Finite element solution of single-phase flow 212
      10.3.1. Treatment of initial conditions 213
      10.3.2. The finite element method 213
      10.3.3. Practical issues 217
10.4. Exercises 220

11. Other Finite Element Methods   221
11.1. The CVFE method 221
      11.1.1. The basic CVFE method 222
      11.1.2. Positive transmissibilities 225
      11.1.3. The CVFE grid construction 226
      11.1.4. Flux continuity 228

11.2. Multipoint flux approximations 230
    11.2.1. Definition of MPFA 230
    11.2.2. A-orthogonal grids 233
11.3. The nonconforming finite element method 235
    11.3.1. Second-order partial differential problems 236
    11.3.2. Nonconforming finite elements on triangles 236
11.4. The mixed finite element method 240
    11.4.1. A one-dimensional model problem 241
    11.4.2. A two-dimensional model problem 247
    11.4.3. Extension to boundary conditions
        of other kinds 250
    11.4.4. Mixed finite element spaces 253
11.5. The discontinuous finite element method 258
    11.5.1. DG methods 258
    11.5.2. Stabilized DG methods 263
11.6. The characteristic finite element method 264
    11.6.1. The modified method of characteristics 266
    11.6.2. The Eulerian–Lagrangian localized
        adjoint method 274
11.7. The adaptive finite element method 278
    11.7.1. Local grid refinement in space 279
    11.7.2. Data structures 284
    11.7.3 *A posteriori* error estimates 285
11.8. The multiscale finite element method 293
    11.8.1. The multiscale finite element method 294
    11.8.2. Boundary conditions of basis functions 295
11.9. Exercises 296

Bibliography   **309**

Index   **319**

List of Figures

1.1. An elastic bar 2
1.2. A metallic rod 2
1.3. A one-dimensional porous medium 3
1.4. An illustration of a function $v \in V_h$ 5
1.5. A basis function in one dimension 7
1.6. Basis functions at the end points 10
1.7. Local coordinate transformation 14
2.1. An elastic membrane 28
2.2. A metallic rectangular plate 28
2.3. A finite element partition in two dimensions 32
2.4. A typical triangle 33
2.5. A basis function in two dimensions 35
2.6. An example of a triangulation 36
2.7. A five-point stencil scheme 37
2.8. Uniform refinement 43
2.9. Nonuniform refinement 43
2.10. Node and triangle enumeration 44
2.11. Local coordinate transformation 46
2.12. A flowchart 49
2.13. An example of enumeration 50
2.14. An illustrative triangulation for the computer program . . . 52
2.15. A triangle and its inscribed circle 59
2.16. The support of a basis function at node x_i 60
2.17. A triangulation example 60
2.18. The reference rectangle 61
2.19. A triangulation 62
3.1. A local coordinate transformation 75
3.2. An elastic plate 77

xvii

4.1. A local coordinate transformation   84
5.1. A rectangle K   90
5.2. Bilinear element   92
5.3. The reference element   92
5.4. The first three serendipity elements   93
5.5. Construction of quadrilateral elements by triangles   96
5.6. The de Veubeke quadrilateral element   96
5.7. The Clough and Felippa quadrilateral element   97
5.8. A local coordinate transformation from the
      reference square   97
5.9. Natural coordinates in two dimensions   98
5.10. Location of T_r nodes on a triangle 101
5.11. A linear triangular element 101
5.12. A quadratic triangular element 102
5.13. A cubic triangular element 102
5.14. A cubic Hermitian triangular element 105
5.15. A quintic Hermitian element 107
5.16. Argyris' triangle 108
5.17. Bell's triangle 109
5.18. Biquadratic element 109
6.1. Trilinear element 112
6.2. The first three members of three-dimensional serendipity
      elements 113
6.3. The volume coordinates 115
6.4. The first three Lagrangian tetrahedral elements 118
6.5. The cubic Hermitian tetrahedral elements 120
6.6. Pentahedral elements 121
6.7. Isoparametric elements 123
6.8. Linear approximation to a curved boundary Γ 125
6.9. Mapping from a regular triangle to a curved triangle 125
7.1. The extrapolation approach 147
8.1. The free body diagram of an infinitesimal element 153
8.2. Stratified material 156
8.3. An elastic body Ω 157
8.4. A three-dimensional solid element 162
8.5. An axisymmetric solid 165
8.6. A ring element 168
9.1. Approximate characteristics 176
9.2. An illustration of the normal direction on triangles 191
9.3. A numerical cavity problem 196
10.1. A flux function f_w 210

10.2.   A tetrahedron 218
11.1.   A control volume 222
11.2.   A base triangle for CVFE 223
11.3.   Two adjacent triangles 225
11.4.   An edge swap 227
11.5.   An addition of a new boundary node 228
11.6.   Two adjacent cells 228
11.7.   A base triangle for MPFA 231
11.8.   A triangle with a circumscribed a-ellipse 233
11.9.   A two-point flux in an a-orthogonal polygonal grid 234
11.10.   The degrees of freedom for the Crouzeix–Raviart element . . 237
11.11.   A nonconforming basis function in two dimensions 239
11.12.   A basis function φ_i in one dimension 244
11.13.   An illustration of the unit normal ν 249
11.14.   The triangular RT 256
11.15.   The rectangular RT 257
11.16.   An illustration of ∂K_- and ∂K_+ 259
11.17.   An ordering of computation for DG 260
11.18.   Adjoining rectangles 262
11.19.   An illustration of the definition $\check{x}_n$ 268
11.20.   An illustration of the definition $\check{x}_n$ 272
11.21.   An illustration of $\mathcal{K}^n$ 276
11.22.   Examples of regular and irregular vertices 281
11.23.   A coarse grid (solid lines) and a refinement (dotted lines) . . 282
11.24.   A local refinement and the corresponding tree structure . . . 284
11.25.   An illustration of ν 288
11.26.   An illustration of Ω_K 291
11.27.   Uniform (left) and adaptive (right) triangulations 292
11.28.   The support of a basis function at node x_i 296

List of Tables

5.1. Basis functions for serendipity elements   93
6.1. Integration points and weights in the one-dimensional
      Gauss quadrature   129
6.2. Integration points and weights over the reference triangle . . .   131
6.3. Integration points and weights over the reference
      tetrahedron   131
10.1. Customary and metric units   199
11.1. A comparison of uniform and adaptive refinements   292

Chapter 1

One-Dimensional Model Problems

The finite element method for a one-dimensional model problem is introduced. The exposition in this chapter has two purposes: (1) to introduce the terminology and (2) to summarize the basic ingredients that are required for the development of the finite element method. In Sec. 1.1, we describe three one-dimensional physical problems to motivate a model problem considered. Then, in Sec. 1.2, we discuss the application of the finite element method to this model problem. The development of a linear system of algebraic equations arising from this method is also considered. In Sec. 1.3, we extend the finite element method to boundary conditions of different types. Section 1.4 is devoted to the construction of basis (shape) functions and system matrices using different approaches. In Sec. 1.5, basic components in programming a finite element code are addressed, and a computer program is stated. In Sec. 1.6, we briefly touch on the equivalence between the Ritz and Galerkin finite element methods and the convergence issues. Finally, in Sec. 1.7, exercise problems are given. These exercises are closely related to the literature of this chapter. Below, d/dx and d^2/dx^2 indicate the first and second derivatives, respectively.

1.1. Examples of one-dimensional problems

As an introduction, three stationary problems in one dimension are described.

Example 1.1 (elastic bar). The first problem is of the simple form

$$-\frac{d^2p}{dx^2} = f(x), \quad 0 < x < 1,$$

$$p(0) = p(1) = 0,$$

(1.1)

where f is a given real-valued bounded function. Note that Eq. (1.1) is a two-point boundary value problem for the unknown variable p, with a *homogeneous boundary condition* (zero boundary condition). A number of problems in physics and mechanics arise in the form (1.1). For example, consider an elastic bar with tension one, fixed at both ends ($x = 0, 1$) and subject to a transversal load of intensity f (Fig. 1.1). Under the assumption of a small displacement, the transversal displacement p satisfies problem (1.1) (Exercise 1.1).

Example 1.2 (heat conduction). Two ends of a metallic rod are held at constant temperatures T_1 and T_2, respectively, with $T_2 > T_1$ (Fig. 1.2). There is a loss of heat through convection and radiation into the surrounding medium that has a given temperature $T_0 < T_2$. The loss through convection is modeled by a dissipation coefficient β proportional to $T_1 + T_2 - 2T_0$. The temperature T of this rod satisfies

$$-\kappa \frac{d^2T}{dx^2} + \beta T = 0, \quad a < x < b,$$

$$T(a) = T_1, \qquad T(b) = T_2,$$

(1.2)

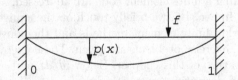

Figure 1.1   An elastic bar.

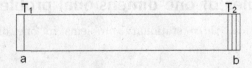

Figure 1.2   A metallic rod.

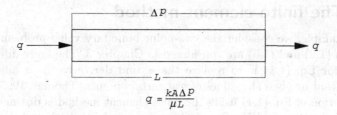

Figure 1.3 A one-dimensional porous medium.

where κ is the thermal diffusion coefficient of the rod and a and b indicate the locations of its two ends. Equation (1.2) is also a two-point boundary value problem for temperature T, with a nonhomogeneous boundary condition (nonzero boundary condition).

Example 1.3 (single phase flow in porous media). A single fluid flows through a porous medium. Assume that the properties of the fluid and the medium are homogeneous in two of the coordinate directions, which results in a single phase flow in one dimension (Fig. 1.3, where A is the cross-sectional area). In the stationary case, mass conservation is

$$\frac{d}{dx}(\rho u) = q, \tag{1.3}$$

where ρ and u are the fluid density and velocity, respectively, and q is the (mass) flow rate. The velocity u is given by Darcy's law (Chen *et al.*, 2006):

$$u = -\frac{k}{\mu}\frac{dp}{dx}, \tag{1.4}$$

where k is the permeability of the medium and μ and p are the fluid viscosity and pressure, respectively. Substituting Eq. (1.4) into Eq. (1.3) gives

$$-\frac{d}{dx}\left(\frac{\rho k}{\mu}\frac{dp}{dx}\right) = q. \tag{1.5}$$

The boundary condition is taken as

$$p(a) = p_a, \quad -\left(\frac{\rho k}{\mu}\frac{dp}{dx}\right)(b) = g, \tag{1.6}$$

where a and b indicate the locations of the two ends of the porous medium, p_a is a given pressure at one end, and g is a given flux at the other end. Equation (1.5) is a two-point boundary value problem for pressure p, with a variable coefficient $\rho k/\mu$. The boundary condition at one end a is of Dirichlet type, while it is of Neumann type (flux type) at the other end b.

1.2. The finite element method

As an example, we consider the two-point boundary value problem (1.1); problems (1.2) and (1.5) are considered in Chapter 3. The finite difference method for Eq. (1.1) is to replace the second derivative by a difference quotient that involves the values of p at certain points (Thomas, 1995). The discretization of Eq. (1.1) using the finite element method is distinct. This method starts by rewriting Eq. (1.1) in an equivalent variational (integral) formulation. For this variational formulation to make sense, we can only use a certain class of functions in the formulation of problem (1.1). This class is referred to as an *admissible space* of functions. For the one-dimensional problem (1.1), the admissible space V consists of the following real-valued functions:

$$V = \left\{ \text{Functions } v \colon \text{ Each } v \text{ is continuous on } [0,1], \right.$$

$$\text{its first derivative } \frac{dv}{dx} \text{ is piecewise continuous}$$

$$\left. \text{and bounded on } (0,1), \text{ and } v(0) = v(1) = 0 \right\}.$$

A function being piecewise continuous on an interval means that this interval can be divided into a finite number of disjoint subintervals such that the function is continuous on each of these subintervals. We define a *functional* F on the space V:

$$F(v) = \frac{1}{2} \int_0^1 \left(\frac{dv}{dx} \right)^2 dx - \int_0^1 fv\,dx, \quad v \in V.$$

The two integrals in this functional are well defined from calculus because of the definition of the admissible space V. It is shown at the end of this chapter that problem (1.1) is equivalent to the minimization problem

$$\text{Find } p \in V \text{ such that } F(p) \leq F(v) \quad \forall v \in V. \tag{1.7}$$

Problem (1.7) is called a *Ritz variational formulation* of Eq. (1.1).

In mechanics, the quantity $\frac{1}{2} \int_0^1 (\frac{dv}{dx})^2 dx$ is the internal elastic energy, $\int_0^1 fv\,dx$ is the load potential, and the functional value $F(v)$ represents the total potential energy associated with the displacement $v \in V$. Therefore, problem (1.7) corresponds to the *fundamental principle of minimum potential energy* in mechanics.

In terms of computations, Eq. (1.1) can be expressed in a more useful, direct formulation. When multiplying the first equation of Eq. (1.1) by any

$v \in V$ (called a *test function*) and integrating over $(0,1)$, we see that

$$-\int_0^1 \frac{d^2p}{dx^2} v \, dx = \int_0^1 fv \, dx.$$

Application of integration by parts to this equation yields

$$\int_0^1 \frac{dp}{dx} \frac{dv}{dx} \, dx = \int_0^1 fv \, dx, \tag{1.8}$$

where we use the fact that $v(0) = v(1) = 0$ from the definition of the space V. Equation (1.8) is called a *Galerkin variational* or *weak formulation* of Eq. (1.1). It corresponds to the *principle of virtual work* in mechanics. If p is a solution to Eq. (1.1), then it also satisfies Eq. (1.8). The converse also holds if the second derivative d^2p/dx^2 exists and is piecewise continuous and bounded in $(0,1)$ (see Exercise 1.2). It can be seen that Eqs. (1.7) and (1.8) are equivalent (see Sec. 1.6).

We now introduce the finite element method for solving Eq. (1.1). Toward that end, for a positive integer M, let $0 = x_0 < x_1 < \cdots < x_M < x_{M+1} = 1$ be a partition of $(0,1)$ into a set of subintervals $I_i = (x_{i-1}, x_i)$, with length $h_i = x_i - x_{i-1}$, $i = 1, 2, \ldots, M + 1$. Set $h = \max\{h_i : i = 1, 2, \ldots, M + 1\}$. The step size h measures how fine the partition is.

On each subinterval I_i, the simplest approximation to the solution p is a linear function, so we define the *finite element space*

$$V_h = \Big\{ \text{Functions } v \colon v \text{ is a continuous function on } [0,1], \, v \text{ is linear}$$
$$\text{on each subinterval } I_i, \text{ and } v(0) = v(1) = 0,$$
$$i = 1, 2, \ldots, M + 1 \Big\}.$$

See Fig. 1.4 for an illustration of a function $v \in V_h$. Note that $V_h \subset V$ (i.e., V_h is a subspace of V).

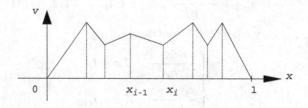

Figure 1.4 An illustration of a function $v \in V_h$.

The discrete version of Eq. (1.7) is

$$\text{Find } p_h \in V_h \text{ such that } F(p_h) \leq F(v) \quad \forall v \in V_h. \tag{1.9}$$

Method (1.9) is referred to as the *Ritz finite element method*. In the same manner as for Eq. (1.8) (see the end of this section), Eq. (1.9) is equivalent to the problem

$$\text{Find } p_h \in V_h \text{ such that } \int_0^1 \frac{dp_h}{dx} \frac{dv}{dx} dx = \int_0^1 fv \, dx \quad \forall v \in V_h. \tag{1.10}$$

This is usually termed the *Galerkin finite element method*, which we study in this book.

It is easy to see that the solution of problem (1.10) is unique. In fact, let $f = 0$, and take $v = p_h$ in Eq. (1.10) to give

$$\int_0^1 \left(\frac{dp_h}{dx} \right)^2 dx = 0,$$

so $dp_h/dx = 0$; i.e., p_h is a constant. It follows from the boundary condition in V_h that $p_h = 0$.

1.2.1. Basis (shape) functions

On each subinterval I_i, an approximate solution p_h (trial function) in V_h has the representation

$$p_h(x) = a_i + b_i x, \quad x \in I_i, \ i = 1, 2, \ldots, M + 1,$$

where the constants a_i and b_i can be determined from the nodal conditions

$$p_h(x_{i-1}) = p_{i-1}, \quad p_h(x_i) = p_i.$$

That is

$$a_i = \frac{x_i p_{i-1} - x_{i-1} p_i}{h_i}, \quad b_i = \frac{p_i - p_{i-1}}{h_i}.$$

Hence, we see that

$$p_h(x) = \frac{x_i - x}{h_i} p_{i-1} + \frac{x - x_{i-1}}{h_i} p_i, \quad x \in I_i, \ i = 1, 2, \ldots, M + 1.$$

For $i = 1, 2, \ldots, M$, we define the functions

$$\varphi_i(x) = \begin{cases} \dfrac{x - x_{i-1}}{h_i} & \text{if } x \in I_i, \\[2mm] \dfrac{x_{i+1} - x}{h_{i+1}} & \text{if } x \in I_{i+1}, \\[2mm] 0 & \text{elsewhere.} \end{cases}$$

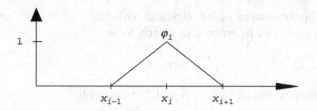

Figure 1.5 A basis function in one dimension.

Note that, for $i, j = 1, 2, \ldots, M$,

$$\varphi_i(x_j) = \begin{cases} 1 & \text{if } i = j, \\ 0 & \text{if } i \neq j. \end{cases}$$

That is, φ_i is a continuous piecewise linear function on $(0, 1)$ such that its value is unity at *node* x_i and zero at other nodes (Fig. 1.5). It is called a *hat* or *chapeau* function. The functions $\{\varphi_1, \varphi_2, \ldots, \varphi_M\}$ can be shown to be linearly independent and form a basis (or *shape*) of the finite element space V_h.

1.2.2. Linear systems

Since the functions $\{\varphi_1, \varphi_2, \ldots, \varphi_M\}$ are a basis of the space V_h, any function $u \in V_h$ can be (uniquely) expressed in a linear combination of them:

$$u(x) = \sum_{i=1}^{M} u_i \varphi_i(x), \quad 0 \leq x \leq 1,$$

where $u_i = u(x_i)$. For each j, take $v = \varphi_j$ in Eq. (1.10) to see that

$$\int_0^1 \frac{dp_h}{dx} \frac{d\varphi_j}{dx} dx = \int_0^1 f\varphi_j dx, \quad j = 1, 2, \ldots, M. \tag{1.11}$$

Set the trial function p_h as

$$p_h(x) = \sum_{i=1}^{M} p_i \varphi_i(x), \quad p_i = p_h(x_i),$$

and substitute it into Eq. (1.11) to give

$$\sum_{i=1}^{M} \int_0^1 \frac{d\varphi_i}{dx} \frac{d\varphi_j}{dx} dx \, p_i = \int_0^1 f\varphi_j dx, \quad j = 1, 2, \ldots, M. \tag{1.12}$$

This is a linear system of M algebraic equations in the M unknowns $p_1, p_2, \ldots, p_M$. It can be written in matrix form

$$\mathbf{A}\mathbf{p} = \mathbf{f}, \qquad (1.13)$$

where the matrix $\mathbf{A}$ and vectors $\mathbf{p}$ and $\mathbf{f}$ are given by

$$\mathbf{A} = \begin{pmatrix} a_{1,1} & a_{1,2} & \cdots & a_{1,M} \\ a_{2,1} & a_{2,2} & \cdots & a_{2,M} \\ \vdots & \vdots & \ddots & \vdots \\ a_{M,1} & a_{M,2} & \cdots & a_{M,M} \end{pmatrix}, \quad \mathbf{p} = \begin{pmatrix} p_1 \\ p_2 \\ \vdots \\ p_M \end{pmatrix}, \quad \mathbf{f} = \begin{pmatrix} f_1 \\ f_2 \\ \vdots \\ f_M \end{pmatrix},$$

with

$$a_{i,j} = \int_0^1 \frac{d\varphi_i}{dx} \frac{d\varphi_j}{dx} dx, \quad f_j = \int_0^1 f\varphi_j dx, \quad i,j = 1, 2, \ldots, M.$$

The matrix $\mathbf{A}$ is referred to as the *stiffness matrix* and $\mathbf{f}$ is the *load vector*. By the definition of the basis functions, we observe that

$$\int_0^1 \frac{d\varphi_i}{dx} \frac{d\varphi_j}{dx} \, dx = 0 \quad \text{if } |i - j| \geq 2,$$

so $\mathbf{A}$ is *tridiagonal*; i.e., only the entries on the main diagonal and the adjacent diagonals may be nonzero. In fact, the entries $a_{i,j}$ can be calculated

$$a_{i,i} = \frac{1}{h_i} + \frac{1}{h_{i+1}}, \quad a_{i-1,i} = -\frac{1}{h_i}, \quad a_{i,i+1} = -\frac{1}{h_{i+1}}.$$

Also, it can be seen that $\mathbf{A}$ is *symmetric*: $a_{i,j} = a_{j,i}$ and *positive definite*

$$\boldsymbol{\eta}^T \mathbf{A} \boldsymbol{\eta} = \sum_{i,j=1}^{M} \eta_i a_{i,j} \eta_j > 0 \quad \text{for all nonzero } \boldsymbol{\eta} \in \mathbb{R}^M,$$

where $\boldsymbol{\eta}^T = (\eta_1, \eta_2, \ldots, \eta_M)$ denotes the transpose of the column vector $\boldsymbol{\eta}$ (i.e., $\boldsymbol{\eta}^T$ is a row vector) and $\mathbb{R}^M$ denotes the set of column vectors with M components (i.e., the M-dimensional real Euclidean space). A positive definite matrix is nonsingular, and the linear system (1.13) thus has a unique solution. Consequently, we have shown that Eq. (1.10) has a unique solution $p_h \in V_h$ in a different way.

The symmetry of $\mathbf{A}$ can be seen from the definition of $a_{i,j}$. The positive definiteness can be checked as follows: with

$$\eta = \sum_{i=1}^{M} \eta_i \varphi_i \in V_h, \quad \boldsymbol{\eta}^T = (\eta_1, \eta_2, \ldots, \eta_M),$$

we see that

$$\sum_{i,j=1}^{M} \eta_i a_{i,j} \eta_j = \sum_{i,j=1}^{M} \eta_i \int_0^1 \frac{d\varphi_i}{dx} \frac{d\varphi_j}{dx} dx \eta_j$$

$$= \int_0^1 \sum_{i=1}^{M} \eta_i \frac{d\varphi_i}{dx} \sum_{j=1}^{M} \eta_j \frac{d\varphi_j}{dx} dx = \int_0^1 \left(\frac{d\eta}{dx}\right)^2 dx \geq 0, \tag{1.14}$$

so, as for Eq. (1.10), the equality holds only for $\eta \equiv 0$, since a constant function η in the space V_h must be zero because of the boundary condition.

We remark that $\mathbf{A}$ is sparse; i.e., only a few entries in each row of $\mathbf{A}$ are nonzero. In the present one-dimensional case, it is tridiagonal. The sparsity of $\mathbf{A}$ depends on the fact that a basis function in V_h is different from zero only on a few intervals; i.e., it has compact support. Thus, it interferes only with a few other basis functions. That basis functions can be chosen in this manner is an important distinctive property of the finite element method.

In the case where the partition is uniform, i.e., $h = h_i$, $i = 1, 2, \ldots, M + 1$, the stiffness matrix $\mathbf{A}$ takes the form

$$\mathbf{A} = \frac{1}{h} \begin{pmatrix} 2 & -1 & 0 & \cdots & 0 & 0 \\ -1 & 2 & -1 & \cdots & 0 & 0 \\ 0 & -1 & 2 & \cdots & 0 & 0 \\ \vdots & \vdots & \vdots & \ddots & \vdots & \vdots \\ 0 & 0 & 0 & \cdots & 2 & -1 \\ 0 & 0 & 0 & \cdots & -1 & 2 \end{pmatrix}.$$

With division by h in $\mathbf{A}$, Eq. (1.10) can be thought of as a variant of the *central difference scheme* where the right-hand side consists of mean values of $f\varphi_j$ over the interval (x_{j-1}, x_{j+1}).

1.3. Boundary conditions

In problem (1.1) only a homogeneous *Dirichlet boundary condition* was considered. Here, we extend the finite element method to a nonhomogeneous Dirichlet boundary condition and other types of boundary conditions.

1.3.1. Nonhomogeneous Dirichlet boundary conditions

In the nonhomogeneous boundary condition, problem (1.1) becomes

$$-\frac{d^2p}{dx^2} = f(x), \quad 0 < x < 1,$$

$$p(0) = p_a, \qquad p(1) = p_b,$$

$$(1.15)$$

where p_a and p_b are given. In this case, the admissible function space V and the finite element space V_h defined earlier remain the same. Now, at the boundary nodes x_0 and x_{M+1}, we introduce their respective basis functions (Fig. 1.6):

$$\varphi_0(x) = \begin{cases} \dfrac{x_1 - x}{h_1} & \text{if } x \in I_1, \\ 0 & \text{elsewhere,} \end{cases}$$

and

$$\varphi_{M+1}(x) = \begin{cases} \dfrac{x - x_M}{h_{M+1}} & \text{if } x \in I_{M+1}, \\ 0 & \text{elsewhere.} \end{cases}$$

Consequently, the approximate solution p_h is represented by

$$p_h(x) = p_a\varphi_0(x) + \sum_{i=1}^{M} p_i\varphi_i(x) + p_b\varphi_{M+1}(x), \quad p_i = p_h(x_i), \ x \in [0,1].$$

Substituting it into Eq. (1.11) gives

$$\sum_{i=1}^{M} \int_0^1 \frac{d\varphi_i}{dx}\frac{d\varphi_j}{dx}\, dx\, p_i = \int_0^1 f\varphi_j\,dx - \int_0^1 \frac{d\varphi_0}{dx}\frac{d\varphi_j}{dx}\,dx\,p_a$$

$$- \int_0^1 \frac{d\varphi_{M+1}}{dx}\frac{d\varphi_j}{dx}\,dx\,p_b, \quad j = 1, 2, \ldots, M.$$

$$(1.16)$$

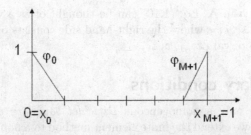

Figure 1.6 Basis functions at the end points.

Therefore, we see that the system matrix $\mathbf{A}$ remains the same, and only the first and last entries of the load vector $\mathbf{f}$ need to be modified because of the definition of the basis functions $\{\varphi_0, \varphi_1, \ldots, \varphi_{M+1}\}$.

An alternative approach is to use all the basis functions φ_0, φ_1, ..., φ_{M+1} to form a larger system of equations, i.e., an $(M+1) \times (M+1)$ system. The procedure for inserting the Dirichlet condition into the system matrix equation is: enter zeros in the first and $(M+1)$th rows of the system matrix $\mathbf{A}$ except for unity in the major diagonal positions of these two rows, and enter p_a and p_b in the first and $(M+1)$th rows of the vector $\mathbf{f}$, respectively. To retain the symmetry of $\mathbf{A}$, the corresponding columns need to be modified as well, which is addressed in Chapter 2 for two-dimensional problems.

1.3.2. General boundary conditions

Instead of imposing a Dirichlet boundary condition, problem (1.1) is now equipped with a general boundary condition:

$$-\frac{d^2p}{dx^2} = f(x), \quad 0 < x < 1,$$

$$p(0) = p_a, \qquad \gamma p(1) + \frac{dp}{dx}(1) = g,$$

(1.17)

where γ and g are given numbers. The boundary condition at the end point $x = 1$ is called a *third, mixed, Robin,* or *Dankwerts* boundary condition. It can be also imposed at the end point $x = 0$. When $\gamma = 0$, the boundary condition is a *second* or *Neumann* condition. When γ is infinite, the boundary condition reduces to a *first* or *Dirichlet condition*, which was considered earlier. A *fourth type* of boundary condition is a *periodic* boundary condition (see Chapter 11). In this section, we treat the case where γ is a finite number.

With the present boundary condition, the *admissible function space V* is modified to

$$V = \left\{ \text{Functions } v \colon \text{Each } v \text{ is continuous on } [0,1], \text{ its first derivative } \frac{dv}{dx} \right.$$

$$\left. \text{ is piecewise continuous and bounded on } (0,1); \text{ and } v(0) = 0 \right\}.$$

Accordingly, the *Galerkin variational formulation* of Eq. (1.17) is

$$\int_0^1 \frac{dp}{dx}\frac{dv}{dx}\,dx + \gamma p(1)v(1) = \int_0^1 fv\,dx + gv(1), \quad v \in V, \qquad (1.18)$$

with $p(0) = p_a$ (see Exercise 1.3).

Note that the boundary condition at the end point $x = 1$ is not imposed in the definition of V. It appears implicitly in Eq. (1.18). A boundary condition that need not be imposed is called a *natural condition*. The pure Neumann boundary condition is natural. The Dirichlet boundary condition has been imposed explicitly in V, and is termed an *essential condition*.

The finite element space V_h is now

$V_h = \{$Functions v: v is a continuous function on $[0,1]$, v is linear

on each subinterval I_i, and $v(0) = 0\}$,

and the piecewise linear approximation p_h to the solution p satisfies

$$\int_0^1 \frac{dp_h}{dx}\frac{dv}{dx}\,dx + \gamma p_h(1)v(1) = \int_0^1 fv\,dx + gv(1), \quad v \in V_h, \qquad (1.19)$$

with $p_h(0) = p_a$. As in the previous section, Eq. (1.19) can be formulated in matrix form (see Exercise 1.4).

1.4. Local coordinate formulation

In Sec. 1.2, the system matrix $\mathbf{A}$ was obtained using a *node-oriented* approach (i.e., looping over all nodes). Another approach is based on *element-oriented* (i.e., looping over all elements, subintervals in the present case). Computational experience shows that this approach is better than the *node-oriented* approach, particularly for multiple dimensions; the node-oriented approach wastes too much time in repeated computations of $\mathbf{A}$ and $\mathbf{f}$.

1.4.1. Element matrices

Note that

$$a_{i,j} = \int_0^1 \frac{d\varphi_i}{dx}\frac{d\varphi_j}{dx}\,dx = \sum_{k=1}^{M+1} \int_{I_k} \frac{d\varphi_i}{dx}\frac{d\varphi_j}{dx}\,dx.$$

Set

$$a_{i,j}^{(k)} = \int_{I_k} \frac{d\varphi_i}{dx}\frac{d\varphi_j}{dx}\,dx, \qquad (1.20)$$

so we see that

$$a_{i,j} = \sum_{k=1}^{M+1} a_{i,j}^{(k)}.$$

The entry $a_{i,j}^{(k)} \neq 0$ only when x_i and x_j are both end points of the kth subinterval or the kth element I_k, i.e., when $i,j \in \{k-1, k\}$. For the kth element, we compute the *element matrix*

$$\mathbf{A}^{(k)} = \begin{pmatrix} a_{k-1,k-1}^{(k)} & a_{k-1,k}^{(k)} \\ a_{k,k-1}^{(k)} & a_{k,k}^{(k)} \end{pmatrix}.$$

In addition, for $k = 1, 2, \ldots, M+1$, we calculate the local right-hand side (load) vector $(f_{k-1}^{(k)}, f_k^{(k)})$, where

$$f_j^{(k)} = \int_{I_k} f\varphi_j \, dx, \quad j = k-1, \ k.$$

To assemble the global matrix $\mathbf{A} = (a_{i,j})$ and the right-hand side vector $\mathbf{f}$, one loops over all subintervals I_k and successively adds the contributions from two adjacent subintervals at each node. That is, for $k = 1, 2, \ldots, M+1$

$$\begin{aligned}
a_{k-1,k-1} &= a_{k-1,k-1} + a_{k-1,k-1}^{(k)}, \\
a_{k-1,k} &= a_{k-1,k} + a_{k-1,k}^{(k)}, \\
a_{k,k-1} &= a_{k,k-1} + a_{k,k-1}^{(k)}, \\
a_{k,k} &= a_{k,k} + a_{k,k}^{(k)},
\end{aligned} \tag{1.21}$$

and

$$f_{k-1} = f_{k-1} + f_{k-1}^{(k)}, \quad f_k = f_k + f_k^{(k)}. \tag{1.22}$$

1.4.2. Local coordinate transformation

In the previous sections, the computation of both the system matrix $\mathbf{A}$ and the right-hand side vector $\mathbf{f}$ was based on a fixed coordinate system as shown in Fig. 1.4, and the basis (shape) function representation for each element used the same reference frame (Fig. 1.5). Such a common system of reference is referred to as a *global coordinate system*. An alternative approach that allows for a more concise formulation utilizes a *local coordinate transformation* specific to each element when the element matrix equations are established. This approach is particularly useful and effective for multiple dimensions, as shown in the subsequent chapters.

Consider the kth subinterval I_k, $k = 1, 2, \ldots, M+1$. The local coordinate transformation from the *reference element* $[0,1]$ to I_k is (Fig. 1.7)

$$x = F_k(\xi) = x_{k-1} + h_k\xi, \quad k = 1, 2, \ldots, M+1. \tag{1.23}$$

On the reference element $[0, 1]$, the approximate solution p_h (trial function) is of the form

$$p_h(\xi) = a + b\xi, \quad 0 \le \xi \le 1, \tag{1.24}$$

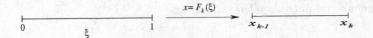

Figure 1.7   Local coordinate transformation.

where the constants a and b are determined by

$$p_h(0) = p_{k-1}, \quad p_h(1) = p_k. \tag{1.25}$$

The *nodal values* of a function remain unchanged during both a translation and a rotation of coordinate axes. While *nodal derivatives* remain unchanged for a translation, they do change for a rotation. Using Eqs. (1.24) and (1.25), we see that

$$p_h(\xi) = (1 - \xi)p_{k-1} + \xi p_k, \quad 0 \le \xi \le 1.$$

We define the *basis* (shape) functions at the end points 0 and 1, respectively:

$$\psi_1(\xi) = 1 - \xi, \quad \psi_2(\xi) = \xi. \tag{1.26}$$

Now, the local basis functions on the kth subinterval I_k are defined by

$$\varphi_{k-1}^{(k)} = \psi_1\big(F_k^{-1}(x)\big), \quad \varphi_k^{(k)} = \psi_2\big(F_k^{-1}(x)\big), \quad x \in I_k, \tag{1.27}$$

for $k = 1, 2, \ldots, M + 1$.

It follows from Eq. (1.20) that

$$a_{i,j}^{(k)} = \int_{I_k} \frac{d\varphi_i^{(k)}}{dx} \frac{d\varphi_j^{(k)}}{dx} dx, \quad i, j \in \{k - 1, k\};$$

consequently, application of the chain rule gives

$$a_{i,j}^{(k)} = \int_0^1 \left(\frac{d\psi_m}{d\xi} \frac{d\xi}{dx} \right) \left(\frac{d\psi_n}{d\xi} \frac{d\xi}{dx} \right) \frac{dF_k}{d\xi} \, d\xi, \quad m, n \in \{1, 2\}, \tag{1.28}$$

where i and j correspond to m and n, respectively. It follows from Eq. (1.23) that $dF_k/d\xi = h_k$ on each subinterval I_k. Similarly, the local right-hand side vector $(f_{k-1}^{(k)}, f_k^{(k)})$ is evaluated as

$$f_i^{(k)} = \int_{I_k} f(x)\varphi_i^{(k)}(x)dx = \int_0^1 f(F(\xi))\psi_m(\xi)\frac{dF_k}{d\xi}d\xi, \quad i = k - 1, \ k,$$

where $m \in \{1, 2\}$ corresponds to i. It can be shown that system (1.13) remains the same with the local coordinate transformation approach (see Exercise 1.5).

1.5. Computer programming considerations

The essential components of a typical computer program implementing the one-dimensional finite element method are summarized as follows:

- input of data such as the computational interval, the right-hand side function, the boundary data, and the coefficients that may appear in a differential problem;

- construction of the partition K_h of the interval;

- computation and assembly of the global stiffness matrix $\mathbf{A}$ and the right-hand side vector $\mathbf{f}$;

- solution of the linear system of algebraic equations $\mathbf{Ap} = \mathbf{f}$; and

- output of computational results.

The data input can be easily implemented in a small subroutine, and the result output depends on the computer system and software the user has in hand. Here, we briefly discuss three other parts of the one-dimensional problem considered in Sec. 1.2.

The partition K_h can be easily constructed from a successive refinement of the original computational interval; fine subintervals can be obtained by dividing each coarse interval into two subintervals. A sequence of uniform refinements will lead to quasi-uniform grids where the subintervals in K_h essentially have the same size. In practical applications, it is often necessary to use subintervals in K_h that vary considerably in size in different regions of the computational interval. For example, one utilizes smaller subintervals in places where the exact solution has a fast variation or where its certain derivatives are large.

After the partition K_h is constructed, one computes the element stiffness matrices with entries $a_{i,j}^{(k)}$ given by Eq. (1.20). We recall that $a_{i,j}^{(k)} = 0$ unless nodes x_i and x_j are both end points of the subinterval I_k in K_h.

In general, the solution of the linear system $\mathbf{Ap} = \mathbf{f}$ can be performed via a direct method (Gaussian elimination) or an iterative method (e.g., the conjugate gradient method). For the tridiagonal system which we currently have, the Thomas method is the best choice (Chen, 2005). The solution of a linear system is not be addressed in this book, and can be performed by one of the standard library solution programs available on most computers.

In the next Fortran computer program for the problem considered in Sec. 1.2, the procedure to obtain the system matrix $\mathbf{A}$ is based on assembly by elements. For this illustrative example, the right-hand side vector $\mathbf{f}$ is taken to be zero (Exercise 1.6 to be referred for a constant right-hand side f). In this program, no attempt is made to optimize the code. A counterpart C++ program is given at the end of this chapter.

```
c        Computer program
c
c        Finite element method for 1D problem
c        Solution of equation (1.15) with f = 0
c        The linear system is solved using a standard
c             library solution program LEQT1F
c
c        The following symbols are used:
c
c        nnode - Total number of nodes
c        nelem - Total number of elements
c        pa, pb - Boundary data
c        x(i) - x-coordinate of node i = 1, 2, ..., nnode
c        sysm(i, j) - Global system matrix A, i, j = 1, 2, ..., nnode
c        sysme(m, n) - Local matrix A, m, n = 1, 2
c        rhs(i, 1) - Right-hand side vector
c        The right-hand side has two subscripts to satisfy
c             the requirement of the program LEQT1F,
c             which also needs a work space wksp(i)
c        The linear system is solved as follows:
c             sysm * solution =rhs
c        The solution is also stored in rhs
c
c        This is the main program
c
         PROGRAM FEM
         call FEM1D
         stop
         end
c
         subroutine FEM1D
         implicit double precision (a - h, o - z)
         parameter (NMAX = 1000)
         dimension x(NMAX), sysm(NMAX, NMAX), sysme(2,2), rhs(NMAX,1)
         dimension wksp (NMAX)
c
         write(6,*) 'enter the number of elements'
         write(6,*) 'suggestion: positive integer'
         read(5,*) nelem
c
```

```
      c       The total number of nodes
              nnode=nelem+1
      c
              write(6,*) 'enter the boundary data'
              read(5,*) pa,pb
      c
              write(6,*) 'enter the x-coordinates of nodes'
              do 11 i = 1,nnode
      11         read(*,10) x(i)
      10      format(f10.3)
      c
      c       The system matrix and right-hand side are initially set zero
      c
              do 30 i = 1,nnode
                do 20 j = 1,nnode
      20        sysm(i,j) = 0.0
      30      rhs(i,1) = 0.0
      c
      c       The element matrices are obtained and assembled
      c
              do 40 i = 1,nelem
                step=x(i+1) − x(i)
                sysme(1,1)=1.0/step
                sysme(1,2)=-sysme(1,1)
                sysme(2,2)=sysme(1,1)
                sysme(2,1)=sysme(1,2)
                sysm(i,i)=sysm(i,i)+sysme(1,1)
                sysm(i,i+1)=sysm(i,i+1)+sysme(1,2)
                sysm(i+1,i)=sysm(i+1,i)+sysme(2,1)
      40      sysm(i+1,i+1)=sysm(i+1,i+1)+sysme(2,2)
      c
      c       The Dirichlet boundary condition is inserted in the
      c          right-hand side and the system matrix is corrected
      c
      c       at the left-hand end i = 1
      c
              rhs(1,1)=pa
              sysm(1,1)=1.0
              do 50 j = 2,nnode
      50      sysm(1,j) = 0.0
      c
      c       at the right-hand end i = nnode
      c
              rhs(nnode,1)=pb
              sysm(nnode,nnode)=1.0
```

```
         do 60 j = 1,nnode-1
60    sysm(nnode,j) = 0.0
c

      do 80 i = 1,nnode
80    write(6,*) (sysm(i,j),j = 1,nnode)
c
c     Solution of the linear system using
c         a standard library solution program
c

      call LEQT1F(sysm,1,nnode,NMAX,rhs,0,wksp,ier)
c
c     The solution has been obtained and printed
c

    · write(6,*) 'Solution has been obtained'
      write(6,*) 'Nodes and solutions at the nodes'
c

      do 71 i = 1,nnode
71    write(6,70) i,rhs(i,1)
70    format((1x,i4,f10.3))
c

      end
```

As a simple example, the input data in the above computer program are entered by the user. An alternative approach is to enter them from an input file, which is addressed in the next chapter. To show an example using this program, the numerical solution for six unequally spaced elements is found in the next example.

Example 1.4. The input data are: The endpoints are 0 and 1, the number of elements is six, the boundary data are $p_a = -1$ and $p_b = 1$, and the nodes and their x-coordinates are $x_1 = 0.000$, $x_2 = 0.100$, $x_3 = 0.300$, $x_4 = 0.500$, $x_5 = 0.700$, $x_6 = 0.900$, and $x_7 = 1.000$. The values of the numerical solution obtained at these nodes are $p_1 = -1.000$, $p_2 = -0.800$, $p_3 = -0.400$, $p_4 = 0.000$, $p_5 = 0.400$, $p_6 = 0.800$, and $p_7 = 1.000$.

1.6. Equivalence and error estimates

We end this chapter with two remarks. The first remark is on the equivalence between Eqs. (1.7) and (1.8). Let p be a solution of Eq. (1.7). Then, for any function $v \in V$ (the admissible space) and any real number ϵ, we have

$$F(p) \le F(p + \epsilon v).$$

With the definition

$$G(\epsilon) = F(p + \epsilon v)$$

$$= \frac{1}{2} \int_0^1 \left(\frac{dp}{dx} \right)^2 dx + \epsilon \int_0^1 \frac{dp}{dx} \frac{dv}{dx} dx + \frac{\epsilon^2}{2} \int_0^1 \left(\frac{dv}{dx} \right)^2 dx$$

$$- \epsilon \int_0^1 fv \, dx - \int_0^1 fp \, dx,$$

we see that G has a minimum at $\epsilon = 0$, so $\dfrac{dG}{d\epsilon}(0) = 0$. Since

$$\frac{dG}{d\epsilon}(0) = \int_0^1 \frac{dp}{dx} \frac{dv}{dx} dx - \int_0^1 fv \, dx,$$

p is a solution of Eq. (1.8). Conversely, suppose that p is a solution of Eq. (1.8). With any $v \in V$, set $w = v - p \in V$; we find that

$$F(v) = F(p + w) = \frac{1}{2} \int_0^1 \left(\frac{d(p + w)}{dx} \right)^2 dx - \int_0^1 f(p + w) \, dx$$

$$= \frac{1}{2} \int_0^1 \left(\frac{dp}{dx} \right)^2 dx - \int_0^1 fp \, dx + \int_0^1 \frac{dp}{dx} \frac{dw}{dx} dx$$

$$- \int_0^1 fw \, dx + \frac{1}{2} \int_0^1 \left(\frac{dw}{dx} \right)^2 dx$$

$$= \frac{1}{2} \int_0^1 \left(\frac{dp}{dx} \right)^2 dx - \int_0^1 fp \, dx + \frac{1}{2} \int_0^1 \left(\frac{dw}{dx} \right)^2 dx \geq F(p),$$

which implies that p is a solution of Eq. (1.7). Because of the equivalence between problems (1.1) and (1.8), Eq. (1.7) is also equivalent to Eq. (1.1).

The second remark is on an error estimate for problem (1.10). In general, the derivation of an error estimate for the finite element method is very technical. Here, we briefly indicate how to obtain an estimate in one dimension. Subtract Eq. (1.10) from Eq. (1.8) to get

$$\int_0^1 \left(\frac{dp}{dx} - \frac{dp_h}{dx} \right) \frac{dv}{dx} dx = 0 \quad \forall v \in V_h. \tag{1.29}$$

We introduce the notation (called a norm)

$$\|v\| = \left(\int_0^1 v^2(x) \, dx \right)^{1/2}.$$

We use the Cauchy or (Cauchy-Schwartz) inequality (Brenner and Scott, 1994; Chen, 2005)

$$\left| \int_0^1 v(x)w(x)\,dx \right| \le \|v\|\,\|w\|. \tag{1.30}$$

Note that, using Eq. (1.29), with $v \in V_h$ (the finite element space) we see that

$$\left\| \frac{dp}{dx} - \frac{dp_h}{dx} \right\|^2 = \int_0^1 \left(\frac{dp}{dx} - \frac{dp_h}{dx} \right)^2 \, dx$$

$$= \int_0^1 \left(\frac{dp}{dx} - \frac{dp_h}{dx} \right) \left(\left[\frac{dp}{dx} - \frac{dv}{dx} \right] + \left[\frac{dv}{dx} - \frac{dp_h}{dx} \right] \right) \, dx$$

$$= \int_0^1 \left(\frac{dp}{dx} - \frac{dp_h}{dx} \right) \left(\frac{dp}{dx} - \frac{dv}{dx} \right) \, dx,$$

so, by Eq. (1.30),

$$\left\| \frac{dp}{dx} - \frac{dp_h}{dx} \right\| \le \left\| \frac{dp}{dx} - \frac{dv}{dx} \right\| \quad \forall\, v \in V_h. \tag{1.31}$$

This equation implies that p_h is the best possible approximation of p in V_h in terms of the norm in Eq. (1.31).

To obtain an error bound, we take v in Eq. (1.31) to be the interpolant $\tilde{p}_h \in V_h$ of p; i.e., $\tilde{p}_h$ is defined by

$$\tilde{p}_h(x_i) = p(x_i), \quad i = 0, 1, \ldots, M + 1. \tag{1.32}$$

It is an easy exercise (see Exercise 1.7) to see that, for $x \in [0, 1]$,

$$|(p - \tilde{p}_h)(x)| \le \frac{h^2}{8} \max_{y \in [0,1]} \left| \frac{d^2 p(y)}{dx^2} \right|,$$

$$\left| \left(\frac{dp}{dx} - \frac{d\tilde{p}_h}{dx} \right)(x) \right| \le h \max_{y \in [0,1]} \left| \frac{d^2 p(y)}{dx^2} \right|. \tag{1.33}$$

With $v = \tilde{p}_h$ in Eq. (1.31) and the second equation of Eq. (1.33), we obtain

$$\left\| \frac{dp}{dx} - \frac{dp_h}{dx} \right\| \le h \max_{y \in [0,1]} \left| \frac{d^2 p(y)}{dx^2} \right|. \tag{1.34}$$

Using the fact that $p(0) - p_h(0) = 0$, we have

$$p(x) - p_h(x) = \int_0^x \left(\frac{dp}{dx} - \frac{dp_h}{dx} \right)(y)\,dy, \quad x \in [0, 1],$$

which, together with Eq. (1.34), implies

$$|p(x) - p_h(x)| \le h \max_{y \in [0,1]} \left| \frac{d^2 p(y)}{dx^2} \right|, \quad x \in [0,1]. \tag{1.35}$$

Note that Eq. (1.35) is less sharp in h than the first estimate in Eq. (1.33) for the interpolation error. With a more delicate analysis, we can show that the first error estimate in Eq. (1.33) holds for p_h as well as $\tilde{p}_h$. In fact, it can be shown that $p_h = \tilde{p}_h$ (see Exercise 1.8), which is true only for one dimension.

In summary, we have obtained the quantitative estimates in Eqs. (1.34) and (1.35), which show that the approximation solution of Eq. (1.10) approaches the exact solution of problem (1.1) as h goes to zero. This implies convergence of the finite element method (1.10).

1.7. Exercises

1.1. Consider an elastic bar with tension one, fixed at both ends ($x = 0, 1$) and subject to a transversal load of intensity f (Fig. 1.1). Under the assumption of small displacements, show that the transversal displacement p satisfies problem (1.1).

1.2. Let the admissible space V be defined as in Sec. 1.2. Show that if $p \in V$ satisfies Eq. (1.8) and p is twice continuously differentiable, then p also satisfies Eq. (1.1).

1.3. Let the admissible space V be defined as in Sec. 1.3.2. Show that if the solution $p \in V$ satisfies Eq. (1.17), then it also satisfies Eq. (1.18).

1.4. Let the finite element space V_h be defined as in Sec. 1.3.2. Write Eq. (1.19) in matrix form.

1.5. Show that system (1.13) remains the same with the local coordinate transformation approach introduced in Sec. 1.4.2.

1.6. Modify the computer program in Sec. 1.5 so that it can handle the case where the right-hand side function f in Eq. (1.15) is any arbitrary constant.

1.7. Derive the estimates in system (1.33).

1.8. Referring to Sec. 1.6, show that the interpolant $\tilde{p}_h \in V_h$ of p defined in Eq. (1.32) equals the finite element solution p_h generated by Eq. (1.10).

1.9. Redo Example 1.4 for the following two cases: (a) When the number of elements becomes 12 and the nodes and their x-coordinates become $x_1 = 0.000$, $x_2 = 0.050$, $x_3 = 0.100$, $x_4 = 0.200$, $x_5 = 0.300$, $x_6 = 0.400$, $x_7 = 0.500$, $x_8 = 0.600$, $x_9 = 0.700$, $x_{10} = 0.800$, $x_{11} = 0.900$, $x_{12} = 0.950$, and $x_{13} = 1.000$, find the corresponding numerical solution at these nodes; (b) When the number of elements becomes 24 and the nodes and their x-coordinates become $x_1 = 0.000$, $x_2 = 0.025$, $x_3 = 0.050$, $x_4 = 0.075$, $x_5 = 0.100$, $x_6 = 0.150$, $x_7 = 0.200$, $x_8 = 0.250$, $x_9 = 0.300$, $x_{10} = 0.350$, $x_{11} = 0.400$, $x_{12} = 0.450$, $x_{13} = 0.500$, $x_{14} = 0.550$, $x_{15} = 0.600$, $x_{16} = 0.650$, $x_{17} = 0.700$, $x_{18} = 0.750$, $x_{19} = 0.800$, $x_{20} = 0.850$, $x_{21} = 0.900$, $x_{22} = 0.925$, $x_{23} = 0.950$, $x_{24} = 0.975$, and $x_{25} = 1.000$, find the corresponding numerical solution at these nodes.

1.10. For problem (1.1), (A) define the finite element method by using piecewise quadratic trial and test functions (referring to Sec. 1.2); (B) Find the basis (shape) function at each node by using nodal function values; (C) Formulate the discrete Galerkin formulation in the matrix form $\mathbf{Ap} = \mathbf{f}$ and find the system matrix $\mathbf{A}$ when the partition of the interval $(0, 1)$ is uniform.

Appendix

A C++ program corresponding to the Fortran program in Sec. 1.5 is given.

/* Computer program

    Finite element method for 1D problem
    Solution of equation (1.15) with $f = 0$
    Linear system is solved by a standard algorithm for tri diagonal system
    The following symbols are used:
    nnode - Total number of nodes
    nelem - Total number of elements
    pa, pb - Boundary data
    $x[i]$ - x-coordinate of node i
    sysm_l[i] - lower diagonal part of global system matrix $\mathbf{A}$,
       $i = 1, \ldots, $ nnode-1
    sysm_d[i] - diagonal part of global system matrix $\mathbf{A}$, $i = 1, 2, \ldots, $ nnode
    sysm_u[i] - upper diagonal part of global system matrix $\mathbf{A}$,
       $i = 1, \ldots, $ nnode-1
    sysme[m][n] - Local matrix $\mathbf{A}$, $m, n = 1, 2$
    rhs[i] - Right-hand side vector

```
*/
#include <iostream>
#include <stdio.h>
#include <string.h>
#include <stdlib.h>
#include <vector>

void solveTriDiagonal(int dim,double *x, const double* b,
        const double *diag_low,const double* diag,const double* diag_upper);

int main(int argc,char** argv)
{
        int nelem=0, nnode=0, i;
        double pa, pb;
        double *x, *sysm_l,*sysm_d,*sysm_u, sysme[2][2], *rhs;

        std::cout << "Enter the number of elements\n";
        std::cin >> nelem;

        nnode = nelem + 1;
        x = new double[nnode*5];
        if (!x)
        {
           std::cout<<"Could not allocate memory\n";
           exit(1);
        }          rhs = x + nnode;
        sysm_l = rhs + nnode;
        sysm_d = sysm_l + nnode;
        sysm_u = sysm_d + nnode;

        for (i = 0;i < nnode*5;++i) x[i] = 0.0;

        std::cout << "Enter boundary data pa pb\n";
        std::cin >> pa >> pb;

        std::cout << "Enter x-coordinates of nodes\n";
        for (i = 0;i < nnode;++i)
        {
           std::cin >> x[i];
        }

        for (i = 0;i < nelem;++i)
        {
           double step = x[i+1] - x[i];
           sysme[0][0] = 1.0/step;
```

```
        sysme[0][1] = - sysme[0][0];
        sysme[1][1] = sysme[0][0];
        sysme[1][0] = sysme[0][1];
        sysm_d[i] += sysme[0][0];
        sysm_u[i] += sysme[0][1];
        sysm_l[i] += sysme[1][0];
        sysm_d[i + 1] += sysme[1][1];
    }
    // The Dirichlet boundary condition is inserted in the right hand
    // side and the system is corrected
    rhs[0] = pa;
    sysm_d[0] = 1.0;
    sysm_u[0] = 0.0;
    rhs[nnode-1] = pb;
    sysm_d[nnode-1] = 1.0;
    sysm_l[nnode-2] = 0.0;

    solveTriDiagonal(nnode,x,rhs,sysm_l,sysm_d,sysm_u);

    char fname_sol[] = "solution1.dat";
    FILE *fp_sol=fopen(fname_sol,"w");
    if (!fp_sol)
    {
        std::cout<<"Could not open file " << fname_sol<<"\n";
        exit(1);
    }
    fprintf(fp_sol,"node_id\tsolution\n");
    for (i = 0;i < nnode;++i)
    {
        fprintf(fp_sol,"%d\t%lf\n",i + 1,x[i]);
    }
    fclose(fp_sol);
    std::cout<<"solution has been written in file " << fname_sol<<"\n";
}


void solveTriDiagonal(int dim,double *x, const double* b,
        const double *diag_low,const double* diag,const double* diag_upper)

    int i;
    std::vector<double> w(dim),v(dim),z(dim);

    for (i = 0;i<dim;++i)
    {
        if (i==0)
```

```
    {
        w[i] = diag[i];
        v[i] = diag_upper[i]/w[i];
        z[i] = b[i]/w[i];
    }
    else
    {
        w[i] = diag[i] - diag_low[i-1]*v[i-1];
        if (i<dim-1) v[i] = diag_upper[i]/w[i];
        z[i] = (b[i]-diag_low[i-1]*z[i-1])/w[i];
    }
}
for (i=dim-1;i> = 0;-i)
{
    if (i==dim-1)
    {
        x[i] = z[i];
    }
    else
    {
        x[i] = z[i]-v[i]*x[i + 1];
    }
}
}
```

Chapter 2

Two-Dimensional
Model Problems

In Chapter 1, a one-dimensional example illustrating the basic concepts of the finite element method was presented. In this chapter, this method is generalized to a two-dimensional differential problem. In Sec. 2.1, we first introduce four two-dimensional physical problems to motivate the model problem considered. Then, in Sec. 2.2, we describe the finite element method for this model problem. The development of a linear system of algebraic equations arising from this method is also discussed. In Sec. 2.3, we discuss the finite element method for boundary conditions of different types. Sec. 2.4 is devoted to the construction of basis (shape) functions and system matrices using different approaches. In Sec. 2.5, basic components in programming a two-dimensional finite element code are addressed, and an illustrative computer program is written. In Sec. 2.6, error estimates are stated. Finally, in Sec. 2.7, exercise problems are given. Below, partial derivatives are indicated by $\partial/\partial x_1$, $\partial/\partial x_2$, $\partial^2/\partial x_1^2$, and $\partial^2/\partial x_2^2$, where $x = (x_1, x_2)$ are the Cartesian coordinates.

2.1. Two-dimensional differential problems

The two-dimensional differential problems described are the counterparts of the corresponding one-dimensional problems discussed in Chapter 1.

Example 2.1 (Elastic membrane). We consider a stationary (i.e., time independent) problem in two dimensions for the unknown p

$$
\begin{aligned}
-\Delta p &= f &&\text{in } \Omega, \\
p &= 0 &&\text{on } \Gamma,
\end{aligned}
\tag{2.1}
$$

27

where Ω is a bounded domain in the plane with boundary Γ, f is a given real-valued bounded function in Ω, and the Laplacian operator Δ is defined by

$$\Delta p = \frac{\partial^2 p}{\partial x_1^2} + \frac{\partial^2 p}{\partial x_2^2}.$$

The notation ∇^2 is also used for this operator in the engineering literature. Corresponding to the one-dimensional problem (1.1) for an elastic bar, consider an elastic membrane fixed at its boundary and subject to a transversal load of intensity f (Fig. 2.1). Under the assumption of small displacements, we can check that the transversal displacement p satisfies Eq. (2.1), which is the Poisson equation, with a homogeneous Dirichlet boundary condition.

Example 2.2 (Heat conduction). The one-dimensional metallic rod problem considered in Example 1.2 is now extended to a rectangular plate. We seek the temperature distribution in this plate, both of whose ends are held at constant temperatures T_1 and T_2, respectively, with $T_2 > T_1$ (Fig. 2.2). There is a loss of heat through convection and radiation into the surrounding medium that is held at a given temperature $T_0 < T_2$. This heat loss is modeled by a dissipation coefficient β proportional to $T_1 + T_2 - 2T_0$. The

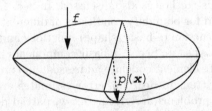

Figure 2.1   An elastic membrane.

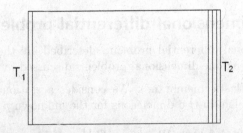

Figure 2.2   A metallic rectangular plate.

temperature T of the plate satisfies

$$-\kappa \Delta T + \beta T = 0 \qquad \text{in } \Omega,$$

$$T|_{\Gamma_1} = T_1, \quad T|_{\Gamma_2} = T_2, \quad \left.\frac{\partial T}{\partial \nu}\right|_{\Gamma_3} = 0, \tag{2.2}$$

where κ is the thermal diffusion coefficient of the plate, Γ is the boundary of the plate Ω, Γ_1 and Γ_2 are the vertical ends, $\Gamma_3 = \Gamma \backslash \{\Gamma_1 \cup \Gamma_2\}$ are the horizontal side edges, and the normal derivative $\partial T / \partial \nu$ is expressed by

$$\frac{\partial T}{\partial \nu} = \frac{\partial T}{\partial x_1}\nu_1 + \frac{\partial T}{\partial x_2}\nu_2,$$

with $\nu = (\nu_1, \nu_2)$ the outward unit normal to Γ. On the vertical ends, the boundary condition is of nonhomogeneous Dirichlet type (given temperature), and on the horizontal sides, it is of homogeneous Neumann type (no heat transfer through these sides).

Example 2.3 (Heat conduction with a variable coefficient). Consider a nonhomogeneous plate that has a variable thermal diffusion coefficient κ; i.e., its property depends on the space location. In this case, the first equation of problem (2.2) becomes

$$-\nabla \cdot (\kappa \nabla T) + \beta T = 0 \qquad \text{in } \Omega, \tag{2.3}$$

where the gradient and divergence operators are

$$\nabla v = \left(\frac{\partial v}{\partial x_1}, \frac{\partial v}{\partial x_2}\right), \quad \nabla \cdot \mathbf{b} = \frac{\partial b_1}{\partial x_1} + \frac{\partial b_2}{\partial x_2},$$

for any differentiable scalar function v and vector function $\mathbf{b} = (b_1, b_2)$.

Example 2.4 (Single phase flow). As the final example in this chapter, we consider a two-dimensional counterpart of the one-dimensional single phase flow problem described in Example 1.3. The mass conservation equation becomes

$$\nabla \cdot (\rho u) = q, \tag{2.4}$$

where ρ and u are the fluid density and velocity, respectively, and q is the (mass) flow rate. Darcy's law is

$$u = -\frac{k}{\mu}\nabla p, \tag{2.5}$$

where k is the permeability of the medium and μ and p are the fluid viscosity and pressure, respectively. Substitution of Eq. (2.5) into Eq. (2.4) gives

$$-\nabla \cdot \left(\frac{\rho k}{\mu} \nabla p \right) = q. \tag{2.6}$$

The pressure-specified boundary condition is

$$p = g_1 \qquad \text{on } \Gamma,$$

where g_1 is a prescribed pressure. A prescribed mass flux boundary condition is

$$-\frac{\rho k}{\mu} \nabla p \cdot \boldsymbol{\nu} = g_2 \qquad \text{on } \Gamma.$$

These two boundary conditions can be also imposed on separate parts of the boundary Γ of the domain Ω.

2.2. The finite element method

As an example, we describe the finite element method for the model problem (2.1); problems (2.2), (2.3), and (2.6) are treated in Chapter 3. For the two-dimensional problem (2.1), the admissible function space V is composed of the following real functions:

$$V = \left\{ \text{Functions } v \colon v \text{ is a continuous function on } \Omega, \frac{\partial v}{\partial x_1} \text{ and } \frac{\partial v}{\partial x_2} \right.$$
$$\left. \text{are piecewise continuous and bounded on } \Omega, \text{ and } v = 0 \text{ on } \Gamma \right\}.$$

This is the two-dimensional counterpart of the admissible space introduced in Chapter 1 for one dimension.

2.2.1. Green's formula

To write Eq. (2.1) in an equivalent variational formulation, we need to use Green's formula, which is an extension to multiple dimensions of integration by parts in one dimension. This important formula is used throughout this book, so we briefly derive it. For a vector-valued function $\mathbf{b} = (b_1, b_2)$, the *divergence theorem* reads

$$\int_\Omega \nabla \cdot \mathbf{b} \, d\mathbf{x} = \int_\Gamma \mathbf{b} \cdot \boldsymbol{\nu} \, d\ell, \tag{2.7}$$

where we recall that the divergence operator is given by

$$\nabla \cdot \mathbf{b} = \frac{\partial b_1}{\partial x_1} + \frac{\partial b_2}{\partial x_2},$$

$\boldsymbol{\nu}$ is the outward unit normal to Γ, the dot product $\mathbf{b} \cdot \boldsymbol{\nu}$ is defined by

$$\mathbf{b} \cdot \boldsymbol{\nu} = b_1 \nu_1 + b_2 \nu_2,$$

and the right-hand side integral in Eq. (2.7) is an integral along the boundary curve Γ. With v, $w \in V$ (the above admissible space), we take $\mathbf{b} = \left(\frac{\partial v}{\partial x_1} w, 0 \right)$ and $\mathbf{b} = \left(0, \frac{\partial v}{\partial x_2} w \right)$ in Eq. (2.7), respectively, to see that

$$\int_\Omega \frac{\partial^2 v}{\partial x_i^2} w \, d\mathbf{x} + \int_\Omega \frac{\partial v}{\partial x_i} \frac{\partial w}{\partial x_i} \, d\mathbf{x} = \int_\Gamma \frac{\partial v}{\partial x_i} w \nu_i \, d\ell, \quad i = 1, 2. \tag{2.8}$$

Recalling the definition of the gradient operator, i.e.,

$$\nabla v = \left(\frac{\partial v}{\partial x_1}, \frac{\partial v}{\partial x_2} \right),$$

we sum (2.8) over $i = 1, 2$ to obtain

$$\int_\Omega \Delta v \, w \, d\mathbf{x} = \int_\Gamma \frac{\partial v}{\partial \boldsymbol{\nu}} w \, d\ell - \int_\Omega \nabla v \cdot \nabla w \, d\mathbf{x}, \tag{2.9}$$

where the *normal derivative* is expressed by

$$\frac{\partial v}{\partial \boldsymbol{\nu}} = \frac{\partial v}{\partial x_1} \nu_1 + \frac{\partial v}{\partial x_2} \nu_2.$$

Relation (2.9) is Green's formula, and it also holds in three dimensions (see Exercise 2.1).

2.2.2. Variational formulation

We define the functional F on the space V

$$F(v) = \frac{1}{2} \int_\Omega |\nabla v|^2 \, d\mathbf{x} - \int_\Omega fv \, d\mathbf{x}, \qquad v \in V,$$

where $|\nabla v|^2 = \nabla v \cdot \nabla v$. As in one dimension, Eq. (2.1) can be formulated as the minimization problem

Find $p \in V$ such that $F(p) \le F(v) \quad \forall v \in V,$

which is the Ritz variational formulation. This problem is equivalent to
the variational problem (2.10) below, using exactly the same proof as for
Eqs. (1.7) and (1.8).

Multiplying the first equation of problem (2.1) by $v \in V$ and integrating
over Ω, we see that

$$-\int_\Omega \Delta p \, v \, d\mathbf{x} = \int_\Omega f v \, d\mathbf{x}.$$

Applying Green's formula (2.9) to this equation and using the homogeneous boundary condition in the space V lead to the Galerkin variational
formulation

$$\text{Find } p \in V \text{ such that } \int_\Omega \nabla p \cdot \nabla v \, d\mathbf{x} = \int_\Omega f v \, d\mathbf{x} \qquad \forall v \in V. \qquad (2.10)$$

We now construct the finite element method for Eq. (2.1). For simplicity, in this section, we assume that Ω is a polygonal domain. A curved
domain Ω is handled in Sec. 6.6. Let K_h be a partition, called a *triangulation*, of Ω into non overlapping (open) triangles K_i (Fig. 2.3):

$$\bar{\Omega} = \bar{K}_1 \cup \bar{K}_2 \cup \cdots \cup \bar{K}_M,$$

such that no vertex of one triangle lies in the interior of an edge of another
triangle, where $\bar{\Omega}$ represents the closure of Ω (i.e., $\bar{\Omega} = \Omega \cup \Gamma$) and a similar
meaning holds for each K_i.

For (open) triangles $K \in K_h$, we define the *mesh parameters*

$$h_K = \text{diam}(K) = \text{the longest edge of } \bar{K} \text{ and } h = \max_{K \in K_h} h_K.$$

Now, we introduce the simplest finite element space in two dimensions

$$V_h = \{\text{Functions } v: v \text{ is a continuous function on } \Omega, v \text{ is}$$
$$\text{linear on each triangle } K \in K_h, \text{ and } v = 0 \text{ on } \Gamma\}.$$

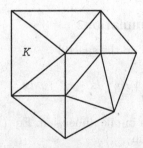

Figure 2.3 A finite element partition in two dimensions.

Again, this space is an extension of the corresponding one-dimensional finite element space to two dimensions. Note that $V_h \subset V$. The finite element method for problem (2.1) is formulated as

$$\text{Find } p_h \in V_h \text{ such that } \int_\Omega \nabla p_h \cdot \nabla v \, d\mathbf{x} = \int_\Omega f v \, d\mathbf{x} \qquad \forall v \in V_h. \quad (2.11)$$

Existence and uniqueness of a solution to Eq. (2.11) can be shown in the same manner as for Eq. (1.10). Also, in the same fashion as in the proof of the equivalence between Eqs. (1.7) and (1.8), one can check that Eq. (2.11) is equivalent to a discrete minimization problem:

$$\text{Find } p_h \in V_h \text{ such that } F(p_h) \leq F(v) \qquad \forall v \in V_h.$$

2.2.3. Basis (shape) functions

Consider a typical triangle (element) K, whose vertices (nodes) $\mathbf{x}_i$, $\mathbf{x}_j$, and $\mathbf{x}_m$ are numbered in the counterclockwise direction (Fig. 2.4) and have the coordinates

$$\mathbf{x}_i = (x_{i,1}, x_{i,2}), \quad \mathbf{x}_j = (x_{j,1}, x_{j,2}), \quad \mathbf{x}_m = (x_{m,1}, x_{m,2}).$$

On the element K, a two-dimensional linear approximation p_h (*trial function*) to the solution p is

$$p_h(x_1, x_2) = a + bx_1 + cx_2, \quad \mathbf{x} = (x_1, x_2) \in K. \quad (2.12)$$

The constants a–c are determined by

$$p_h(x_{i,1}, x_{i,2}) = p_i, \quad p_h(x_{j,1}, x_{j,2}) = p_j, \quad p_h(x_{m,1}, x_{m,2}) = p_m.$$

That is

$$\begin{aligned} a + bx_{i,1} + cx_{i,2} &= p_i, \\ a + bx_{j,1} + cx_{j,2} &= p_j, \\ a + bx_{m,1} + cx_{m,2} &= p_m. \end{aligned} \quad (2.13)$$

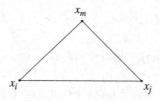

Figure 2.4 A typical triangle.

The system of Eqs. (2.13) will produce a unique solution a, b, and c if the determinant of the coefficient matrix does not vanish:

$$2\Delta_K = \begin{vmatrix} 1 & x_{i,1} & x_{i,2} \\ 1 & x_{j,1} & x_{j,2} \\ 1 & x_{m,1} & x_{m,2} \end{vmatrix} \neq 0. \tag{2.14}$$

It can simply be shown by trigonometry that this determinant equals twice the area Δ_K of the triangle K, as illustrated in Eq. (2.14). Because the area of a nontrivial triangle never equals zero, the solution for the constants a–c exists and is unique. Solving system (2.13) gives

$$a = \frac{1}{2\Delta_K}(a_i p_i + a_j p_j + a_m p_m),$$

$$b = \frac{1}{2\Delta_K}(b_i p_i + b_j p_j + b_m p_m), \tag{2.15}$$

$$c = \frac{1}{2\Delta_K}(c_i p_i + c_j p_j + c_m p_m),$$

where

$$a_i = x_{j,1}x_{m,2} - x_{m,1}x_{j,2}, \quad b_i = x_{j,2} - x_{m,2}, \quad c_i = -(x_{j,1} - x_{m,1}), \tag{2.16}$$

and a_j, a_m, b_j, b_m, c_j, and c_m can be obtained by cyclic permutation of the indices $\{i, j, m\}$. Substituting expression (2.15) into Eq. (2.12) yields

$$p_h(x_1, x_2) = \frac{1}{2\Delta_K}\{(a_i + b_i x_1 + c_i x_2)p_i + (a_j + b_j x_1 + c_j x_2)p_j$$

$$+ (a_m + b_m x_1 + c_m x_2)p_m\}, \quad (x_1, x_2) \in K. \tag{2.17}$$

We define the local two-dimensional *basis functions* (*shape functions*) on the element K as follows:

$$\varphi_i(x_1, x_2) = \frac{1}{2\Delta_K}(a_i + b_i x_1 + c_i x_2),$$

$$\varphi_j(x_1, x_2) = \frac{1}{2\Delta_K}(a_j + b_j x_1 + c_j x_2), \tag{2.18}$$

$$\varphi_m(x_1, x_2) = \frac{1}{2\Delta_K}(a_m + b_m x_1 + c_m x_2), \quad (x_1, x_2) \in K.$$

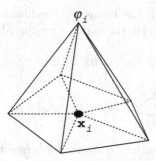

Figure 2.5 A basis function in two dimensions.

Note that these functions satisfy

$$\varphi_i(x_{i,1}, x_{i,2}) = 1, \quad \varphi_i(x_{j,1}, x_{j,2}) = 0, \quad \varphi_i(x_{m,1}, x_{m,2}) = 0,$$
$$\varphi_j(x_{i,1}, x_{i,2}) = 0, \quad \varphi_j(x_{j,1}, x_{j,2}) = 1, \quad \varphi_j(x_{m,1}, x_{m,2}) = 0,$$
$$\varphi_m(x_{i,1}, x_{i,2}) = 0, \quad \varphi_m(x_{j,1}, x_{j,2}) = 0, \quad \varphi_m(x_{m,1}, x_{m,2}) = 1.$$

Now, denote all the vertices (*nodes*) of the triangles in K_h by $\mathbf{x}_1, \mathbf{x}_2, \ldots,$ $\mathbf{x}_{\tilde{M}}$ (including the vertices on the boundary Γ). The global basis (shape) functions φ_i, $i = 1, 2, \ldots, \tilde{M}$, are then required to satisfy

$$\varphi_i(\mathbf{x}_j) = \begin{cases} 1 & \text{if } i = j, \\ 0 & \text{if } i \neq j, \end{cases} \quad j = 1, 2, \ldots, \tilde{M}.$$

The *support* of φ_i, i.e., the set where $\varphi_i(\mathbf{x}) \neq 0$, consists of the triangles with the common node $\mathbf{x}_i$ (Fig. 2.5). The function φ_i is a two-dimensional hat (chapeau) function. On each element K, these functions become the local shape functions defined in system (2.18).

2.2.4. Linear systems

Let $M < \tilde{M}$ be the number of interior vertices in K_h; for convenience, let the first M vertices be the interior ones. As in Chapter 1, any function $v \in V_h$ has the unique representation

$$v(\mathbf{x}) = \sum_{i=1}^{M} v_i \varphi_i(\mathbf{x}), \qquad \mathbf{x} \in \Omega,$$

where $v_i = v(\mathbf{x}_i)$. Due to the boundary condition imposed in the finite element space V_h, we exclude the vertices on the boundary of Ω (i.e., the boundary nodes).

In the same way as for Eq. (1.10), with

$$p_h(\mathbf{x}) = \sum_{i=1}^{M} p_i \varphi_i(\mathbf{x}), \quad p_i = p_h(\mathbf{x}_i), \quad \mathbf{x} \in \Omega, \qquad (2.19)$$

Equation (2.11) can be written in matrix form (see Exercise 2.2)

$$\mathbf{A}\mathbf{p} = \mathbf{f}, \qquad (2.20)$$

where, as before, the matrix $\mathbf{A}$ and vectors $\mathbf{p}$ and $\mathbf{f}$ are given by

$$\mathbf{A} = (a_{i,j}), \quad \mathbf{p} = (p_j), \quad \mathbf{f} = (f_j),$$

with

$$a_{i,j} = \int_{\Omega} \nabla \varphi_i \cdot \nabla \varphi_j \, d\mathbf{x}, \quad f_j = \int_{\Omega} f \varphi_j \, d\mathbf{x}, \quad i, j = 1, 2, \ldots, M.$$

As in one dimension, it can be checked that the stiffness matrix $\mathbf{A}$ is symmetric and positive definite. In particular, it is nonsingular. Consequently, Eqs. (2.20) and thus (2.11) have a unique solution.

Note that the matrix $\mathbf{A}$ is sparse from the construction of the basis functions. As an example, we consider the case where the domain is the unit square $\Omega = (0, 1) \times (0, 1)$ and K_h is a uniform triangulation of Ω as illustrated in Fig. 2.6 with the indicated *enumeration* of nodes. In this case,

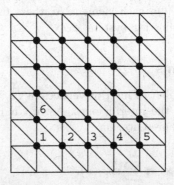

Figure 2.6 An example of a triangulation.

the matrix $\mathbf{A}$ has the form (see Exercise 2.3)

$$\mathbf{A} = \begin{pmatrix}
4 & -1 & 0 & 0 & \cdots & 0 & -1 & 0 & \cdots & 0 & 0 \\
-1 & 4 & -1 & 0 & \cdots & 0 & 0 & -1 & \cdots & 0 & 0 \\
0 & -1 & 4 & -1 & \cdots & 0 & 0 & 0 & \cdots & -1 & 0 \\
0 & 0 & -1 & 4 & \cdots & 0 & 0 & 0 & \cdots & 0 & -1 \\
\vdots & \vdots & \vdots & \vdots & \ddots & \vdots & \vdots & \vdots & \ddots & \vdots & \vdots \\
0 & 0 & 0 & 0 & \cdots & 4 & -1 & 0 & \cdots & 0 & 0 \\
-1 & 0 & 0 & 0 & \cdots & -1 & 4 & -1 & \cdots & 0 & 0 \\
0 & -1 & 0 & 0 & \cdots & 0 & -1 & 4 & \cdots & 0 & 0 \\
\vdots & \vdots & \vdots & \vdots & \ddots & \vdots & \vdots & \vdots & \ddots & \vdots & \vdots \\
0 & 0 & -1 & 0 & \cdots & 0 & 0 & 0 & \cdots & 4 & -1 \\
0 & 0 & 0 & -1 & \cdots & 0 & 0 & 0 & \cdots & -1 & 4
\end{pmatrix}.$$

Associated with the four corner nodes (e.g., node 1), there are only three nonzeros per row; the adjacent diagonal entry for such a node (e.g., node 5) may be zero. For other nodes adjacent to the boundary (e.g., node 2), there are solely four nonzeros per row. From the form of $\mathbf{A}$ shown, the left-hand side of the ith equation in Eq. (2.20) is a linear combination of the values of p_h at most at the five nodes illustrated in Fig. 2.7. After division by h^2, system (2.20) can be treated as a linear system generated by a *five-point difference stencil scheme* for problem (2.1) (Thomas, 1995).

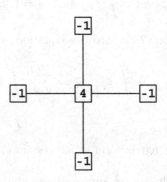

Figure 2.7 A five-point stencil scheme.

2.3. Extensions to general boundary conditions

In the previous section, a homogeneous Dirichlet boundary condition was considered. We now extend the finite element method to the *stationary Poisson problem* with other types of boundary conditions.

2.3.1. Nonhomogeneous Dirichlet boundary conditions

With a nonhomogeneous Dirichlet boundary condition, problem (2.1) becomes

$$\begin{aligned} -\Delta p &= f && \text{in } \Omega, \\ p &= g && \text{on } \Gamma, \end{aligned} \qquad (2.21)$$

where the function g is given on the boundary Γ. This nonhomogeneous boundary value problem can be handled in the same fashion as in the one dimension outlined in Chapter 1.

Approach A. The definition of the admissible function space V remains unchanged. Now, the approximate solution has the representation

$$p_h(\mathbf{x}) = \sum_{i=1}^{M} p_i \varphi_i(\mathbf{x}) + \sum_{i=M+1}^{\tilde{M}} g_i \varphi_i(\mathbf{x}), \quad \mathbf{x} \in \Omega, \qquad (2.22)$$

where

$$p_i = p_h(\mathbf{x}_i), \quad g_i = g(\mathbf{x}_i).$$

With $v = \varphi_j$ in Eq. (2.11), we see that

$$\int_\Omega \nabla p_h \cdot \nabla \varphi_j \, d\mathbf{x} = \int_\Omega f \varphi_j \, d\mathbf{x}, \quad j = 1, 2, \ldots, M. \qquad (2.23)$$

Substitution of expression (2.22) into this equation gives

$$\sum_{i=1}^{M} \int_\Omega \nabla \varphi_i \cdot \nabla \varphi_j \, d\mathbf{x} \, p_i = \int_\Omega f \varphi_j \, d\mathbf{x} - \sum_{i=M+1}^{\tilde{M}} \int_\Omega \nabla \varphi_i \cdot \nabla \varphi_j \, d\mathbf{x} \, g_i,$$

$$j = 1, 2, \ldots, M, \quad (2.24)$$

which can be written in matrix form as for system (2.20) with a modified right-hand side vector $\mathbf{f}$.

Approach B. An alternative approach uses all the basis functions φ_1, φ_2, ..., $\varphi_{\tilde{M}}$ to form a larger system of equations, i.e., an $\tilde{M} \times \tilde{M}$ system.

If $\mathbf{x}_i$ is a node for which the nodal value is specified explicitly as g_i, then the procedure for inserting this Dirichlet condition into the system matrix equation is: enter zeros in the ith row of the system matrix $\mathbf{A}$ except for unity in the major diagonal position and enter g_i in the ith row of the vector $\mathbf{f}$. To retain the symmetry of $\mathbf{A}$, the corresponding columns need to be modified as well. This procedure can be most simply presented through an example.

Consider the system of equations

$$
\begin{pmatrix}
a_{1,1} & a_{1,2} & a_{1,3} & a_{1,4} & a_{1,5} \\
a_{2,1} & a_{2,2} & a_{2,3} & a_{2,4} & a_{2,5} \\
a_{3,1} & a_{3,2} & a_{3,3} & a_{3,4} & a_{3,5} \\
a_{4,1} & a_{4,2} & a_{4,3} & a_{4,4} & a_{4,5} \\
a_{5,1} & a_{5,2} & a_{5,3} & a_{5,4} & a_{5,5}
\end{pmatrix}
\begin{pmatrix}
p_1 \\ p_2 \\ p_3 \\ p_4 \\ p_5
\end{pmatrix}
=
\begin{pmatrix}
f_1 \\ f_2 \\ f_3 \\ f_4 \\ f_5
\end{pmatrix}.
\tag{2.25}
$$

Suppose that the boundary condition is

$$
p_1 = g_1, \quad p_4 = g_4.
$$

Values p_1 and p_4 are specified so they must be replaced in system (2.25). The substitution disturbs the symmetry of matrix $\mathbf{A}$, but this can be restored by inserting $p_1 = g_1$ and $p_4 = g_4$ in each of the other component equations and modifying the right-hand sides accordingly. For example, the second component equation in Eq. (2.25) is modified as follows:

$$
a_{2,2}p_2 + a_{2,3}p_3 + a_{2,5}p_5 = f_2 - a_{2,1}p_1 - a_{2,4}p_4.
$$

Using this modification for the third and fifth component equations in Eq. (2.25), the resulting system becomes

$$
\begin{pmatrix}
1 & 0 & 0 & 0 & 0 \\
0 & a_{2,2} & a_{2,3} & 0 & a_{2,5} \\
0 & a_{3,2} & a_{3,3} & 0 & a_{3,5} \\
0 & 0 & 0 & 1 & 0 \\
0 & a_{5,2} & a_{5,3} & 0 & a_{5,5}
\end{pmatrix}
\begin{pmatrix}
p_1 \\ p_2 \\ p_3 \\ p_4 \\ p_5
\end{pmatrix}
=
\begin{pmatrix}
g_1 \\
f_2 - a_{2,1}p_1 - a_{2,4}p_4 \\
f_3 - a_{3,1}p_1 - a_{3,4}p_4 \\
g_4 \\
f_5 - a_{5,1}p_1 - a_{5,4}p_4
\end{pmatrix},
\tag{2.26}
$$

which is symmetric. This approach can be summarized: if a nodal variable p_i is given as $p_i = g_i$, one first replaces f_i by the given value and $a_{i,i}$ by unity, next overwrites the remainder of the ith row and the ith column of $\mathbf{A}$ with zeros, and finally, for each nonprescribed f_j, subtracts $a_{j,i}p_i$ from f_j.

In a large system of equations with many prescribed Dirichlet conditions, this approach reduces the size of the final system. For example, system (2.26) can be written as

$$
\begin{pmatrix} a_{2,2} & a_{2,3} & a_{2,5} \\ a_{3,2} & a_{3,3} & a_{3,5} \\ a_{5,2} & a_{5,3} & a_{5,5} \end{pmatrix} \begin{pmatrix} p_2 \\ p_3 \\ p_5 \end{pmatrix} = \begin{pmatrix} f_2 - a_{2,1}p_1 - a_{2,4}p_4 \\ f_3 - a_{3,1}p_1 - a_{3,4}p_4 \\ f_5 - a_{5,1}p_1 - a_{5,4}p_4 \end{pmatrix}. \tag{2.27}
$$

Both of the above two approaches (Approaches A and B) for handling the nonhomogeneous Dirichlet condition produce the same final system of equations (see Exercise 2.4).

2.3.2. General boundary conditions

We now extend the finite element method to the stationary Poisson problem with another type of boundary condition:

$$
\begin{aligned}
-\Delta p &= f && \text{in } \Omega, \\
\gamma p + \frac{\partial p}{\partial \boldsymbol{\nu}} &= g && \text{on } \Gamma,
\end{aligned} \tag{2.28}
$$

where γ and g are given functions and we recall that $\partial p/\partial \boldsymbol{\nu}$ is the normal derivative. Recall that this type of boundary condition is a *third*, *mixed*, *Robin*, or *Dankwerts* boundary condition. When $\gamma = 0$, the boundary condition is a *second* or *Neumann* condition. When γ is infinite, the boundary condition reduces to a *first* or *Dirichlet condition*, which was considered in the previous section. A *fourth type* of boundary condition (i.e., a *periodic* boundary condition) is considered in Chapter 11. In this section, we treat the case where γ is bounded.

Note that if $\gamma = 0$ on the boundary Γ, Green's formula Eq. (2.9) and Eq. (2.28) imply that (see Exercise 2.5)

$$
\int_\Omega f \, d\mathbf{x} + \int_\Gamma g \, d\ell = 0. \tag{2.29}
$$

For Eq. (2.28) to have a solution, the *compatibility condition* (2.29) must be satisfied. In this case, p is unique only up to an additive constant.

Now, we introduce the admissible function space

$$
V = \left\{ \text{Functions } v \colon v \text{ is a continuous function on } \Omega, \text{ and } \frac{\partial v}{\partial x_1} \text{ and } \right.
$$

$$
\left. \frac{\partial v}{\partial x_2} \text{ are piecewise continuous and bounded on } \Omega \right\}.
$$

Then, as in the previous section, Eq. (2.28) can be written as (see Exercise 2.6): find $p \in V$ such that

$$\int_{\Omega} \nabla p \cdot \nabla v \, d\mathbf{x} + \int_{\Gamma} \gamma p v \, d\ell = \int_{\Omega} f v \, d\mathbf{x} + \int_{\Gamma} g v \, d\ell \qquad \forall v \in V. \quad (2.30)$$

Note that the boundary condition in Eq. (2.28) is not imposed in the definition of the space V. It appears implicitly in Eq. (2.30). A boundary condition that need not be imposed is called a *natural condition*. The pure Neumann boundary condition is natural. The Dirichlet boundary condition has been imposed explicitly in V in Secs. 2.2 and 2.3.1, and is termed an *essential condition*.

If $\gamma \equiv 0$, the definition of V needs to be modified to take into account the up-to a constant uniqueness of the solution to Eq. (2.28). The space V can be modified to, say

$$V = \left\{ \text{Functions } v: v \text{ is a continuous function on } \Omega, \frac{\partial v}{\partial x_1} \text{ and } \frac{\partial v}{\partial x_2} \right.$$

$$\left. \text{are piecewise continuous and bounded on } \Omega, \text{ and } \int_{\Omega} v \, d\mathbf{x} = 0 \right\}.$$

That is, the zero mean value on the domain Ω is imposed for each function in the space V.

To construct the finite element method for problem (2.28), let K_h be a triangulation of Ω as in the previous section. The finite element space V_h is defined by

$$V_h = \{ \text{Functions } v: v \text{ is a continuous function on } \Omega \text{ and}$$

$$\text{linear on each triangle } K \in K_h \}. \quad (2.31)$$

Note that the functions in V_h are not required to satisfy any boundary condition. Now, the finite element solution $p_h \in V_h$ satisfies

$$\int_{\Omega} \nabla p_h \cdot \nabla v \, d\mathbf{x} + \int_{\Gamma} \gamma p_h v \, d\ell = \int_{\Omega} f v \, d\mathbf{x} + \int_{\Gamma} g v \, d\ell \qquad \forall v \in V_h. \quad (2.32)$$

For the pure Neumann boundary condition, V_h needs to be modified to

$$V_h = \left\{ \text{Functions } v: v \text{ is a continuous function on } \Omega \text{ and linear} \right.$$

$$\left. \text{on each triangle } K \in K_h, \text{ and } \int_{\Omega} v \, d\mathbf{x} = 0 \right\}.$$

Note that the zero mean value condition is imposed for the space V_h; however, the standard nodal basis functions defined in Sec. 2.2.3 do not satisfy it. Instead of imposing it explicitly, the space V_h can be defined in the usual way (2.31), and the approximate solution is subtracted from its mean at the end of calculations so that it satisfies the condition. Alternatively, because the solution is unique up to a constant, one can remove this condition simply by fixing a value for the solution at a node or any point on the domain Ω. Another note is that as in the previous section, Eq. (2.32) can be formulated in matrix form.

2.4. Local coordinate formulations

Two-dimensional basis (shape) functions can be also obtained using a local coordinate system transformation as in one dimension.

2.4.1. Local element matrices

In practical computations (Sec. 2.2.4), the entries $a_{i,j}$ in $\mathbf{A}$ are obtained by summing the contributions from different triangles $K \in K_h$:

$$a_{i,j} = \sum_{K \in K_h} a_{i,j}^K,$$

where

$$a_{i,j}^K = \int_K \nabla \varphi_i \cdot \nabla \varphi_j \, d\mathbf{x}. \tag{2.33}$$

Using the definition of the basis functions, we see that $a^K(\varphi_i, \varphi_j) = 0$ unless nodes $\mathbf{x}_i$ and $\mathbf{x}_j$ are both vertices of K. Thus, for the triangle K with vertices $\mathbf{x}_i$, $\mathbf{x}_j$, and $\mathbf{x}_m$ (Fig. 2.4), the local *element matrix* on K is

$$\mathbf{A}^K = \begin{pmatrix} a_{i,i}^K & a_{i,j}^K & a_{i,m}^K \\ a_{j,i}^K & a_{j,j}^K & a_{j,m}^K \\ a_{m,i}^K & a_{m,j}^K & a_{m,m}^K \end{pmatrix}, \tag{2.34}$$

where

$$a_{l,n}^K = \int_K \nabla \varphi_l \cdot \nabla \varphi_n \, d\mathbf{x}, \quad l, \, n \in \{i, j, m\}.$$

The (linear) basis functions φ_i, φ_j, and φ_m on K are given by system (2.18), so

$$a_{l,n}^K = \frac{1}{4\Delta_K} \left(b_l b_n + c_l c_n \right), \quad l, \, n \in \{i, j, m\}, \tag{2.35}$$

where b_l and c_l are given in Eq. (2.16).

The right-hand side on K is locally computed similarly by

$$f_l^K = \int_K f\varphi_l \, d\mathbf{x}, \quad l = i, j, m. \tag{2.36}$$

2.4.2. Construction of triangulations

To assemble the global matrix $\mathbf{A}$ and right-hand side vector $\mathbf{f}$, we need to consider the construction of the triangulation K_h. It can be constructed from a successive refinement of an initial coarse partition of the domain Ω; fine triangles can be obtained by connecting the midpoints of edges of coarse triangles, for example. A sequence of uniform refinements will lead to *quasi-uniform* grids where the triangles in K_h essentially have the same size in all regions of Ω (Fig. 2.8). If the boundary Γ of Ω is a curve, special care needs to be taken of near Γ (Sec. 6.6).

In practice, it is often necessary to use triangles in K_h that vary considerably in size in different regions of Ω. For example, one utilizes smaller triangles in regions where the exact solution has a fast variation or where its certain derivatives are large (Fig. 2.9, where a *local refinement* strategy

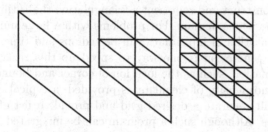

Figure 2.8 Uniform refinement.

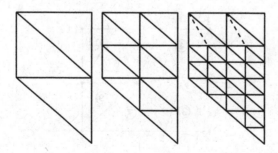

Figure 2.9 Nonuniform refinement.

is carried out). In this strategy, proper care is taken in the transition zone between regions with triangles of different sizes so that a so-called *regular* local refinement results (i.e., no vertex of one triangle lies in the interior of an edge of another triangle). Methods that automatically refine grids where needed are called *adaptive methods* (see Sec. 11.7).

Let a triangulation K_h have M nodes and $\mathcal{M}$ triangles. The triangulation can be represented by two arrays $\mathbf{Z}(M, 2)$ and $\mathbf{IZ}(\mathcal{M}, 3)$, where $\mathbf{Z}(i, j)$ ($j = 1, 2$) indicates the coordinates of the ith node, $1 = 1, 2, \ldots, M$, and $\mathbf{IZ}(k, j)$ ($j = 1, 2, 3$) enumerates the nodes of the kth triangle, $k = 1, 2, \ldots, \mathcal{M}$. An example is demonstrated in Fig. 2.10, where the triangle numbers are denoted in circles. For this example, the array $\mathbf{IZ}(\mathcal{M}, 3)$ is of the form, where $M = \mathcal{M} = 11$:

$$
\mathbf{IZ} = \begin{pmatrix}
1 & 1 & 2 & 3 & 4 & 4 & 5 & 6 & 7 & 7 & 8 \\
2 & 4 & 5 & 4 & 5 & 7 & 9 & 7 & 9 & 10 & 10 \\
4 & 3 & 4 & 6 & 7 & 6 & 7 & 8 & 10 & 8 & 11
\end{pmatrix}.
$$

In general, when local refinement is involved in a triangulation K_h, it is very difficult to enumerate nodes and triangles; some strategies are given in Sec. 11.7. For a simple domain Ω (e.g., a convex polygonal Ω), it is rather easy to construct and represent a triangulation that utilizes uniform refinement in the whole domain. For problems with a large number of elements, the preparation of input data can be tedious and time consuming. Automatic grid generation is one way to speed up this task. When certain basic information, such as the location of corner and boundary nodes, element type, and density of elements, is provided, a typical grid generation program will generate a desired grid and provide a list of nodes and their coordinates. Although such a program can be integrated into a finite element solution code, it often runs separately so that the output can be

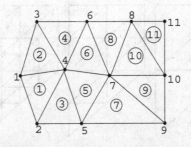

Figure 2.10   Node and triangle enumeration.

checked. For further information on grid generation, see Thompson *et al.* (1998).

2.4.3. Assembly of stiffness matrices

After the triangulation K_h is constructed, one computes the *element stiffness matrices* with entries $a_{i,j}^K$ given by Eq. (2.33). We recall that $a_{i,j}^K = 0$ unless nodes $\mathbf{x}_i$ and $\mathbf{x}_j$ are both vertices of the triangle $K \in K_h$.

For a kth triangle $K^{(k)}$, $\mathbf{IZ}(k, m)$ ($m = 1, 2, 3$) are the numbers of the vertices (nodes) of $K^{(k)}$, and the element stiffness matrix $\mathbf{A}^{K^{(k)}} = \left(a_{m,n}^{(k)} \right)_{m,n=1}^{3}$ is now calculated as follows:

$$a_{m,n}^{(k)} = \int_{K^{(k)}} \nabla\varphi_m \cdot \nabla\varphi_n \, d\mathbf{x}, \quad m, \, n = 1, 2, 3,$$

where the (linear) basis function φ_m on the triangle $K^{(k)}$ satisfies

$$\varphi_m\left(\mathbf{x}_{\mathbf{IZ}(k,n)}\right) = \begin{cases} 1 & \text{if } m = n, \\ 0 & \text{if } m \neq n. \end{cases}$$

The right-hand side on $K^{(k)}$ is computed by

$$f_m^{(k)} = \int_{K^{(k)}} f\varphi_m \, d\mathbf{x}, \quad m = 1, 2, 3.$$

Note that m and n are the local numbers of the three vertices of $K^{(k)}$, while i and j used in Eq. (2.33) are the global numbers of the vertices (nodes) in K_h.

To assemble the global matrix $\mathbf{A} = (a_{i,j})$ and the right-hand side vector $\mathbf{f} = (f_j)$, one loops over all triangles $K^{(k)}$ and successively adds the contributions from different $K^{(k)}$'s:

For $k = 1, 2, \ldots, \mathcal{M}$, compute
$$a_{\mathbf{IZ}(k,m),\mathbf{IZ}(k,n)} = a_{\mathbf{IZ}(k,m),\mathbf{IZ}(k,n)} + a_{m,n}^{(k)},$$
$$f_{\mathbf{IZ}(k,m)} = f_{\mathbf{IZ}(k,m)} + f_m^{(k)}, \quad m, \, n = 1, 2, 3.$$

The approach used is *element oriented*; i.e., we loop over elements (i.e., triangles). As noted in Chapter 1, this approach is better than the *node-oriented* approach (i.e., looping over all nodes); the latter approach wastes too much time in repeated computations of $\mathbf{A}$ and $\mathbf{f}$, as noted in Chapter 1.

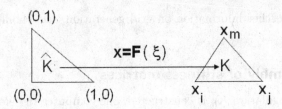

Figure 2.11 Local coordinate transformation.

2.4.4. Local coordinate transformation

As in one dimension, the local stiffness matrix $\mathbf{A}^K$ and right-hand side vector $\mathbf{f}^K$ can be obtained using local coordinate system transformations. For a typical triangle K in the global coordinate system with vertices $\mathbf{x}_i$, $\mathbf{x}_j$, and $\mathbf{x}_m$, suppose that $\mathbf{F}$ is a one-to-one mapping from the reference triangle $\hat{K}$ in the $\xi_1\xi_2$-coordinate system, with vertices $(0,0)$, $(1,0)$, and $(0,1)$, onto K (Fig. 2.11); these vertices correspond to the node identifers 1, 2, and 3, respectively. On the reference element $\hat{K}$, the approximate solution p_h (trial function) is of the form

$$p_h(\xi_1, \xi_2) = a + b\xi_1 + c\xi_2, \quad 0 \le \xi_1, \xi_2, \xi_1 + \xi_2 \le 1, \tag{2.37}$$

where the constants a–c are determined by

$$p_h(0,0) = p_i, \quad p_h(1,0) = p_j, \quad p_h(0,1) = p_m. \tag{2.38}$$

That is

$$a = p_i, \quad b = p_j - p_i, \quad c = p_m - p_i. \tag{2.39}$$

Substitution of these values into Eq. (2.37) gives

$$p_h(\xi_1, \xi_2) = (1 - \xi_1 - \xi_2)p_i + \xi_1 p_j + \xi_2 p_m. \tag{2.40}$$

We define

$$\psi_1(\xi_1, \xi_2) = 1 - \xi_1 - \xi_2, \quad \psi_2(\xi_1, \xi_2) = \xi_1, \quad \psi_3(\xi_1, \xi_2) = \xi_2. \tag{2.41}$$

These are the basis (shape) functions on the reference triangle $\hat{K}$. Now, the transformation $\mathbf{F}$ is given by

$$\mathbf{x} = \mathbf{F}(\xi) = \mathbf{F}(\xi_1, \xi_2) = \mathbf{x}_i\psi_1 + \mathbf{x}_j\psi_2 + \mathbf{x}_m\psi_3. \tag{2.42}$$

Its inverse is

$$\xi_1 = \frac{1}{2\Delta_K}\left(a_j + b_j x_1 + c_j x_2\right),$$

$$\xi_2 = \frac{1}{2\Delta_K}\left(a_m + b_m x_1 + c_m x_2\right), \tag{2.43}$$

where Δ_K is the area of the triangle K and the coefficients a_j, a_m, b_j, b_m, c_j, and c_m are defined in Eq. (2.16).

We define the local basis functions on the element K:

$$\varphi_i(\mathbf{x}) = \psi_1\big(\mathbf{F}^{-1}(\mathbf{x})\big), \quad \varphi_j(\mathbf{x}) = \psi_2\big(\mathbf{F}^{-1}(\mathbf{x})\big),$$

$$\varphi_m(\mathbf{x}) = \psi_3\big(\mathbf{F}^{-1}(\mathbf{x})\big), \quad \mathbf{x} \in K. \tag{2.44}$$

We now consider the computation of the local stiffness matrix $\mathbf{A}^K$ in Eq. (2.34) when the local coordinate transformation is used to obtain the basis functions. As an example, we consider to compute the entry

$$a_{i,j}^K = \int_K \nabla\varphi_i \cdot \nabla\varphi_j \, d\mathbf{x}. \tag{2.45}$$

It follows from the chain rule that

$$\frac{\partial\varphi_i}{\partial x_k} = \frac{\partial}{\partial x_k}\left(\psi_1\big(\mathbf{F}^{-1}(\mathbf{x})\big)\right) = \frac{\partial\psi_1}{\partial\xi_1}\frac{\partial\xi_1}{\partial x_k} + \frac{\partial\psi_1}{\partial\xi_2}\frac{\partial\xi_2}{\partial x_k},$$

for $k = 1, 2$. Consequently, we see that

$$\nabla\varphi_i = \mathbf{G}^{-T}\nabla\psi_1,$$

where $\mathbf{G}^{-T}$ is the transpose of the Jacobian of $\mathbf{F}^{-1}$

$$\mathbf{G}^{-T} = \begin{pmatrix} \dfrac{\partial\xi_1}{\partial x_1} & \dfrac{\partial\xi_2}{\partial x_1} \\[2mm] \dfrac{\partial\xi_1}{\partial x_2} & \dfrac{\partial\xi_2}{\partial x_2} \end{pmatrix}.$$

Similarly, we see that

$$\nabla\varphi_j = \mathbf{G}^{-T}\nabla\psi_2.$$

When we apply the change of variable $\mathbf{F} : \hat{K} \to K$ to Eq. (2.45), we have

$$a_{i,j}^K = \int_{\hat{K}} (\mathbf{G}^{-T}\nabla\psi_1) \cdot (\mathbf{G}^{-T}\nabla\psi_2)|\det\mathbf{G}| \, d\boldsymbol{\xi}, \tag{2.46}$$

where $|\det \mathbf{G}|$ is the absolute value of the determinant of the Jacobian $\mathbf{G}$

$$\mathbf{G} = \begin{pmatrix} \dfrac{\partial x_1}{\partial \xi_1} & \dfrac{\partial x_1}{\partial \xi_2} \\[2mm] \dfrac{\partial x_2}{\partial \xi_1} & \dfrac{\partial x_2}{\partial \xi_2} \end{pmatrix}.$$

Applying an algebraic computation, we see that

$$\mathbf{G}^{-T} = \left(\mathbf{G}^{-1}\right)^{T} = \frac{1}{\det \mathbf{G}} \mathbf{G}',$$

where

$$\mathbf{G}' = \begin{pmatrix} \dfrac{\partial x_2}{\partial \xi_2} & -\dfrac{\partial x_2}{\partial \xi_1} \\[2mm] -\dfrac{\partial x_1}{\partial \xi_2} & \dfrac{\partial x_1}{\partial \xi_1} \end{pmatrix}.$$

Hence, Eq. (2.46) becomes

$$a_{i,j}^{K} = \int_{\hat{K}} (\mathbf{G}' \nabla \psi_1) \cdot (\mathbf{G}' \nabla \psi_2) \frac{1}{|\det \mathbf{G}|} \, d\boldsymbol{\xi}. \tag{2.47}$$

Therefore, the matrix entry $a_{i,j}^{K}$ on K can be calculated by either Eq. (2.46) or Eq. (2.47). In general, it is difficult to evaluate these two integrals analytically. However, they can be relatively easily evaluated using a numerical integration formula (or a *quadrature rule*; see Sec. 6.7).

The local right-hand side (f_i^K, f_j^K, f_m^K) on element K can be similarly computed as follows:

$$f_l^K = \int_K f \varphi_l \, d\mathbf{x} = \int_{\hat{K}} f(\mathbf{F}(\boldsymbol{\xi})) \psi_n \, |\det \mathbf{G}| \, d\boldsymbol{\xi}, \quad l = i,j,m, \quad n = 1,2,3, \tag{2.48}$$

where i, j, m correspond to $1, 2, 3$, respectively.

2.5. Programming considerations

The essential components of a typical computer program implementing the finite element method were stated in Sec. 1.5 and are repeated as follows:

- input of data such as the domain Ω, the right-hand side function f, the boundary data γ and g (system (2.28)), and the coefficients that may appear in a differential problem;

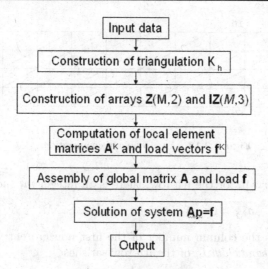

Figure 2.12 A flowchart.

- construction of the triangulation K_h;

- computation and assembly of the stiffness matrix $\mathbf{A}$ and the right-hand vector $\mathbf{f}$;

- solution of the linear system of algebraic equations $\mathbf{Ap} = \mathbf{f}$; and

- output of computational results.

A flowchart for the computer program is shown in Fig. 2.12. The data input can be easily implemented in a small subroutine, and the result output depends on the computer system and software the user has in hand. The second and third parts were discussed in Secs. 2.4.2 and 2.4.3, respectively. The solution of the linear system $\mathbf{Ap} = \mathbf{f}$ can be performed via a direct method (Gaussian elimination) or an iterative method (e.g., the conjugate gradient method), which is discussed in linear solver books (Chen, 2005; Saad, 2004). Below, we briefly mention the numbering of nodes in K_h and the matrix storage.

2.5.1. Numbering of nodes

If a direct method (Gaussian elimination) is employed to solve the linear system $\mathbf{Ap} = \mathbf{f}$, the nodes should be enumerated in such a way that the *bandwidth* of each row in $\mathbf{A}$ is as small as possible. The matrix $\mathbf{A} = (a_{i,j})_{M \times M}$

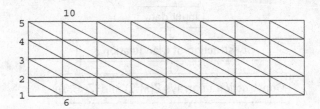

Figure 2.13 An example of enumeration.

is a *banded matrix* if, for the ith row, there is an integer m_i such that

$$a_{i,j} = 0 \quad \text{if } j < m_i, \quad i = 1, 2, \ldots, M.$$

Note that m_i is the column number of the first nonzero entry in the ith row. Then, the *bandwidth* L_i of the ith row satisfies

$$L_i = i - m_i, \qquad i = 1, 2, \ldots, M.$$

It should be noted that $2L_i + 1$ is sometimes called the bandwidth.

Set $L_m = \max\limits_{1 \le i \le M} L_i$. Note that

$$a_{i,j} = \int_\Omega \nabla \varphi_i \cdot \nabla \varphi_j \, d\mathbf{x}, \qquad i, \, j = 1, 2, \ldots, M,$$

where $\{\varphi_i\}_{i=1}^M$ is a basis of the finite element space V_h. Then, we see that

$$L_m = \max\{|i - j| : \varphi_i \text{ and } \varphi_j \text{ correspond to degrees of}$$
$$\text{freedom belonging to the same element}\}.$$

Consequently, the bandwidth depends on the enumeration of nodes. For example, with a vertical enumeration of nodes in Fig. 2.13, L_m is 6 (assuming that one degree of freedom is associated with each node). With a horizontal enumeration, L_m will be 11. In addition to the *natural ordering* of the nodes considered, there are other ordering methods designed to save computational time and computer storage (Chen *et al.*, 2006).

It is generally expensive to solve a large system of algebraic equations via a direct method like Gaussian elimination. On the other hand, we recall that the *condition number* of a *symmetric* matrix $\mathbf{A}$ is defined by

$$\text{cond}(\mathbf{A}) = \frac{\text{the largest eigenvalue of } \mathbf{A}}{\text{the smallest eigenvalue of } \mathbf{A}}.$$

For the matrix $\mathbf{A}$ in system (2.20) (for second-order partial differential problems), it has a condition number proportional to h^{-2} (Chen, 2005).

The efficiency of *iterative methods* degrades on problems with a large condition number. The common technique to obtain the solution of a large-scale system is to use a *preconditioned* iterative approach (Saad, 2004).

2.5.2. Matrix storage

In use of direct or iterative methods, it is not necessary to exploit an array $\mathbf{A}(M, M)$ to store the entire stiffness matrix $\mathbf{A}$. Instead, if $\mathbf{A}$ is a sparse symmetric banded matrix, computer storage can be greatly reduced by storing only the elements within the band. This can be completed by storing the major diagonal elements with either the codiagonal elements to the left or to the right; these are known as subdiagonal or superdiagonal storage modes, respectively. This technique is illustrated by using the following example for the case of subdiagonal storage of the symmetric matrix $\mathbf{A}$ with band width $L_m = 2$:

$$\mathbf{A} = \begin{pmatrix} a_{1,1} & a_{1,2} & a_{1,3} & 0 & 0 \\ a_{2,1} & a_{2,2} & a_{2,3} & a_{2,4} & 0 \\ a_{3,1} & a_{3,2} & a_{3,3} & a_{3,4} & a_{3,5} \\ 0 & a_{4,2} & a_{4,3} & a_{4,4} & a_{4,5} \\ 0 & 0 & a_{5,3} & a_{5,4} & a_{5,5} \end{pmatrix} . \tag{2.49}$$

The entries to be retained for the subdiagonal storage are

$$\begin{pmatrix} a_{1,1} & & \\ a_{2,1} & a_{2,2} & \\ a_{3,1} & a_{3,2} & a_{3,3} \\ a_{4,2} & a_{4,3} & a_{4,4} \\ a_{5,3} & a_{5,4} & a_{5,5} \end{pmatrix} . \tag{2.50}$$

It is often desirable to store a matrix or portion as a vector of elements, i.e., as a string of elements. The band symmetric storage mode can be converted to the vector storage mode either by storing successive rows or successive columns. There are also matrix-free linear solvers that do not require matrix storage but perform the action of matrix and vector multiplication (Saad, 2004).

2.5.3. Computer program

An illustrative Fortran computer program for solving problem (2.28) is considered for the type of partition shown in Fig. 2.14, where the boundary

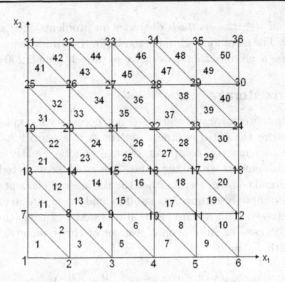

Figure 2.14 An illustrative triangulation for the computer program.

Γ is split into the union of a solution-prescribed part and a nonprescribed part. The procedure based on *assembly by elements* is used to obtain the system matrix **A**. As an example, the right-hand side vector **f** is taken to be zero. As in one dimension, no attempt is made to optimize the code. The corresponding C++ program is given at the end of this chapter.

```
c      Computer program
c
c      Finite element method for a 2D problem
c      Solution of equation (2.28) with f = 0
c      The linear system is solved using a standard
c          library solution program LEQT1F
c
c      The following symbols are used:
c
c      nnode - Total number of nodes
c      nelem - Total number of elements
c      npres - Total number of nodes where
c          the solution is prescribed
c      np(i) - The number of node where
c          the solution is prescribed, i = 1, 2, . . . , npres
c      pres(i) - The prescribed value of solution at
```

```
c        node np(i), i = 1, 2, . . . , npres
c     x(i), y(i) - x and y coordinates of node i = 1, 2, . . . , nnode
c     iz(i, j) - The node numbers of three nodes of each
c        element, j = 1, 2, 3, i = 1, 2, . . . , nnode
c     sysm(i, j) - Global system matrix A, i, j = 1, 2, . . . , nnode
c     rhs(i, 1) - Right-hand side vector
c     The right-hand side has two subscripts to satisfy
c        the requirement of the program LEQT1F,
c        which also needs a work space wksp(i)
c     The linear system is solved as follows:
c          sysm * solution =rhs
c     The solution is also stored in rhs
c     The following symbols are "working variables":
c     xx(i), yy(i) - x and y coordinates of the three nodes
c        of each local element, i = 1, 2, 3
c     b(i), c(i) - Parameters defined in equation (2.16), i = 1, 2, 3
c     Delta - Area of each triangle
c     coef - Local matrix A
c
c     This is the main program
c

      PROGRAM FEM
      call FEM2D
      stop
      end
c

      subroutine FEM2D
      implicit double precision (a-h,o-z)
      parameter (NMAX = 1000)
      dimension x(NMAX),y(NMAX),sysm(NMAX,NMAX),rhs(NMAX,1)
      dimension iz(NMAX,3),np(NMAX),pres(NMAX),wksp(NMAX)
      Dimension xx(3),yy(3),b(3),c(3)
c
c     The format of input file:
c     nelem nnode npres
c     x(i) y(i)i = 1, . . . , nnode
c     iz(i, 1) iz(i, 2) iz(i, 3) i = 1, . . . , nelem
c     np(i) pres(i) i = 1, . . . , npres
c
      iu = 7
      open(unit=iu,file="input.dat",status="old",err=999)
```

```
c
      read(iu,*) nelem, nnode, npres
c
      do 10 i = 1,nnode
10    read(iu,*) x(i),y(i)
c
      do 20 i = 1,nelem
20    read(iu,*) iz(i,1),iz(i,2),iz(i,3)
c
      do 30 i = 1,npres
30    read(iu,*) np(i),pres(i)
c
      close(iu)
c
c     The system matrix and right-hand side are initially set zero
c
      do 50 i = 1,nnode
        do 40 j = 1,nnode
40        sysm(i,j) = 0.0
50    rhs(i,1) = 0.0
c
c     Element matrices are obtained and assembled immediately
c
      do 120 i = 1,nelem
        do 60 j = 1,3
        ij = iz(i,j)
        xx(j) = x(ij)
60      yy(j) = y(ij)
        do 100 j = 1,3
        ij = j+1
        jj = j+2
        if (ij - 3) 90,80,70
70      ij = 1
        jj = 2
        go to 90
80      jj = 1
90      b(j) = yy(ij) - yy(jj)
100       c(j) = xx(jj) - xx(ij)
        delta=(c(3)*b(2) - c(2)*b(3))*0.5
        do 110 m = 1,3
        do 110 n = 1,3
```

```
           mm = iz(i, m)
           nn = iz(i, n)
           coef=(b(m)*b(n) + c(m)*c(n))/(4.0*delta)
  110      sysm(mm, nn) = sysm(mm, nn) + coef
  120   continue
c
c       The prescribed boundary conditions are inserted
c
        do 140 i = 1, npres
           ii = np(i)
           do 130 j = 1, nnode
  130      sysm(ii, j) = 0.0
           sysm(ii, ii) = 1.0
  140   rhs(ii, 1) = pres(i)
c
        do 300 i = 1, nnode
  300   write(*,*) (sysm(i, j), j = 1, nnode)
c
c       Solution of the linear system using
c           a standard library solution program
c
        call LEQT1F(sysm,1,nnode,NMAX,rhs,0,wksp,ier)
c
c       The solution has been obtained and printed
c
        write(6,*) "Solution has been obtained"
        write(6,*) "Nodes and solutions at the nodes"
c
        do 400 i = 1, nnode
  400   write(*,401) i,rhs(i,1)
  401   format((1x,i4,f10.3))
        goto 888
c
  999   write(*,*)"could not open the grid file"
  888   continue
        end
```

To illustrate the use of this computer program, the numerical solution for 50 equal triangles is sought on the square $[0, 10] \times [0, 10]$ (Fig. 2.14).

Example 2.5. The input data are: the number of elements is 50, the number of nodes is 36, and the number of the prescribed nodes is 11. The

Dirichlet boundary condition is

$$p = -20x_1 + 200, \quad 0 \leq x_1 \leq 10, \, x_2 = 10,$$
$$p = -20x_2 + 200, \quad 0 \leq x_2 \leq 10, \, x_1 = 10,$$

and the remaining boundary of the domain is equipped with the homogeneous Neumann boundary condition. For Fig. 2.14, the input.dat has the form

50 36 11
0.0 0.0
2.0 0.0
4.0 0.0
6.0 0.0
8.0 0.0
10.0 0.0
0.0 2.0
2.0 2.0
4.0 2.0
6.0 2.0
8.0 2.0
10.0 2.0
0.0 4.0
2.0 4.0
4.0 4.0
6.0 4.0
8.0 4.0
10.0 4.0
0.0 6.0
2.0 6.0
4.0 6.0
6.0 6.0
8.0 6.0
10.0 6.0
0.0 8.0
2.0 8.0
4.0 8.0
6.0 8.0
8.0 8.0
10.0 8.0
0.0 10.0
2.0 10.0
4.0 10.0

6.0 10.0
8.0 10.0
10.0 10.0
1 2 7
2 8 7
2 3 8
3 9 8
3 4 9
4 10 9
4 5 10
5 11 10
5 6 11
6 12 11
7 8 13
8 14 13
8 9 14
9 15 14
9 10 15
10 16 15
10 11 16
11 17 16
11 12 17
12 18 17
13 14 19
14 20 19
14 15 20
15 21 20
15 16 21
16 22 21
16 17 22
17 23 22
17 18 23
18 24 23
19 20 25
20 26 25
20 21 26
21 27 26
21 22 27
22 28 27
22 23 28
23 29 28
23 24 29

24 30 29
25 26 31
26 32 31
26 27 32
27 33 32
27 28 33
28 34 33
28 29 34
29 35 34
29 30 35
30 36 35
31 200.0
32 160.0
33 120.0
34 80.0
35 40.0
36 0.0
30 40.0
24 80.0
18 120.0
12 160.0
6 200.0

The values of the numerical solution obtained at the 36 nodes are

Node	Sol	Node	Sol	Node	Sol	Node	Sol
1	132.826	2	132.826	3	133.925	4	139.417
5	156.020	6	200.000	7	132.826	8	132.277
9	131.728	10	133.861	11	142.331	12	160.000
13	133.925	14	131.728	15	126.849	16	121.969
17	119.444	18	120.000	19	139.417	20	133.861
21	121.969	22	107.723	23	93.476	24	80.000
25	156.020	26	142.331	27	119.444	28	93.476
29	66.738	30	40.000	31	200.000	32	160.000
33	120.000	34	80.000	35	40.000	36	0.000

2.6. Error estimates

We end this chapter with a remark on an error estimate. As noted earlier, the derivation of an estimate is very delicate; all error estimates for multi-dimensional problems will just be stated without proof. By using the same argument as for inequality (1.31), we can show

$$\|\nabla p - \nabla p_h\| \le \|\nabla p - \nabla v\| \qquad \forall v \in V_h,$$

where p and p_h are the respective solutions of Eqs. (2.10) and (2.11), and we recall that $\|\cdot\|$ is the norm in terms of the first partial derivatives:

$$\|\nabla p\| = \left(\int_\Omega \left(\left(\frac{\partial p}{\partial x_1}\right)^2 + \left(\frac{\partial p}{\partial x_2}\right)^2 \right) d\mathbf{x} \right)^{1/2}.$$

This means that the finite element solution p_h is the best possible approximation of p in V_h in terms of this norm. Applying the approximation theory (Ciarlet, 1978; Chen, 2005), it holds that

$$\|p - p_h\| + h\|\nabla p - \nabla p_h\| \le Ch^2, \tag{2.51}$$

where the constant C depends on the second partial derivatives of p and the smallest angle of the triangles $K \in K_h$, but does not depend on h. The error estimate (2.51) indicates that if the solution is sufficiently smooth, the approximate solution p_h converges to the exact solution p in the norm $\|\cdot\|$ as h approaches zero.

For $h > 0$, let K_h be a triangulation of Ω into triangles, as defined above. For $K \in K_h$, as previously, we define the mesh parameters

$$h_K = \text{diam}(K) = \text{the longest edge of } \bar{K}, \quad h = \max_{K \in K_h} h_K.$$

We also define the quantity (Fig. 2.15)

$$\rho_K = \text{the diameter of the largest circle inscribed in } K.$$

We say that a triangulation is *shape regular* if there is a constant $\beta > 0$, independent of h, such that

$$\frac{h_K}{\rho_K} \le \beta \qquad \forall K \in K_h. \tag{2.52}$$

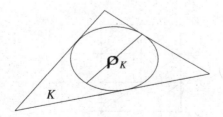

Figure 2.15 A triangle and its inscribed circle.

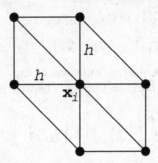

Figure 2.16 The support of a basis function at node $\mathbf{x}_i$.

This condition says that the triangles in K_h are not arbitrarily thin, or equivalently, the angles of the triangles are not arbitrarily small. The constant β is a measure of the smallest angle over all $K \in K_h$. The constant C in estimate (2.51) depends on this constant (Fig. 2.16).

2.7. Exercises

2.1. Prove Green's formula (2.9) in three space dimensions.

2.2. Carry out the derivation of system (2.20) in detail.

2.3. For Fig. 2.16, construct the linear basis function at node $\mathbf{x}_i$ according to the definition in Sec. 2.2.3 (i.e., φ_i equals unity at $\mathbf{x}_i$ and zero at all other nodes). Then, use this result to show that the stiffness matrix $\mathbf{A}$ in Eq. (2.20) for the uniform partition of the unit square $(0,1) \times (0,1)$ given in Fig. 2.6 is determined as in Sec. 2.2.4.

2.4. For problem (2.21) and Fig. 2.17, show that the two approaches of handling the nonhomogeneous Dirichlet boundary conditions discussed in Sec.   2.3.1   produce   the   same   final   system   of   algebraic equations.

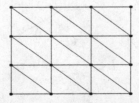

Figure 2.17 A triangulation example.

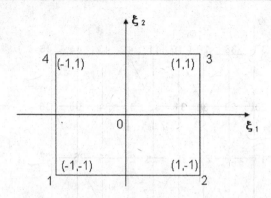

Figure 2.18   The reference rectangle.

2.5. Use Green's formula (2.9) to show Eq. (2.29) from problem (2.28) with $\gamma = 0$.

2.6. Derive Eq. (2.30) from problem (2.28) in detail.

2.7. Modify the computer program in Sec. 2.5.3 so that it can handle the case where the right-hand side function f in Eq. (2.28) is any arbitrary constant (use the same triangulation as in Fig. 2.14).

2.8. Redo Example 2.5 with the same input data and triangulation except that the Dirichlet boundary condition is changed to

$$p = 20x_1, \quad 0 \leq x_1 \leq 10, \; x_2 = 0,$$
$$p = 20x_2, \quad 0 \leq x_2 \leq 10, \; x_1 = 0,$$

and the remaining two sides of the domain is equipped with the homogeneous Neumann boundary condition.

2.9. For the nonhomogeneous problem on the unit square

$$-\Delta p = 0 \qquad \text{in } (0,1) \times (0,1),$$
$$p = g \qquad \text{on } \Gamma,$$

where the function g on the boundary Γ is given by

$$g(x_1, x_2) = 3 - 2x_2, \qquad 0 \leq x_2 \leq 1, \; x_1 = 0,$$
$$g(x_1, x_2) = 0, \qquad \text{elsewhere,}$$

and for the triangulation given in Fig. 2.19, find the input.dat file in the computer program defined in Sec. 2.5.3 and compute the numerical solution at the 45 nodes.

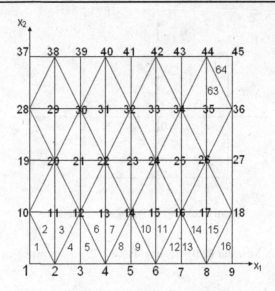

Figure 2.19   A triangulation.

2.10. The vertices of the reference rectangle in Fig. 2.18 have node identifiers 1, 2, 3, and 4, respectively. The bilinear (linear in each direction) basis (shape) function at each node has the property that its value is unity at that node and zero at the other three nodes. Find all four basis functions satisfying such a property.

Appendix

A C++ program corresponding to the Fortran program in Sec. 2.5 is presented.

```
/*      Computer program
C
C       Finite element method for 2D problem
C       Solution of equation (2.28) with f=0
C       The linear system is solved using a standard library solution
C
C       The following symbols are used:
C
C       nnode - Total number of nodes
C       nelem - Total number of elements
C       npres - Total number of nodes where the solution is prescribed
C       np(i) - The number of node where the solution is prescribed
```

```
C              i = 1, 2, . . . , npres
C         pres(i) - The prescribed value of solution at node np(i)
C         x(i), y(i) - x and y coordinates of nodes i = 1, . . . , nnodes
C         iz(i, j) - The node number of three nodes of each element
C              j = 1, 2, 3, i = 1, 2, . . . , nelem
C         sysm(i, j) - Global system matrix A, i, j = 1, 2, . . . , nnode
C         rhs(i) - Right-hand side vector
C
C         xx(i), yy(i)- x and y coordinates of three nodes of a local element
C         b(i), c(i) - Parameters defined in equation (2.16), i = 1, 2, 3
C         Delta - Area of each triangle
C         coef - local matrix A
C
C
C         format of input file:
C         nelem nnode npres
C         x(i) y(i) i = 1, . . . , nnode
C         iz(i, 1) iz(i, 2) iz(i, 3) i = 1, . . . , nelem
C         np(i) pres(i) i = 1, . . . , npres
C
*/
#include <iostream>
#include <stdio.h>
#include <string.h>
#include <stdlib.h>
#include <math.h>
#include <vector>

int CG(const std::vector<std::vector<double> > & A, std::vector<double> & x,
       const std:: vector<double> & b, int &max_iter, double &tol);

int main(int argc,char **argv)
{
       int nnode,nelem, npres,i, j;
       std::vector<double> x, y,pres;
       std::vector<int> iz[3],np;
       int max_iter=100000;
       double tol=1e-6;

       std::vector<std::vector<double> > sysm;
       std::vector<double> rhs,sol,xx(3),yy(3),b(3),c(3);

       char fname_input[] = "input.dat";
       FILE *fp=fopen(fname_input,"r");
       if (!fp)
```

```
{
  std::cout<<"Error open input file"<<fname_input <<"\n";
  exit(0);
}
fscanf(fp,"%d %d %d",&nelem,&nnode,&npres);
x.resize(nnode);
y.resize(nnode);
pres.resize(npres);

iz[0].resize(nelem);
iz[1].resize(nelem);
iz[2].resize(nelem);
np.resize(npres);

sysm.resize(nnode);
for (i = 0;i < nnode;++i) sysm[i].resize(nnode);
rhs.resize(nnode);
sol.resize(nnode);

for (i = 0;i < nnode;++i) fscanf(fp,"%lf %lf",&x[i],&y[i]);
for (i = 0;i < nelem;++i) fscanf(fp,"%d %d %d",&(iz[0][i]),
    &(iz[1][i]),&(iz[2][i]));
for (i = 0;i < npres;++i) fscanf(fp,"%d %lf",&np[i],&pres[i]);
fclose(fp);
std::cout<<"Got input data from file " << fname_input<<"\n";

for (j = 0;j < nnode;++j)
{
  for (i = 0;i < nnode;++i)
  {
    sysm[i][j] = 0.0;
  }
  rhs[j] = 0.0;
  sol[j] = 0.0;
}
for (i = 0;i < nelem;++i)
{
  for (j = 0;j < 3;++j)
  {
    int ij = iz[j][i] - 1;
    xx[j] = x[ij];
    yy[j] = y[ij];
  }

  b[0] = yy[1] - yy[2];
```

```
c[0] = xx[2] - xx[1];
b[1] = yy[2] - yy[0];
c[1] = xx[0] - xx[2];
b[2] = yy[0] - yy[1];
c[2] = xx[1] - xx[0];
double delta = 0.5*(c[2]*b[1] - c[1]*b[2]);
for (int m = 0;m < 3;++m)
{
    int mm = iz[m][i] - 1;
    for (int n = 0;n < 3;++n)
    {
        int nn = iz[n][i] - 1;
        double coef = (b[m]*b[n] + c[m]*c[n])/(4.0*delta);
        sysm[mm][nn] += coef;
    }
}

}
for (i = 0;i < npres;++i)
{
    int ii = np[i] - 1;
    for (j = 0;j < nnode;++j)
    {
        sysm[ii][j] = 0.0;
    }

    for (j = 0;j < nnode;++j)
    {
        rhs[j]- = sysm[j][ii]*pres[i];
        sysm[j][ii] = 0.0;
    }
    sysm[ii][ii] = 1.0;
    rhs[ii] = pres[i];
}

if (CG(sysm,sol,rhs,max_iter,tol)!=0)
{
    std::cout<<"convergent tolerance not achieved or maximum number of
        iterations reached\n";
}

std::cout << "Number of CG-iterations performed: " <<max_iter<<"\n";
std::cout << "L2-norm of residual: "<<tol<<"\n";
```

```
char fname_sol[]="solution2.dat";
FILE *fp_sol=fopen(fname_sol,"w");
if (fp_sol)
{
    fprintf(fp_sol,"node_id solution\n");
    for (i = 0;i < nnode;++i)
    {
        fprintf(fp_sol,"%d %lf\n",i + 1,sol[i]);
    }
    std::cout<<"Solution is written in file "<<fname_sol<<"\n";
}
}

void matrixTimesVector(std::vector<double>& result,const std::::
    vector<double>& vec,
    const std::vector<std::vector<double>> & A )
{
    int i,j,n = A.size();
    for (i = 0;i < n;++i)
    {
        result[i] = 0.0;
        for (j = 0;j < n;++j)
        {
            if (A[i][j]! = 0.0) result[i]+ = A[i][j]*vec[j];
        }
    }
}


int CG(const std::vector<std::vector<double>>& A, std::vector<double>&
    x, const std::vector<double> & b, int &max_iter, double &tol)
{
    double resid;
    int i, num = A.size();
    std::vector<double> r(num), p(num), z(num), q(num), tmp(num);
    double alpha, beta, rho, rho_1,dpq;

    double normb = 0.0;
    for (i = 0;i < num;++i) normb + = b[i]*b[i];
    normb = sqrt(normb);

    matrixTimesVector(tmp,x,A);
    for (i = 0;i < num;++i)
    {
        r[i] = b[i] − tmp[i];
    }
```

```
if (normb == 0.0) normb = 1;
resid = 0.0;
for (i = 0;i < num;++i) resid += r[i]*r[i];
resid = sqrt(resid);

if ((resid / normb) <= tol)
{
  tol = resid;
  max_iter = 0;
  return 0;
}

for (int jj = 1; jj <= max_iter; jj++)
{
  // preconditioning is disabled
  for (i = 0;i < num;++i) z[i] = r[i];
  rho = 0.0;
  for (i = 0;i < num;++i) rho += r[i]*z[i];

  if (jj == 1)
  {
    for (i = 0;i < num;++i) p[i] = z[i];
  }
  else
  {
    beta = rho / rho_1;
    for (i = 0;i < num;++i) p[i] = z[i]+ beta*p[i];
  }

  matrixTimesVector(q,p,A);
  dpq = 0.0;
  for (i = 0;i < num;++i) dpq += p[i]*q[i];
  alpha = rho / dpq;

  for (i = 0;i < num;++i) x[i]+= alpha*p[i];
  for (i = 0;i < num;++i) r[i]-= alpha*q[i];

  resid = 0.0;
  for (i = 0;i < num;++i) resid +=r[i]*r[i];
  resid = sqrt(resid);
  if ((resid / normb) <= tol)
  {
    tol = resid;
    max_iter = jj;
    return 0;
```

```
      }

   rho_1 = rho;
   }

   tol = resid;
   return 1;
}
```

Chapter 3

General Variational Formulation

The introductory finite element methods discussed in the previous two chapters are written in a more general formulation in this chapter. This general formulation will result in an application of the finite element method to more differential problems. We first provide this formulation and its simple analysis, and then give several concrete examples. More examples are given as exercise problems in this chapter and given in subsequent chapters as well.

3.1. Continuous variational formulation

Let V be a given admissible real function space, and $\mathbb{R}$ be the set of real numbers. Suppose that $a(\cdot, \cdot) : V \times V \to \mathbb{R}$ is a *bilinear form* in the sense

$$a(u, \alpha v + \beta w) = \alpha a(u, v) + \beta a(u, w),$$

$$a(\alpha u + \beta v, w) = \alpha a(u, w) + \beta a(v, w),$$

for $\alpha, \beta \in \mathbb{R}$, $u, v, w \in V$. The bilinear form $a(\cdot, \cdot)$ is *symmetric* if

$$a(u, v) = a(v, u) \quad \forall u, \ v \in V. \tag{3.1}$$

Also, assume that we are given a linear form (functional) $L : V \to \mathbb{R}$:

$$L(\alpha u + \beta v) = \alpha L(u) + \beta L(v), \quad \alpha, \beta \in \mathbb{R}, \quad u, v \in V.$$

We define the functional $F : V \to \mathbb{R}$ by

$$F(v) = \frac{1}{2} a(v, v) - L(v), \quad v \in V.$$

69

We now consider the general (Ritz) *minimization problem*

$$\text{Find } p \in V \text{ such that } F(p) \leq F(v) \quad \forall v \in V, \tag{3.2}$$

and the general (Galerkin) *variational problem*

$$\text{Find } p \in V \text{ such that } a(p,v) = L(v) \quad \forall v \in V. \tag{3.3}$$

Under certain conditions on the bilinear form $a(\cdot,\cdot)$ and the linear form $L(\cdot)$, problems (3.2) and (3.3) have the same unique solution $p \in V$ (Chen, 2005). Concrete examples for the space V, the bilinear form $a(\cdot,\cdot)$, and the linear form $L(\cdot)$ are given in Sec. 3.3.

3.2. The finite element method

Suppose that V_h is a finite element (finite dimensional) subspace of V. The respective discrete counterparts of Eqs. (3.2) and (3.3) are

$$\text{Find } p_h \in V_h \text{ such that } F(p_h) \leq F(v) \quad \forall v \in V_h, \tag{3.4}$$

and

$$\text{Find } p_h \in V_h \text{ such that } a(p_h,v) = L(v) \quad \forall v \in V_h. \tag{3.5}$$

Let $\{\varphi_i\}_{i=1}^M$ be a basis of V_h, where M is the dimension of V_h. We choose $v = \varphi_j$ in Eq. (3.5) to give

$$a(p_h, \varphi_j) = L(\varphi_j), \quad j = 1, 2, \ldots, M. \tag{3.6}$$

Represent the trial function p_h as

$$p_h = \sum_{i=1}^M p_i \varphi_i, \quad p_i \in \mathbb{R},$$

and substitute it into Eq. (3.6) to give

$$\sum_{i=1}^M a(\varphi_i, \varphi_j) p_i = L(\varphi_j), \quad j = 1, 2, \ldots, M.$$

In matrix form, it is

$$\mathbf{Ap} = \mathbf{L}, \tag{3.7}$$

where

$$\mathbf{A} = (a_{i,j}), \quad \mathbf{p} = (p_i), \quad \mathbf{L} = (L_i),$$

$$a_{i,j} = a(\varphi_i, \varphi_j), \quad L_i = L(\varphi_i), \quad i, j = 1, 2, \ldots, M.$$

If the bilinear form $a(\cdot, \cdot)$ is symmetric, so is the system matrix $\mathbf{A}$. Under a V_h-ellipticity or coercivity on this bilinear form (i.e., it is positive in the norm of V_h), it can be shown that this matrix is also *positive definite* (Eq. (1.14)). For more information, refer to Chen (2005).

3.3. Examples

We now state several examples for the choice of the admissible function space V, the bilinear form $a(\cdot, \cdot)$, the linear form $L(\cdot)$, and the finite element space V_h. More examples are found in the next eight chapters.

Example 3.1 Let us return to the one-dimensional problem (1.1). The admissible function space V is

$$V = \Big\{ \text{Functions } v\text{: Each } v \text{ is continuous on } [0,1],$$

$$\text{its first derivative } \frac{dv}{dx} \text{ is piecewise continuous}$$

$$\text{and bounded on } (0,1), \text{ and } v(0) = v(1) = 0 \Big\}.$$

In addition, we define

$$a(v,w) = \int_0^1 \frac{dv}{dx}\frac{dw}{dx}\,dx, \quad L(v) = \int_0^1 fv\,dx, \quad v,\ w \in V,$$

where the function f is given on the unit interval $[0,1]$.

For $h > 0$, let K_h be a partition of $(0,1)$ into subintervals as in Sec. 1.2. Associated with K_h, let $V_h \subset V$ be the space of piecewise linear polynomials introduced in Sec. 1.2. The basis (shape) functions $\{\varphi_i\}_{i=1}^M$ of V_h were constructed in Sec. 1.2 as well.

Example 3.2 We now consider the Poisson equation (2.1) in two dimensions. For a two-dimensional polygon $\Omega \subset \mathbb{R}^2$ (the real plane), the admissible space V is

$$V = \Big\{ \text{Functions } v : v \text{ is a continuous function on } \Omega, \frac{\partial v}{\partial x_1} \text{ and } \frac{\partial v}{\partial x_2}$$

$$\text{are piecewise continuous and bounded on } \Omega, \text{ and } v = 0 \text{ on } \Gamma \Big\}.$$

Furthermore, the bilinear form $a(\cdot, \cdot)$ and the linear form $L(\cdot)$ are defined as

$$a(v,w) = \int_\Omega \nabla v \cdot \nabla w\,d\mathbf{x}, \quad L(v) = \int_\Omega fv\,d\mathbf{x}, \quad v,\ w \in V,$$

where the given function f is defined in Ω.

For $h > 0$, let K_h be a triangulation of Ω into triangles, as defined in Sec. 2.2. For $K \in K_h$, as previously, we define the mesh parameters

$$h_K = \text{diam}(K) = \text{the longest edge of } \bar{K}, \quad h = \max_{K \in K_h} h_K.$$

We also define the quantity

$$\rho_K = \text{the diameter of the largest circle inscribed in } K.$$

We recall that a triangulation is *shape regular* if there is a constant $\beta > 0$, independent of h, such that

$$\frac{h_K}{\rho_K} \leq \beta \quad \forall K \in K_h. \tag{3.8}$$

This condition says that the triangles in K_h are not arbitrarily thin, or equivalently, the angles of the triangles are not arbitrarily small (see Fig. 2.14). The constant β is a measure of the smallest angle over all $K \in K_h$. Condition (3.8) is required to carry out an error analysis for the two-dimensional finite element method (Ciarlet, 1978; Chen, 2005).

The finite element space V_h is defined as in Sec. 2.2; i.e.,

$$V_h = \{\text{Functions } v : v \text{ is a continuous function on } \Omega, v \text{ is linear}$$
$$\text{on each triangle } K \in K_h, \text{ and } v = 0 \text{ on } \Gamma\}.$$

The basis (shape) functions $\{\varphi_i\}_{i=1}^M$ of V_h were constructed in Sec. 2.2.

Example 3.3 The study in the previous example can be extended to a more general second-order problem:

$$-\nabla \cdot (\mathbf{a}\nabla p) + \boldsymbol{\beta} \cdot \nabla p + cp = f \quad \text{in } \Omega,$$
$$p = 0 \quad\quad\quad\quad\quad\quad\quad\quad \text{on } \Gamma, \tag{3.9}$$

where $\mathbf{a} = (a_{ij})$ is a 2×2 matrix, $\boldsymbol{\beta}$ is a constant vector, and c is a bounded, nonnegative function. They, together with function f, are given functions. Assume that the coefficient $\mathbf{a}$ satisfies the positive-definiteness and boundedness condition

$$0 < a_* \leq |\boldsymbol{\eta}|^2 \sum_{i,j=1}^2 a_{ij}(\mathbf{x})\eta_i\eta_j \leq a^* < \infty, \quad \mathbf{x} \in \Omega, \boldsymbol{\eta} \neq \mathbf{0} \in \mathbb{R}^2,$$

where $|\boldsymbol{\eta}|^2 = |\eta_1|^2 + |\eta_2|^2$, $\boldsymbol{\eta} = (\eta_1, \eta_2)$. This problem is an example of a *convection–diffusion–reaction problem*; the first term corresponds to diffusion with the diffusion coefficient $\mathbf{a}$, the second term to convection in the direction $\boldsymbol{\beta}$, and the third term to reaction with the coefficient c. We consider the case where the size of $|\boldsymbol{\beta}|$ is moderate. For convection- or advection-dominated problems, the reader should refer to Chapter 11. Many problems arise in form (3.9), e.g., the problems from multiphase flows in porous media (see Chapter 10).

The admissible function space V is defined as in Example 3.2. Multi- plying both sides of the first equation in Eq. (3.9) by a test function $v \in V$ and integrating the resulting equation on the domain Ω give

$$\int_{\Omega} (-\nabla \cdot (\mathbf{a}\nabla p)v + \boldsymbol{\beta} \cdot \nabla p \, v + cpc)dx = \int_{\Omega} fv \, d\mathbf{x}, \quad v \in V.$$

Applying Green's formula (2.9) to the first term in the left-hand side of this equation and noting that $v|_{\Gamma} = 0$, we see that

$$\int_{\Omega} (\mathbf{a}\nabla p \cdot \nabla v + \boldsymbol{\beta} \cdot \nabla pv + c \, p \, v) \, d\mathbf{x} = \int_{\Omega} fv \, d\mathbf{x}, \quad v \in V. \tag{3.10}$$

Consequently, the bilinear form $a(\cdot, \cdot)$ is

$$a(v, w) = \int_{\Omega} (\mathbf{a}\nabla v \cdot \nabla w + \boldsymbol{\beta} \cdot \nabla v \, w + c \, v \, w) \, d\mathbf{x}, \quad v, \, w \in V.$$

The linear form $L(\cdot)$ remains unchanged as in Example 3.2. Note that $a(\cdot, \cdot)$ is not symmetric due to the presence of the term involving $\boldsymbol{\beta}$ (unless $\boldsymbol{\beta} = \mathbf{0}$). Applying Green's formula (2.9), we see that

$$\int_{\Omega} \boldsymbol{\beta} \cdot \nabla v \, v \, d\mathbf{x} = \int_{\Gamma} \boldsymbol{\beta} \cdot \boldsymbol{\nu} \, v^2 \, d\ell - \int_{\Omega} v \, \boldsymbol{\beta} \cdot \nabla v \, d\mathbf{x},$$

so, by the fact that $v|_{\Gamma} = 0$,

$$\int_{\Omega} \boldsymbol{\beta} \cdot \nabla v \, v \, d\mathbf{x} = 0.$$

Hence, we obtain

$$a(v, v) = \int_{\Omega} (\mathbf{a}\nabla v \cdot \nabla v + c \, v \, v) \, d\mathbf{x}, \quad v \in V,$$

which is a very important property in the analysis of the finite element method for problem (3.9) (Chen, 2005). The construction of the finite el- ement space V_h and its basis functions $\{\varphi_i\}_{i=1}^{M}$ remains the same as in Example 3.2.

While we have considered only the Dirichlet boundary condition in these three examples, boundary conditions of other types can be analyzed as in Sec. 2.3 (see Exercises 3.1–3.4).

Example 3.4 In this example, we consider a *fourth-order problem* in one dimension:

$$\frac{d^4 p}{dx^4} = f(x), \quad 0 < x < 1,$$

$$p(0) = p(1) = \frac{dp}{dx}(0) = \frac{dp}{dx}(1) = 0,$$

(3.11)

where f is a given function on the unit interval $I = (0,1)$. For this example, the admissible space V is

$$V = \left\{ \text{Functions } v : v \text{ and } \frac{dv}{dx} \text{ are continuous on } I, \right.$$

$$\text{its second derivative } \frac{d^2 v}{dx^2} \text{ is piecewise continuous and}$$

$$\left. \text{bounded in } I, \text{ and } v(0) = v(1) = \frac{dv}{dx}(0) = \frac{dv}{dx}(1) = 0 \right\}.$$

Multiplication of both sides of the first equation in Eq. (3.11) by a test function $v \in V$ and integration of the resulting equation on the interval I yield

$$\int_0^1 \frac{d^4 p}{dx^4} v \, dx = \int_0^1 f v \, dx.$$

Applying integration by parts to the left-hand side of this equation and noting that $v(0) = v(1) = 0$, we see that

$$-\int_0^1 \frac{d^3 p}{dx^3} \frac{dv}{dx} \, dx = \int_0^1 f v \, dx.$$

Another application of integration by parts and use of the condition that $dv/dx(0) = dv/dx(1) = 0$ give

$$\int_0^1 \frac{d^2 p}{dx^2} \frac{d^2 v}{dx^2} \, dx = \int_0^1 f v \, dx, \quad v \in V.$$

As a result, we define

$$a(v, w) = \int_0^1 \frac{d^2 v}{dx^2} \frac{d^2 w}{dx^2} \, dx, \quad L(v) = \int_0^1 f v \, dx, \quad v, \ w \in V.$$

As in Sec. 1.2, let $K_h : 0 = x_0 < x_1 < \cdots < x_M < x_{M+1} = 1$ be a partition of the interval $I = (0,1)$ into subintervals $I_i = (x_{i-1}, x_i)$, with length $h_i = x_i - x_{i-1}$, $i = 1, 2, \ldots, M + 1$. Set $h = \max\{h_i : i =$

$1, 2, \ldots, M + 1$. We introduce the finite element space

$$V_h = \Big\{ \text{Functions } v : v \text{ and } \frac{dv}{dx} \text{ are continuous on } I, v \text{ is a polynomial of}$$
$$\text{degree 3 on each subinterval } I_i, i = 1, 2, \ldots, M + 1,$$
$$\text{and } v(0) = v(1) = \frac{dv}{dx}(0) = \frac{dv}{dx}(1) = 0 \Big\}.$$

As parameters, or *degrees of freedom*, to describe the functions $v \in V_h$, we can use the values and first derivatives of v at the nodes $\{x_i\}_{i=0}^{M+1}$ of K_h. Note that because $v \in V_h$ and its first derivative are required to be continuous on I, it has at least four degrees of freedom on each subinterval in K_h. Thus, the degree of v must be greater than or equal to three.

To determine a basis in V_h, we consider a local coordinate transformation from the reference interval $(-1, 1)$ to a typical subinterval (x_{i-1}, x_i) (Fig. 3.1):

$$x = F_i(\xi) = x_{i-1/2} + \frac{h_i}{2}\xi,$$

where $h_i = x_i - x_{i-1}$ and $x_{i-1/2} = (x_i + x_{i-1})/2$. On the interval $(-1, 1)$, the approximate solution p_h is of the form

$$p_h(\xi) = a + b\xi + c\xi^2 + d\xi^3, \quad -1 \le \xi \le 1, \tag{3.12}$$

where the constants a–d are determined by

$$p_h(-1) = p_{i-1}, \quad p_h(1) = p_i,$$
$$\frac{dp_h}{d\xi}(-1) = \frac{dp_{i-1}}{d\xi}, \quad \frac{dp_h}{d\xi}(1) = \frac{dp_i}{d\xi}, \tag{3.13}$$

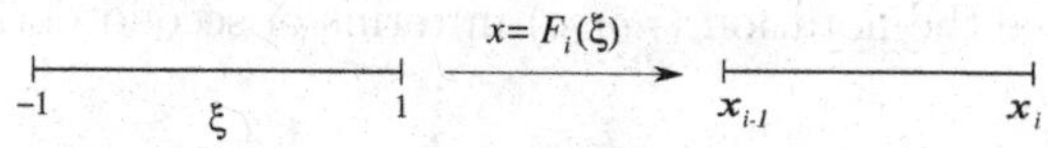

Figure 3.1 A local coordinate transformation.

where $p_{i-1} = p_h(x_{i-1})$, $p_i = p_h(x_i)$, $dp_{i-1}/d\xi = dp_h(x_{i-1})/d\xi$, and $dp_i/d\xi = dp_h(x_i)/d\xi$. That is,

$$a = \frac{p_{i-1} + p_i}{2} - \frac{1}{4}\left(\frac{dp_i}{d\xi} - \frac{dp_{i-1}}{d\xi}\right),$$

$$b = \frac{3}{4}(p_i - p_{i-1}) - \frac{1}{4}\left(\frac{dp_i}{d\xi} + \frac{dp_{i-1}}{d\xi}\right),$$

$$c = \frac{1}{4}\left(\frac{dp_i}{d\xi} - \frac{dp_{i-1}}{d\xi}\right),$$

$$d = -\frac{1}{4}(p_i - p_{i-1}) + \frac{1}{4}\left(\frac{dp_i}{d\xi} + \frac{dp_{i-1}}{d\xi}\right).$$

Substituting these values into Eq. (3.12) gives

$$p_h(\xi) = \frac{1}{4}(\xi - 1)^2(2 + \xi)p_{i-1} + \frac{1}{4}(\xi + 1)^2(2 - \xi)p_i$$
$$+ \frac{1}{4}(\xi - 1)^2(\xi + 1)\frac{dp_{i-1}}{d\xi} + \frac{1}{4}(\xi + 1)^2(\xi - 1)\frac{dp_i}{d\xi}. \tag{3.14}$$

We define the cubic basis (shape) functions on the interval $[-1, 1]$:

$$\psi_{0,1}(\xi) = \frac{1}{4}(\xi - 1)^2(2 + \xi), \quad \psi_{0,2}(\xi) = \frac{1}{4}(\xi + 1)^2(2 - \xi),$$
$$\psi_{1,1}(\xi) = \frac{1}{4}(\xi - 1)^2(\xi + 1), \quad \psi_{1,2}(\xi) = \frac{1}{4}(\xi + 1)^2(\xi - 1). \tag{3.15}$$

They correspond to the degrees of freedom using function values and first derivatives at the end points -1 and 1, respectively. Now, the local basis functions on (x_{i-1}, x_i) are defined through the local coordinate transformation as follows: for $i = 1, 2, \ldots, M+1$,

$$\varphi_{0,i-1}(x) = \psi_{0,1}\big(F_i^{-1}(x)\big), \quad \varphi_{0,i}(x) = \psi_{0,2}\big(F_i^{-1}(x)\big),$$
$$\varphi_{1,i-1}(x) = \psi_{1,1}\big(F_i^{-1}(x)\big), \quad \varphi_{1,i}(x) = \psi_{1,2}\big(F_i^{-1}(x)\big), \quad x \in x_{i-1}, x_i). \tag{3.16}$$

We introduce the notation (*norm*) in terms of second derivatives:

$$|v|_2 = \left(\int_0^1 \left(\frac{d^2v}{dx^2}\right)^2 dx\right)^{1/2}.$$

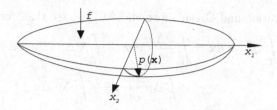

Figure 3.2 An elastic plate.

It can be seen (Chen, 2005) that the difference between the solution p to Eq. (3.11) and the finite element solution p_h using the current finite element space satisfies

$$|p - p_h|_2 \leq Ch,$$

where the constant C depends on the third derivative of p on the interval I.

Example 3.5 In this example, we extend the one-dimensional fourth-order problem to two dimensions, i.e., to the *biharmonic problem*:

$$\Delta^2 p = f \quad \text{in } \Omega,$$
$$p = \frac{\partial p}{\partial \nu} = 0 \quad \text{on } \Gamma, \tag{3.17}$$

where $\Delta^2 = \Delta\Delta$ is the *biharmonic operator* and $\partial p/\partial\nu$ is the normal derivative of the solution p. This problem is a formulation of the Stokes equations in fluid mechanics (see Exercise 9.5). It also models the displacement of a thin elastic plate, *clamped* at its boundary and under a transversal load of intensity f; refer to Fig. 3.2, where the thin elastic plate has a surface given by $\Omega \subset \mathbb{R}^2$ (the real plane). The first boundary condition $p|_\Gamma = 0$ says that the displacement p is held fixed (at the zero height) at the boundary Γ, while the second condition $\partial p/\partial\nu|_\Gamma = 0$ means that the rotation of the plate is also prescribed at Γ. These boundary conditions thus imply that the plate is *clamped*.

We introduce the admissible real function space

$$V = \Big\{ \text{Functions } v : v \text{ and } \nabla v \text{ are continuous on } \Omega,$$

the second partial derivatives of v are piecewise

continuous and bounded in Ω, and $v = \frac{\partial v}{\partial \nu} = 0 \text{ on } \Gamma \Big\}.$

With this definition and Green's formula (2.9), we see that, for $v \in V$

$$\int_\Omega \Delta^2 p \, v \, d\mathbf{x} = \int_\Gamma \frac{\partial \Delta p}{\partial \boldsymbol{\nu}} \, v \, d\ell - \int_\Omega \nabla \Delta p \cdot \nabla v \, d\mathbf{x}$$

$$= -\int_\Gamma \Delta p \, \frac{\partial v}{\partial \boldsymbol{\nu}} \, d\ell + \int_\Omega \Delta p \, \Delta v \, d\mathbf{x}$$

$$= \int_\Omega \Delta p \, \Delta v \, d\mathbf{x},$$

where we used the condition that $v = \dfrac{\partial v}{\partial \boldsymbol{\nu}} = 0$ on Γ. Consequently, we define

$$a(v, w) = \int_\Omega \Delta v \, \Delta w \, d\mathbf{x}, \quad L(v) = \int_\Omega f v \, d\mathbf{x}, \quad v, \ w \in V.$$

Let K_h be a triangulation of Ω into triangles as in Example 3.2. Associated with K_h, we define the finite element space

$$V_h = \Big\{ \text{Functions } v : \ v \text{ and } \nabla v \text{ are continuous on } \Omega, v \text{ is a polynomial}$$

$$\text{of degree 5 on each triangle } K \in K_h$$

$$\text{and } v = \frac{\partial v}{\partial \boldsymbol{\nu}} = 0 \text{ on } \Gamma \Big\}.$$

This space is often known as the *Argyris triangle*. Since the first partial derivatives of the functions in V_h are required to be continuous on $\Omega \subset \mathbb{R}^2$, there are at least six degrees of freedom on each interior edge in K_h. Thus, the polynomial degree must be greater than or equal to five. The degrees of freedom for describing the functions in V_h and the construction of the corresponding basis functions are discussed in Chapter 5. With this V_h, the finite element method (3.5) applies to Eq. (3.17). For the present problem (3.17), we introduce the notation (norm) in terms of second partial derivatives:

$$|v|_2 = \left(\int_\Omega \left[\left(\frac{\partial^2 v}{\partial x_1^2} \right)^2 + \left(\frac{\partial^2 v}{\partial x_1 \partial x_2} \right)^2 + \left(\frac{\partial^2 v}{\partial x_2 \partial x_1} \right)^2 + \left(\frac{\partial^2 v}{\partial x_2^2} \right)^2 \right] d\mathbf{x} \right)^{1/2}.$$

Then, we can show that the error between the solution p to Eq. (3.17) and the finite element solution p_h using the Argyris triangular element satisfies

$$|p - p_h|_2 \leq Ch,$$

where C depends on the third partial derivatives of the solution p on the domain Ω and the constant $\beta > 0$ in condition (3.8) (Ciarlet, 1978; Chen, 2005).

3.4. Exercises

3.1. Consider a one-dimensional problem with an inhomogeneous boundary condition:
$$-\frac{d^2p}{dx^2} = f(x), \quad 0 < x < 1,$$
$$p(0) = p_a, \quad p(1) = p_b,$$
where f is a given real-valued bounded function on $(0,1)$, and p_a and p_b are real numbers. Write this problem in a Galerkin variational formulation, and construct the finite element method using piecewise linear functions. Determine the corresponding linear system of algebraic equations for a uniform partition.

3.2. Consider a one-dimensional problem with a Neumann boundary condition at $x = 1$:
$$-\frac{d^2p}{dx^2} = f(x), \quad 0 < x < 1,$$
$$p(0) = \frac{dp}{dx}(1) = 0.$$
Express this problem in a Galerkin variational formulation, formulate the finite element method using piecewise linear functions, and determine the corresponding linear system of algebraic equations for a uniform partition.

3.3. For the two-dimensional problem with a Neumann boundary condition on the boundary Γ of the domain Ω:
$$-\Delta p + cp = f \quad \text{in } \Omega,$$
$$\frac{\partial p}{\partial \nu} = g \quad \text{on } \Gamma,$$
where $c(\mathbf{x}) \geq c_* > 0$, $\mathbf{x} \in \Omega$, give a Galerkin variational formulation for this problem.

3.4. Give a Galerkin variational formulation for the d-dimensional problem ($d = 2$ or 3):
$$-\nabla \cdot (\mathbf{a}\nabla p) + cp = f \quad \text{in } \Omega,$$
$$p = g_D \quad \text{on } \Gamma_D,$$
$$\gamma p + \mathbf{a}\nabla p \cdot \boldsymbol{\nu} = g_N \quad \text{on } \Gamma_N,$$
where $\mathbf{a}$ is a $d \times d$ matrix ($d = 2$ or 3), c, f, g_D, and g_N are given functions of $\mathbf{x}$, γ is a constant, and the boundary Γ satisfies that $\bar{\Gamma} = \bar{\Gamma}_D \cup \bar{\Gamma}_N$, $\Gamma_D \cap \Gamma_N = \emptyset$ (empty set).

3.5. Consider the Poisson equation (2.1) with an inhomogeneous boundary condition, i.e.,

$$-\Delta p = f \quad \text{in } \Omega,$$
$$p = g \quad \text{on } \Gamma,$$

where Ω is a bounded domain in the plane with boundary Γ, and f and g are given functions. Express this problem in a Galerkin variational formulation, formulate the finite element method using piecewise linear functions, and determine the corresponding linear system of algebraic equations for a uniform partition of $\Omega = (0,1) \times (0,1)$ as given in Fig. 2.6.

3.6. Consider the problem

$$-\Delta p = f \quad \text{in } \Omega,$$
$$p = g_D \quad \text{on } \Gamma_D,$$
$$\frac{\partial p}{\partial \nu} = g_N \quad \text{on } \Gamma_N,$$

where Ω is a bounded domain in the plane with boundary Γ, $\bar{\Gamma} = \bar{\Gamma}_D \cup \bar{\Gamma}_N$, $\Gamma_D \cap \Gamma_N = \emptyset$, and f, g_D, and g_N are given functions. Write down a Galerkin variational formulation for this problem and formulate the finite element method using piecewise linear functions.

Chapter 4

One-Dimensional Elements and Their Properties

In the previous three chapters several different finite element spaces were introduced in the development and application of the finite element method. Each element trial function represents the unknown function as a linear combination of its nodal values. These values are composed of the values of the function at the nodes, together with, in the case of derivative elements the values of its derivatives. In principle, an approximation involving a nonlinear dependency on these nodal values could be utilized, but an improvement in accuracy would be unlikely to outweigh the additional complexity of such a nonlinear approximation.

It was observed in the previous chapters that the trial function must be compatible with the associated element to uniquely determine the coefficients in the trial function representation. More specifically, it was seen that the form of the trial function, the location and number of the element nodes, and the number of unknowns per node cannot be independently specified. Furthermore, the type and order of the differential equation problem and the convergence requirement of the variational method must also be taken into account when choosing proper elements and their trial functions. Subject to these constraints, a variety of acceptable finite elements are developed in this and the next two chapters. This chapter focuses on one-dimensional elements and their properties. After a brief classification of elements in Sec. 4.1, different approaches for deriving basis (shape) functions are discussed in Sec. 4.2. Then, two types of one-dimensional elements, Lagrangian and Hermitian, are addressed in Secs. 4.3 and 4.4, respectively. The chapter is concluded with exercise problems in Sec. 4.5.

4.1. Element classification

The most obvious classification of elements is into one-, two-, and three-dimensional element categories. The treatment in this and subsequent chapters is based on these groupings. The categories may be further subdivided, according to whether the nodal values involve solely the function (*Lagrangian elements*) or include its derivatives as well (*Hermitian elements*).

4.2. Different approaches for deriving basis functions

For a Lagrangian element, there is only one degree of freedom per node, i.e., the value of the function. Thus, the general trial (approximate) function p_h over any element K is represented as

$$p_h(x) = \sum_{i=1}^{s} p_i \varphi_i(x), \quad x \in K, \qquad (4.1)$$

where s is the total number of nodes, φ_i is a basis (shape) function, and p_i is a nodal value on K.

For a Hermitian element, the derivative(s) of the function now appears as variables at the nodes (see Example 3.4). If each of the s nodes of element K has q degrees of freedom, then the trial function p_h over K has the form

$$p_h(x) = \sum_{i=1}^{s} p_{0,i} \varphi_{0,i}(x) + \sum_{i=1}^{s} p_{1,i} \varphi_{1,i}(x) + \cdots + \sum_{i=1}^{s} p_{q-1,i} \varphi_{q-1,i}(x),$$
$$x \in K, \quad (4.2)$$

which is the general representation for a Hermitian element, provided that appropriate basis functions and nodal values are adopted. Note that $s = q = 2$ in Example 3.4.

In the previous three chapters, the basis functions were obtained using the global coordinate x (or coordinates x_1 and x_2) or the local coordinate ξ (or coordinates ξ_1 and ξ_2). It can be shown that basis functions can be always found in these two approaches. Below a third approach, the *interpolation function approach*, is also described.

4.2.1. Global coordinate approach

The global coordinate approach is particularly suitable for simple elements based on polynomials of low order that are complete (i.e., linear, quadratic polynomials, etc.). The derivation of basis functions becomes tedious for more complex elements.

This approach is illustrated by finding the quadratic basis functions φ_{i-1} and φ_i on a subinterval (x_{i-1}, x_i) such that

$$\varphi_{i-1}(x_{i-1}) = 1, \quad \varphi_{i-1}(x_i) = 0, \quad \varphi_i(x_{i-1}) = 0, \quad \varphi_i(x_i) = 1. \qquad (4.3)$$

A quadratic function φ_{i-1} is of the form

$$\varphi_{i-1}(x) = a + bx + cx^2, \quad x \in [x_{i-1}, x_i], \qquad (4.4)$$

where the constants a–c are to be determined. Two equations in Eq. (4.3) can be used in their determination; one more independent equation is thus needed. For this, an additional node, i.e., $x_{i-1/2} = (x_{i-1} + x_i)/2$, is introduced such that

$$\varphi_{i-1}(x_{i-1/2}) = 0. \qquad (4.5)$$

Applying these three conditions, we see that

$$a = \frac{2}{h_i^2}, \quad b = -\frac{1}{h_i}\left(\frac{4x_{i-1/2}}{h_i} + 1\right), \quad c = \frac{x_{i-1/2}}{h_i}\left(\frac{2x_{i-1/2}}{h_i} + 1\right), \qquad (4.6)$$

where $h_i = x_i - x_{i-1}$. Substitution of these values into Eq. (4.4) yields

$$\varphi_{i-1}(x) = \frac{x - x_{i-1/2}}{h_i}\left(\frac{2}{h_i}(x - x_{i-1/2}) - 1\right), \quad x \in [x_{i-1}, x_i].$$

Similarly, using the other two conditions in Eq. (4.3) and the condition that $\varphi_i(x_{i-1/2}) = 0$ for the basis function φ_i, we see that (see Exercise 4.1)

$$\varphi_i(x) = \frac{x - x_{i-1/2}}{h_i}\left(\frac{2}{h_i}(x - x_{i-1/2}) + 1\right), \quad x \in [x_{i-1}, x_i]. \qquad (4.7)$$

An additional node $x_{i-1/2}$ is introduced in the subinterval (x_{i-1}, x_i) so that a corresponding quadratic basis function $\varphi_{i-1/2}$ needs to be defined at this node, which is required to satisfy

$$\varphi_{i-1/2}(x_{i-1}) = 0, \quad \varphi_{i-1/2}(x_{i-1/2}) = 1, \quad \varphi_{i-1/2}(x_i) = 0.$$

An argument similar to that for deriving φ_{i-1} can be used to obtain

$$\varphi_{i-1/2}(x) = 1 - \frac{4}{h_i^2}(x - x_{i-1/2})^2, \quad x \in [x_{i-1}, x_i].$$

Therefore, three nodes and three corresponding basis functions exist on each subinterval (x_{i-1}, x_i), $i = 1, 2, \ldots, M + 1$.

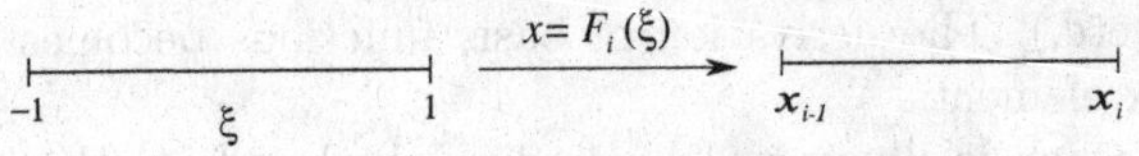

Figure 4.1 A local coordinate transformation.

4.2.2. Local coordinate transformation approach

As in the previous chapters, the quadratic basis functions can be also obtained using a local coordinate transformation (Fig. 4.1):

$$x = F_i(\xi) = x_{i-1/2} + \frac{h_i}{2}\xi, \quad \xi \in [-1, 1], \tag{4.8}$$

where $h_i = x_i - x_{i-1}$ and $[-1, 1]$ is the one-dimensional reference element. A quadratic trial function on $[-1, 1]$ is

$$p_h(\xi) = a + b\xi + c\xi^2, \quad \xi \in [-1, 1]. \tag{4.9}$$

The constants a–c are determined by

$$p_h(-1) = p_{i-1}, \quad p_h(0) = p_{i-1/2}, \quad p_h(1) = p_i.$$

Simple algebraic computations yield

$$a = p_{i-1/2}, \quad b = \frac{p_i - p_{i-1}}{2}, \quad c = \frac{p_i + p_{i-1}}{2} - p_{i-1/2}. \tag{4.10}$$

Substitution of Eq. (4.10) into Eq. (4.9) gives

$$p_h(\xi) = \frac{1}{2}\xi(\xi - 1)p_{i-1} + (1 - \xi^2)p_{i-1/2} + \frac{1}{2}\xi(\xi + 1)p_i, \quad \xi \in [-1, 1]. \tag{4.11}$$

Consequently, we define the basis (shape) functions on the reference element $[-1, 1]$:

$$\psi_1(\xi) = \frac{1}{2}\xi(\xi - 1), \quad \psi_2(\xi) = 1 - \xi^2, \quad \psi_3(\xi) = \frac{1}{2}\xi(\xi + 1), \quad \xi \in [-1, 1]. \tag{4.12}$$

Accordingly, the local basis functions on (x_{i-1}, x_i) are defined as

$$\varphi_{i-1}(x) = \psi_1\big(F_i^{-1}(x)\big), \quad \varphi_{i-1/2}(x) = \psi_2\big(F_i^{-1}(x)\big),$$
$$\varphi_i(x) = \psi_3\big(F_i^{-1}(x)\big), \quad x \in (x_{i-1}, x_i). \tag{4.13}$$

It can be seen that this approach will produce the same basis functions φ_{i-1}, $\varphi_{i-1/2}$, and φ_i as those obtained in the previous section (see Exercise 4.2).

When evaluating an element contribution to the global system matrix, derivatives such as dp/dx and product terms such as $x\,dp/dx$ are often present in an integral. With a local coordinate transformation, element contributions can be generally expressed as a product of nodal values and integrals of the form

$$\int_{x_{i-1}}^{x_i} L_1^m(x) L_2^n(x) \ dx,$$

where m and n are integers, and

$$L_1(x) = \frac{x - x_i}{x_{i-1} - x_i}, \quad L_2(x) = \frac{x - x_{i-1}}{x_i - x_{i-1}}, \quad x \in (x_{i-1}, x_i).$$

This integration can be analytically performed using the formula

$$\int_{x_{i-1}}^{x_i} L_1^m(x) L_2^n(x) \ dx = \frac{m!\,n!}{(m+n+1)!} h_i, \qquad (4.14)$$

where $h_i = x_i - x_{i-1}$.

4.2.3. Interpolation function approach

The interpolation function approach is particularly suitable for polynomials of higher degree. In the subsequent two sections, we obtain both Lagrangian and Hermitian elements of any order.

4.3. Lagrangian elements

In one dimension, a complete polynomial v of degree r is of the form

$$v(x) = v_0 + v_1 x + v_2 x^2 + \cdots + v_r x^r, \quad r = 0, 1, 2, \ldots,$$

where x is the variable and the coefficients $v_0, v_1, \ldots, v_r$ are real numbers. The choices of $r = 0, 1, 2$, and 3 correspond to the constant, linear, quadratic, and cubic functions, respectively.

On an interval K we consider the approximation p_h of the function p by a polynomial of degree r, where the values of p are given as $p_1, p_2, \ldots, p_{r+1}$ at the $r+1$ distinct points $x_1, x_2, \ldots, x_{r+1}$ on K. From numerical analysis we see that the approximation p_h can be represented as the rth order polynomial:

$$p_h(x) = \sum_{i=1}^{r+1} p_i \, \varphi_i(x), \quad x \in K, \qquad (4.15)$$

where φ_i is the Lagrange polynomial

$$\varphi_i(x) = \prod_{j=1, j \neq i}^{r+1} \frac{x - x_j}{x_i - x_j}, \quad i = 1, 2, \ldots, r+1. \qquad (4.16)$$

It is worth noting that the discrete points $x_1, x_2, \ldots, x_{r+1}$ on K need not be equally spaced, although this is often convenient. It follows from the property of Lagrange polynomials that φ_i has the value of unity at node x_i and zero at all other nodes, as required by a basis function. The Lagrange polynomials φ_i are the shape functions at these nodes. The cases $r = 1, 2$ were previously considered, while expression (4.16) applies to any order $r \geq 1$.

4.4. Hermitian elements

The basis functions for Hermitian elements can be obtained in a similar fashion, but using Hermitian polynomials in place of Lagrange polynomials. In addition to the function values at nodes, the nodal values now also contain the derivative values. As an illustration, we consider an s-node one-dimensional element K where the nodes are not necessarily equally spaced. Assume that each node has two degrees of freedom, the function value p and its first derivative dp/dx. As a result, the trial function p_h on element K is represented as

$$p_h(x) = \sum_{i=1}^{s} \left(p_i \varphi_{0,i}(x) + \frac{dp_i}{dx} \varphi_{1,i}(x) \right), \quad x \in K, \qquad (4.17)$$

where the first subscript in each basis function $\varphi_{k,i}$ denotes the order of differentiation of the underlying nodal variable and the second subscript identifies the node. These functions are given by the Hermitian polynomials

$$\varphi_{0,i}(x) = \prod_{j=1, j \neq i}^{s} \frac{(x - x_j)^2}{(x_i - x_j)^2} \left(1 + 2 \sum_{j=1, j \neq i}^{s} \frac{x_i - x}{x_i - x_j} \right),$$

$$\varphi_{1,i}(x) = \prod_{j=1, j \neq i}^{s} \frac{(x - x_j)^2}{(x_i - x_j)^2} (x - x_i), \quad x \in K, i = 1, 2, \ldots, s. \qquad (4.18)$$

The case $s = 2$ was described in Example 3.4. The definition can be extended to including, in addition to the function and its derivative, its higher order derivatives as well.

4.5. Exercises

4.1. Carry out the derivation of Eq. (4.7) in detail.

4.2. Show that the global coordinate and local coordinate transformation approaches described in Secs. 4.2.1 and 4.2.2 produce the same basis functions φ_{i-1}, $\varphi_{i-1/2}$, and φ_i.

4.3. Set $I = (0, 1)$ and

$$P_r(I) = \{v \colon v \text{ is a polynomial of degree at most } r \text{ on } I\},$$

where $r = 0, 1, 2, \ldots$. If v is zero at $r+1$ distinct points on the interval I, then $v \equiv 0$. Hint: If $v \in P_r(I)$ is zero at some point $x_0 \in I$, then $v(x) = (x - x_0)w(x)$, where $w \in P_{r-1}(I)$.

4.4. Find the Hermitian basis (shape) functions on an s-node one-dimensional element K for $s = 3$.

Chapter 5

Two-Dimensional Elements and Their Properties

The one-dimensional finite elements introduced in Chapter 4 are now generalized to two dimensions. The most straightforward generalization is the two-dimensional rectangular elements that use products of one-dimensional basis (shape) functions. Quadrilateral and triangular elements are also considered. Rectangular elements on their own are not well suited to irregular two-dimensional domains, but can be combined with the more adaptive triangular elements. Quadrilateral elements are better adapted for irregular domains, but have not achieved as wide applications as triangular elements. Rectangular and quadrilateral elements are introduced in Sec. 5.1, while triangular elements are considered in Sec. 5.2. Exercise problems are given in Sec. 5.3.

5.1. Rectangular and quadrilateral elements

Even though rectangular elements may not be as flexible as triangular elements in curve fitting, it should be remembered that for certain applications they may be used to advantage. We consider the case where Ω is a rectangular domain and K_h is a partition of Ω into non overlapping rectangles such that the horizontal and vertical edges of these rectangles are parallel to the x_1- and x_2-coordinate axes, respectively. We also require that no vertex of any rectangle lies in the interior of an edge of another rectangle.

The tensor products of one-dimensional polynomials of degree r in the x_1 and x_2 directions lead to the two-dimensional polynomials:

$$v(x_1, x_2) = \sum_{j=0}^{r}\sum_{i=0}^{r} v_{ij} x_1^i x_2^j, \quad r \geq 0, \tag{5.1}$$

where x_1 and x_2 are variables and the coefficients v_{ij} are real numbers. The total number of terms in such a function v is $(r+1)^2$. For $r = 1$, the function v is a *bilinear polynomial*:

$$v(x_1, x_2) = v_{00} + v_{10}x_1 + v_{01}x_2 + v_{11}x_1x_2.$$

For $r = 2$, it becomes a *biquadratic function*

$$\begin{aligned}
v(x_1, x_2) = {} & v_{00} + v_{10}x_1 + v_{01}x_2 + v_{11}x_1x_2 + v_{20}x_1^2 \\
& + v_{21}x_1^2 x_2 + v_{02}x_2^2 + v_{12}x_1x_2^2 + v_{22}x_1^2 x_2^2.
\end{aligned}$$

5.1.1. Lagrangian rectangular elements

Consider a two-dimensional element K (Fig. 5.1) where there are $m + 1$ and $n + 1$ distinct nodes in the x_1- and x_2 directions, respectively. With each node with coordinates $(x_{i,1}, x_{j,2})$, the associated basis (shape) function is defined as the tensor product of two Lagrange polynomials in their respective directions:

$$\begin{aligned}
\varphi_{i,j}(x_1, x_2) = \varphi_i(x_1)\varphi_j(x_2), \quad & i = 1, 2, \ldots, m+1, \\
j = 1, 2, \ldots, n+1, \quad & (x_1, x_2) \in K,
\end{aligned} \tag{5.2}$$

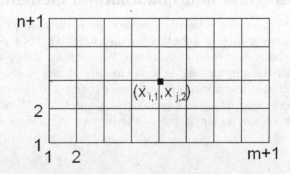

Figure 5.1 A rectangle K.

where the one-dimensional Lagrange polynomials φ_i and φ_j are

$$\varphi_i(x_1) = \prod_{k=1, k \neq i}^{m+1} \frac{x_1 - x_{k,1}}{x_{i,1} - x_{k,1}}, \quad i = 1, 2, \ldots, m+1,$$

and

$$\varphi_j(x_2) = \prod_{k=1, k \neq j}^{n+1} \frac{x_2 - x_{k,2}}{x_{j,2} - x_{k,2}}, \quad j = 1, 2, \ldots, n+1.$$

It follows from the property of Lagrange polynomials that $\varphi_{i,j}$ has the value of unity at node $(x_{i,1}, x_{j,2})$ and zero at all other nodes, as required by a basis function. This particular family of elements is termed the Lagrangian rectangular family due to its direct derivation from the Lagrange polynomials. The trial function p_h on the element K can be represented as

$$p_h(x_1, x_2) = \sum_{i=1}^{m+1} \sum_{j=1}^{n+1} p_{i,j} \varphi_{i,j}(x_1, x_2), \quad (x_1, x_2) \in K, \tag{5.3}$$

where $p_{i,j} = p_h(x_{i,1}, x_{j,2})$.

The commonly used rectangular elements are the case where $m = n$. In addition, the $m + 1$ nodes on the horizontal and vertical edges of each rectangle are equally spaced. In this case, the elements can readily possess the *geometric conformity* where no vertex of any rectangle lies in the interior of an edge of another rectangle. Furthermore, the Lagrangian rectangular elements possess continuity of the trial function across interelement boundaries and thus throughout the entire domain. This can be seen from the fact that the number of nodes on any side of a rectangle is the same as the number of coefficients in the polynomial along that side, hence enabling these coefficients to be determined uniquely. The polynomial on the common side of two adjacent elements is uniquely pegged on the same nodal values; so, there must be continuity of the trial function across the interelement boundaries. This family of continuous functions is *admissible* for the discretization of second-order partial differential equations using the finite element method illustrated in Chapters 2 and 3.

As an example, we consider the case $m = n = 1$, i.e., the bilinear element. In this case, the four basis functions are (Fig. 5.2)

$$\varphi_{i,j}(x_1, x_2) = \frac{x_1 - x_{i+1,1}}{x_{i,1} - x_{i+1,1}} \frac{x_2 - x_{j+1,2}}{x_{j,2} - x_{j+1,2}},$$

$$\varphi_{i+1,j}(x_1, x_2) = \frac{x_1 - x_{i,1}}{x_{i+1,1} - x_{i,1}} \frac{x_2 - x_{j+1,2}}{x_{j,2} - x_{j+1,2}},$$

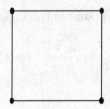

Figure 5.2   Bilinear element.

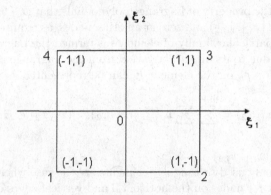

Figure 5.3   The reference element.

$$\varphi_{i+1,j+1}(x_1, x_2) = \frac{x_1 - x_{i,1}}{x_{i+1,1} - x_{i,1}} \frac{x_2 - x_{j,2}}{x_{j+1,2} - x_{j,2}},$$

$$\varphi_{i,j+1}(x_1, x_2) = \frac{x_1 - x_{i+1,1}}{x_{i,1} - x_{i+1,1}} \frac{x_2 - x_{j,2}}{x_{j+1,2} - x_{j,2}}.$$

In terms of the local coordinates (ξ_1, ξ_2) (Fig. 5.3 for the reference rectangle $[-1, 1] \times [-1, 1]$ with node identifies 1, 2, 3, and 4), these basis functions become

$$\psi_1(\xi_1, \xi_2) = \frac{1}{4}(\xi_1 - 1)(\xi_2 - 1), \quad \psi_2(\xi_1, \xi_2) = -\frac{1}{4}(\xi_1 + 1)(\xi_2 - 1),$$

$$\psi_3(\xi_1, \xi_2) = \frac{1}{4}(\xi_1 + 1)(\xi_2 + 1), \quad \psi_4(\xi_1, \xi_2) = -\frac{1}{4}(\xi_1 - 1)(\xi_2 + 1),$$

where the basis functions on the reference element $\hat{K}$ are indicated using the notation ψ_i, $i = 1, 2, 3, 4$, following the notation in Chapters 1–4.

5.1.2. Serendipity elements

Except for the bilinear element, the Lagrangian rectangular elements suffer from the disadvantage of having *internal nodes*. These internal nodes can be eliminated to obtain the *serendipity elements*. The first three members of this family of elements are given in Fig. 5.4, where the description of these elements as linear, quadratic, and cubic refers to the variations of the function in the ξ_1 direction at constant ξ_2 or in the ξ_2 direction at constant ξ_1. The trial functions from these three elements are incomplete quadratic, cubic, and quadric polynomials, respectively, in ξ_1 and ξ_2.

The basis functions for the first three serendipity elements were originally found by inspection and are shown in Table 5.1 in terms of the local coordinates (ξ_1, ξ_2). They can be also obtained using the following

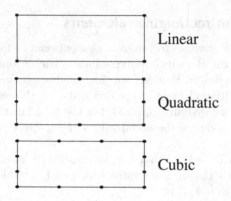

Linear

Quadratic

Cubic

Figure 5.4 The first three serendipity elements.

Table 5.1 Basis functions for serendipity elements.

Type	Nodes	Basis functions
Linear	Vertex $(\xi_{i,1}, \xi_{i,2})$	$\psi_i = \frac{1}{4}(1 + \xi_1\xi_{i,1})(1 + \xi_2\xi_{i,2})$
Quadratic	Vertex $(\xi_{i,1}, \xi_{i,2})$	$\psi_i = \frac{1}{4}(1 + \xi_1\xi_{i,1})(1 + \xi_2\xi_{i,2})(\xi_1\xi_{i,1} + \xi_2\xi_{i,2} - 1)$
	On edge $\xi_{i,1} = 0$	$\psi_i = \frac{1}{2}(1 - \xi_1^2)(1 + \xi_2\xi_{i,2})$
	On edge $\xi_{i,2} = 0$	$\psi_i = \frac{1}{2}(1 + \xi_1\xi_{i,1})(1 - \xi_2^2)$
Cubic	Vertex $(\xi_{i,1}, \xi_{i,2})$	$\psi_i = \frac{1}{32}(1 + \xi_1\xi_{i,1})(1 + \xi_2\xi_{i,2})[9(\xi_1^2 + \xi_2^2) - 10])$
	On edge $\xi_{i,1} = \pm 1$	
	$\quad \xi_{i,2} = \pm\frac{1}{3}$	$\psi_i = \frac{9}{32}(1 + \xi_1\xi_{i,1})(1 - \xi_2^2)(1 + 9\xi_2\xi_{i,2})$
	On edge $\xi_{i,1} = \pm\frac{1}{3}$	
	$\quad \xi_{i,2} = \pm 1$	$\psi_i = \frac{9}{32}(1 + \xi_2\xi_{i,2})(1 - \xi_1^2)(1 + 9\xi_1\xi_{i,1})$

incomplete polynomials:

$$v_{00} + v_{10}\xi_1 + v_{01}\xi_2 + v_{11}\xi_1\xi_2 \quad \text{(linear)},$$

$$v_{00} + v_{10}\xi_1 + v_{01}\xi_2 + v_{11}\xi_1\xi_2 + v_{20}\xi_1^2 + v_{21}\xi_1^2\xi_2$$
$$+ v_{02}\xi_2^2 + v_{12}\xi_1\xi_2^2 \quad \text{(quadratic)}, \tag{5.4}$$

$$v_{00} + v_{10}\xi_1 + v_{01}\xi_2 + v_{11}\xi_1\xi_2 + v_{20}\xi_1^2 + v_{21}\xi_1^2\xi_2$$
$$+ v_{02}\xi_2^2 + v_{12}\xi_1\xi_2^2 + v_{30}\xi_1^3 + v_{31}\xi_1^3\xi_2 + v_{13}\xi_1\xi_2^3 + v_{03}\xi_2^3 \quad \text{(cubic)},$$

where the coefficients v_{ij} use the notation in definition (5.1). That is, symmetric pairs of terms have been omitted from the corresponding complete polynomials to retain geometric conformity. Along element sides, the trial function of a serendipity element is a complete polynomial, and thus there is interelement continuity of this trial function.

5.1.3. Hermitian rectangular elements

Basis functions for Hermitian rectangular elements can be found by forming tensor products of the Hermitian polynomials in each coordinate direction in a manner analogous to that for the Lagrangian rectangular elements. Consider the Hermitian element of the first order for the reference element in the $\xi_1\xi_2$ coordinate system (Fig. 5.3). For the trial function, the degrees of freedom at each node are the nodal values of p_h, $\partial p_h/\partial\xi_1$, $\partial p_h/\partial\xi_2$, and $\partial^2 p_h/\partial\xi_1\partial\xi_2$.

It follows from Eq. (4.17) that for the one-dimensional two-node Hermitian element with the first derivatives as nodal variables, the element approximation p_h on $[-1, 1]$ is

$$p_h(\xi) = \sum_{i=1}^{2} \left(p_i\psi_{0,i}(\xi) + \frac{dp_i}{d\xi}\psi_{1,i}(\xi) \right), \quad \xi \in [-1, 1], \tag{5.5}$$

where (see Eq. (3.15))

$$\psi_{0,1}(\xi) = \frac{1}{4}(\xi - 1)^2(2 + \xi), \quad \psi_{0,2}(\xi) = \frac{1}{4}(\xi + 1)^2(2 - \xi),$$

$$\psi_{1,1}(\xi) = \frac{1}{4}(\xi - 1)^2(\xi + 1), \quad \psi_{1,2}(\xi) = \frac{1}{4}(\xi + 1)^2(\xi - 1). \tag{5.6}$$

For the corresponding Hermitian rectangular element on $[-1, 1] \times [-1, 1]$, the approximation p_h has the form (Fig. 5.3)

$$p_h(\xi_1, \xi_2) = \sum_{i=1}^{4} \left(p_i\psi_{00,i} + \frac{\partial p_i}{\partial\xi_1}\psi_{10,i} + \frac{\partial p_i}{\partial\xi_2}\psi_{01,i} + \frac{\partial^2 p_i}{\partial\xi_1\partial\xi_2}\psi_{11,i} \right), \tag{5.7}$$

where

$$\psi_{00,1}(\xi_1, \xi_2) = \psi_{0,1}(\xi_1)\psi_{0,1}(\xi_2), \quad \psi_{10,1}(\xi_1, \xi_2) = \psi_{1,1}(\xi_1)\psi_{0,1}(\xi_2),$$
$$\psi_{01,1}(\xi_1, \xi_2) = \psi_{0,1}(\xi_1)\psi_{1,1}(\xi_2), \quad \psi_{11,1}(\xi_1, \xi_2) = \psi_{1,1}(\xi_1)\psi_{1,1}(\xi_2),$$

$$\psi_{00,2}(\xi_1, \xi_2) = \psi_{0,2}(\xi_1)\psi_{0,1}(\xi_2), \quad \psi_{10,2}(\xi_1, \xi_2) = \psi_{1,2}(\xi_1)\psi_{0,1}(\xi_2),$$
$$\psi_{01,2}(\xi_1, \xi_2) = \psi_{0,2}(\xi_1)\psi_{1,1}(\xi_2), \quad \psi_{11,2}(\xi_1, \xi_2) = \psi_{1,2}(\xi_1)\psi_{1,1}(\xi_2),$$

$$\psi_{00,3}(\xi_1, \xi_2) = \psi_{0,2}(\xi_1)\psi_{0,2}(\xi_2), \quad \psi_{10,3}(\xi_1, \xi_2) = \psi_{1,2}(\xi_1)\psi_{0,2}(\xi_2),$$
$$\psi_{01,3}(\xi_1, \xi_2) = \psi_{0,2}(\xi_1)\psi_{1,2}(\xi_2), \quad \psi_{11,3}(\xi_1, \xi_2) = \psi_{1,2}(\xi_1)\psi_{1,2}(\xi_2),$$

$$\psi_{00,4}(\xi_1, \xi_2) = \psi_{0,1}(\xi_1)\psi_{0,2}(\xi_2), \quad \psi_{10,4}(\xi_1, \xi_2) = \psi_{1,1}(\xi_1)\psi_{0,2}(\xi_2),$$
$$\psi_{01,4}(\xi_1, \xi_2) = \psi_{0,1}(\xi_1)\psi_{1,2}(\xi_2), \quad \psi_{11,4}(\xi_1, \xi_2) = \psi_{1,1}(\xi_1)\psi_{1,2}(\xi_2).$$

$$(5.8)$$

Using Eq. (5.6), these basis functions at node 1 become (see Exercise 5.2)

$$\psi_{00,1}(\xi_1, \xi_2) = \frac{1}{16}(\xi_1 - 1)^2(2 + \xi_1)(\xi_2 - 1)^2(2 + \xi_2),$$

$$\psi_{10,1}(\xi_1, \xi_2) = \frac{1}{16}(\xi_1 - 1)^2(\xi_1 + 1)(\xi_2 - 1)^2(2 + \xi_2),$$

$$\psi_{01,1}(\xi_1, \xi_2) = \frac{1}{16}(\xi_1 - 1)^2(2 + \xi_1)(\xi_2 - 1)^2(\xi_2 + 1),$$

$$\psi_{11,1}(\xi_1, \xi_2) = \frac{1}{16}(\xi_1 - 1)^2(\xi_1 + 1)(\xi_2 - 1)^2(\xi_2 + 1).$$

The trial function p_h is an incomplete sixth-order polynomial in terms of ξ_1 and ξ_2. It can be shown that it possesses continuity in both the function and its first partial derivatives. This rectangular element is referred to as the *Bogner–Fox–Schmit rectangle*.

A condensation method can be used to eliminate some degrees of freedom in this Hermitian rectangular element. The condensation rule is that the reduced polynomial is of degree three in one variable when the other variable is held to be constant. It can be seen that to eliminate the parameter $\psi_{11,i}$, the terms involving $\xi_1^2\xi_2^2$, $\xi_1^3\xi_2^2$, $\xi_1^2\xi_2^3$, and $\xi_1^3\xi_2^3$ must be omitted from the basis functions $\psi_{00,i}$, $\psi_{10,i}$, and $\psi_{01,i}$, $i = 1, 2, 3, 4$. The resulting Hermitian element is termed the constrained bi-cubic *Adini element*.

Higher order trial functions, based on Hermitian rectangular elements with more than two nodes on each side and using partial derivatives of order greater than two, can be developed, but become rather complex and are seldom used.

5.1.4. Quadrilateral elements

Approach A. One approach to handling quadrilateral elements is to form such elements from triangles and use certain averaging based on these triangles. Figure 5.5 shows how averaging can be used with simple linear triangular elements to obtain a quadrilateral element. First, the quadrilateral is divided by one diagonal, and then by the other diagonal. The element matrix on each divided quadrilateral is obtained from the linear representations with the component triangles, and then the two element matrices are averaged to find the final quadrilateral matrix.

Higher order triangles can be used to form higher order quadrilaterals. Figure 5.6 illustrates the quadrilateral element developed by Fraeijs de Veubeke (1968) from four complete cubic polynomial triangles. There exist 16 degrees of freedom for this element, three at each vertex using p, $\partial p/\partial x_1$, and $\partial p/\partial x_2$, and one at each edge midpoint using the outward normal derivative $\partial p/\partial \nu$.

The Clough and Felippa (1968) quadrilateral element shown in Fig. 5.7 is also obtained from four triangles, each of which is constructed from three triangular subregions. There are 12 degrees of freedom for this element, three at each vertex node.

Approach B. An alternative approach to handling quadrilateral elements is to use a local coordinate transformation to transfer a quadrilateral onto a rectangle (or a square). The relationship between the reference square $[-1, 1] \times [-1, 1]$ in the $\xi_1\xi_2$ coordinate system to a quadrilateral in the x_1x_2

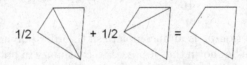

$$1/2 \qquad + 1/2 \qquad =$$

Figure 5.5 Construction of quadrilateral elements by triangles.

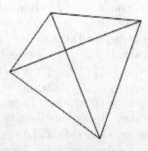

Figure 5.6 The de Veubeke quadrilateral element.

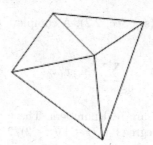

Figure 5.7 The Clough and Felippa quadrilateral element.

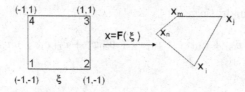

Figure 5.8 A local coordinate transformation from the reference square.

plane is (Fig. 5.8)

$$\mathbf{x} = \mathbf{F}(\boldsymbol{\xi}) = \psi_1(\boldsymbol{\xi})\mathbf{x}_i + \psi_2(\boldsymbol{\xi})\mathbf{x}_j + \psi_3(\boldsymbol{\xi})\mathbf{x}_m + \psi_4(\boldsymbol{\xi})\mathbf{x}_n,$$
$$\boldsymbol{\xi} = (\xi_1, \xi_2) \in [-1, 1] \times [-1, 1], \tag{5.9}$$

where $\mathbf{x}_i$, $\mathbf{x}_j$, $\mathbf{x}_m$, and $\mathbf{x}_n$ are the coordinates of the four vertices of the quadrilateral and

$$\psi_1(\boldsymbol{\xi}) = \frac{1}{4}(1 - \xi_1)(1 - \xi_2), \quad \psi_2(\boldsymbol{\xi}) = \frac{1}{4}(1 + \xi_1)(1 - \xi_2),$$

$$\psi_3(\boldsymbol{\xi}) = \frac{1}{4}(1 + \xi_1)(1 + \xi_2), \quad \psi_4(\boldsymbol{\xi}) = \frac{1}{4}(1 - \xi_1)(1 + \xi_2).$$

The trial function on the reference square $[-1, 1] \times [-1, 1]$ in the $\xi_1 \xi_2$ system can be transformed to that on any quadrilateral in the $x_1 x_2$ system.

5.2. Triangular elements

Triangles are the simplest polygonal figures into which any two-dimensional polygonal domain can be partitioned, and this flexibility in part accounts for the popularity of triangular elements. We now consider the case where a polygonal domain Ω in the plane is divided into non overlapping (open) triangles K such that no vertex of one triangle lies in the interior of an edge of another triangle (see Sec. 2.2). A complete polynomial of degree r in variables x_1 and x_2 is

$$v(\mathbf{x}) = \sum_{0 \le i+j \le r} v_{ij} x_1^i x_2^j, \quad r \ge 0,$$

where the coefficients v_{ij} are real numbers. The number of terms in each complete polynomial of degree r is $(r+1)(r+2)/2$. For $r = 1$, v is a linear function

$$v(\mathbf{x}) = v_{00} + v_{10} x_1 + v_{01} x_2.$$

For $r = 2$, it is quadratic:

$$v(\mathbf{x}) = v_{00} + v_{10} x_1 + v_{01} x_2 + v_{20} x_1^2 + v_{11} x_1 x_2 + v_{02} x_2^2.$$

5.2.1. Natural coordinates in two dimensions

Triangular elements can be conveniently defined using *natural coordinates* (or *area coordinates*). These area coordinates are similar to the length coordinate described in Chapter 4 in one dimension. For any point P on a triangle K with the vertex coordinates $(x_{1,1}, x_{1,2})$, $(x_{2,1}, x_{2,2})$, and $(x_{3,1}, x_{3,2})$ (Fig. 5.9), such coordinates are defined by taking the area subtended by P and its appropriate base (or datum) and dividing this area by the whole area of the triangle K. Hence, as demonstrated in Fig. 5.9, the area coordinates of point P are

$$\lambda_1 = \frac{\Delta_1}{\Delta}, \quad \lambda_2 = \frac{\Delta_2}{\Delta}, \quad \lambda_3 = \frac{\Delta_3}{\Delta}, \quad \quad (5.10)$$

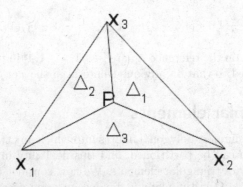

Figure 5.9   Natural coordinates in two dimensions.

where Δ_i is the area of the triangle formed by this point and the λ_i-datum (base), $i = 1, 2, 3$, and Δ is the area of the whole triangle K. These coordinates are also called the *barycentric coordinates* of point P. It is obvious that

$$0 \le \lambda_1, \lambda_2, \lambda_3 \le 1, \quad \lambda_1 + \lambda_2 + \lambda_3 = 1. \tag{5.11}$$

The triangle bases correspond to $\lambda_1 = 0$, $\lambda_2 = 0$, and $\lambda_3 = 0$, and the opposite vertices to $\lambda_1 = 1$, $\lambda_2 = 1$, and $\lambda_3 = 1$, respectively. The midpoints of the three edges opposite to the corresponding vertices have the area coordinates

$$\left(0, \frac{1}{2}, \frac{1}{2}\right), \quad \left(\frac{1}{2}, 0, \frac{1}{2}\right), \quad \left(\frac{1}{2}, \frac{1}{2}, 0\right). \tag{5.12}$$

In addition, the centroid (*center of gravity*) of triangle K has the area coordinates

$$\left(\frac{1}{3}, \frac{1}{3}, \frac{1}{3}\right). \tag{5.13}$$

It follows from simple geometry that the relationship between the global coordinates (x_1, x_2) and the area coordinates is

$$\begin{pmatrix} x_1 \\ x_2 \\ 1 \end{pmatrix} = \begin{pmatrix} x_{1,1} & x_{2,1} & x_{3,1} \\ x_{1,2} & x_{2,2} & x_{3,2} \\ 1 & 1 & 1 \end{pmatrix} \begin{pmatrix} \lambda_1 \\ \lambda_2 \\ \lambda_3 \end{pmatrix}. \tag{5.14}$$

That is,

$$\begin{aligned} x_1 &= \sum_{i=1}^{3} x_{i,1} \lambda_i, \\ x_2 &= \sum_{i=1}^{3} x_{i,2} \lambda_i, \\ \lambda_1 + \lambda_2 + \lambda_3 &= 1. \end{aligned} \tag{5.15}$$

The area coordinates λ_1, λ_2, and λ_3 are analogous to the basis functions in that they have the values of 1 and 0 at the three vertices (nodes). In fact, they are the local linear basis functions of triangle K.

We define

$$a_i = x_{j,1} x_{m,2} - x_{m,1} x_{j,2}, \quad b_i = x_{j,2} - x_{m,2}, \quad c_i = -(x_{j,1} - x_{m,1}), \tag{5.16}$$

and a_j, a_m, b_j, b_m, c_j, and c_m can be obtained by cyclic permutation of the indices $\{i, j, m\} = \{1, 2, 3\}$. Then, system (5.14) can be solved for λ_1,

λ_2, and λ_3 in terms of x_1 and x_2:

$$\begin{pmatrix} \lambda_1 \\ \lambda_2 \\ \lambda_3 \end{pmatrix} = \frac{1}{2\Delta} \begin{pmatrix} a_1 & b_1 & c_1 \\ a_2 & b_2 & c_2 \\ a_3 & b_3 & c_3 \end{pmatrix} \begin{pmatrix} 1 \\ x_1 \\ x_2 \end{pmatrix}. \tag{5.17}$$

An element matrix originally obtained through reference to the global system can be transferred to the natural system of area coordinates by transformation (5.14). Generally, the element contributions then involve integrals $\int_K \lambda_1^l \lambda_2^m \lambda_3^n \, dx_1 dx_2$, which can be analytically computed as follows:

$$\int_K \lambda_1^l \lambda_2^m \lambda_3^n \, dx_1 dx_2 = 2\Delta \frac{l!m!n!}{(l+m+n+2)!}, \tag{5.18}$$

where l, m, and n are nonnegative integers. Note that natural coordinates can also be developed for quadrilaterals.

5.2.2. Lagrangian triangular elements

Lagrangian triangular elements can be simply constructed through selecting a sufficient number of nodes to allow a unique determination of the coefficients in the chosen polynomial trial function. A complete polynomial of order r contains $T_r = (r+1)(r+2)/2$ coefficients, and a triangular element must accordingly contain the same number T_r of nodes. While other possibilities exist for the location of these nodes, the next choice leads to relatively simple basis functions.

On each edge of a triangle K, the trial function is a polynomial of order r, so $r+1$ nodes are needed on this edge, which are composed of two vertices and $r-1$ internal edge nodes. Thus, there are $E_r = 3(r-1)+3 = 3r$ nodes on three edges of triangle K. The internal nodes within K are

$$I_r = \frac{1}{2}(r+1)(r+2) - 3r = \frac{1}{2}(r-1)(r-2), \quad r \geq 1.$$

The triangle K is subdivided into r^2 smaller triangles by drawing $r-1$ equally spaced straight lines parallel to each side of the triangle (Fig. 5.10). The vertices of all these smaller triangles are the nodes on K, which are uniformly distributed throughout this element.

The natural coordinates $(\lambda_{k,1}, \lambda_{k,2}, \lambda_{k,3})$ of these T_r nodes on triangle K are

$$\lambda_{k,i} = \frac{1}{r} \sum_{j=1}^{3} \alpha_j^k \lambda_{i,j}, \quad k = 1, 2, \ldots, T_r, \ i = 1, 2, 3, \tag{5.19}$$

where the nonnegative integers α_1^k, α_2^k, and α_3^k are required to satisfy

$$\alpha_1^k + \alpha_2^k + \alpha_3^k = r, \quad k = 1, 2, \ldots, T_r, \tag{5.20}$$

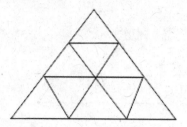

Figure 5.10 Location of T_r nodes on a triangle.

and $(\lambda_{i,1}, \lambda_{i,2}, \lambda_{i,3})$ are the natural coordinates of three vertices, $i = 1, 2, 3$. In general, $\boldsymbol{\alpha} = (\alpha_1, \alpha_2, \alpha_3) \in Z_+^3$ (three copies of the set of nonnegative integers) is a multi index (called a 3-tuple), with $\alpha_1, \alpha_2, \alpha_3$ nonnegative integers, and $|\boldsymbol{\alpha}| = \alpha_1 + \alpha_2 + \alpha_3$ is the length of $\boldsymbol{\alpha}$.

For $r = 1$, $T_1 = 3$ and the indices $(\alpha_1^k, \alpha_2^k, \alpha_3^k)$ $(k = 1, 2, 3)$ satisfying Eq. (5.20) are

$$(1, 0, 0), \quad (1, 0, 0), \quad (1, 0, 0), \tag{5.21}$$

which are inserted into Eq. (5.19) to obtain the natural coordinates of the three nodes

$$(1, 0, 0), \quad (1, 0, 0), \quad (1, 0, 0).$$

They correspond to the three vertices of a triangle (Fig. 5.11). For $r = 2$, $T_1 = 6$ and the indices $(\alpha_1^k, \alpha_2^k, \alpha_3^k)$ $(k = 1, 2, \ldots, 6)$ satisfying Eq. (5.20) are

$$(2, 0, 0), \quad (0, 2, 0), \quad (0, 0, 2), \quad (0, 1, 1), \quad (1, 0, 1), \quad (1, 1, 0). \tag{5.22}$$

Substitution of these indices into Eq. (5.19) gives the natural coordinates of the six nodes

$$(1, 0, 0), \quad (1, 0, 0), \quad (1, 0, 0),$$

$$\left(0, \frac{1}{2}, \frac{1}{2}\right), \quad \left(\frac{1}{2}, 0, \frac{1}{2}\right), \quad \left(\frac{1}{2}, \frac{1}{2}, 0\right),$$

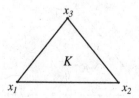

Figure 5.11 A linear triangular element.

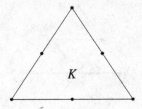

Figure 5.12 A quadratic triangular element.

which correspond to the three vertices and edge midpoints of the triangle
(Fig. 5.12). For $r = 3$, $T_1 = 10$ and the indices $(\alpha_1^k, \alpha_2^k, \alpha_3^k)$ $(k = 1, 2, \ldots, 10)$
satisfying Eq. (5.20) are

$$
\begin{aligned}
&(3,0,0), \quad (0,3,0), \quad (0,0,3), \quad (0,2,1), \quad (0,1,2), \\
&(1,0,2), \quad (2,0,1), \quad (2,1,0), \quad (1,2,0), \quad (1,1,1).
\end{aligned}
\tag{5.23}
$$

Substituting them into Eq. (5.19) yields the natural coordinates of the
10 nodes

$$(1,0,0), \qquad (1,0,0), \qquad (1,0,0),$$

$$\left(0, \frac{2}{3}, \frac{1}{3}\right), \quad \left(0, \frac{1}{3}, \frac{2}{3}\right), \quad \left(\frac{1}{3}, 0, \frac{2}{3}\right),$$

$$\left(\frac{2}{3}, 0, \frac{1}{3}\right), \quad \left(\frac{2}{3}, \frac{1}{3}, 0\right), \quad \left(\frac{1}{3}, \frac{2}{3}, 0\right),$$

$$\left(\frac{1}{3}, \frac{1}{3}, \frac{1}{3}\right).$$

These are the natural coordinates of the three vertices, two points on each
edge, and the centroid of the triangle (Fig. 5.13).

For Lagrangian triangular elements, there is a degree of freedom at
each node. For any $r \geq 1$, the corresponding basis (shape) functions at the
T_r nodes on triangle K can be given in a unified form in terms of natural

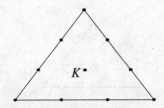

Figure 5.13 A cubic triangular element.

coordinates:

$$\psi_i(\lambda_1, \lambda_2, \lambda_3) = \prod_{j=1}^{3} \left\{ \frac{1}{\alpha_j^i!} \prod_{k=0}^{\alpha_j^i - 1} (r\lambda_j - k) \right\}, \quad i = 1, 2, \ldots, T_r, \qquad (5.24)$$

where the indices $(\alpha_1^k, \alpha_2^k, \alpha_3^k)$ satisfy Eq. (5.20).

For $r = 1$, the indices $(\alpha_1^k, \alpha_2^k, \alpha_3^k)$ $(k = 1, 2, 3)$ satisfying Eq. (5.20) are given in Eq. (5.21), and are inserted into Eq. (5.24) to find the three linear basis functions:

$$\psi_1 = \lambda_1, \quad \psi_2 = \lambda_2, \quad \psi_3 = \lambda_3,$$

which are the basis functions corresponding to the three vertices of a triangle. For $r = 2$, the indices $(\alpha_1^k, \alpha_2^k, \alpha_3^k)$ $(k = 1, 2, \ldots, 6)$ satisfying Eq. (5.20) are shown in Eq. (5.22), and are substituted into Eq. (5.24) to obtain the quadratic basis functions:

$$\psi_i = \lambda_i(2\lambda_i - 1), \quad i = 1, 2, 3,$$
$$\psi_4 = 4\lambda_2\lambda_3, \quad \psi_5 = 4\lambda_1\lambda_3, \quad \psi_6 = 4\lambda_1\lambda_2,$$

which are the basis functions at the three vertices and edge midpoints of the triangle, respectively. For $r = 3$, a similar argument can be used to obtain the cubic basis functions (see Exercise 5.3):

$$\psi_i = \frac{1}{2}\lambda_i(3\lambda_i - 1)(3\lambda_i - 2), \quad i = 1, 2, 3,$$

$$\psi_4 = \frac{9}{2}\lambda_1\lambda_2(3\lambda_1 - 1), \quad \psi_5 = \frac{9}{2}\lambda_1\lambda_2(3\lambda_2 - 1),$$

$$\psi_6 = \frac{9}{2}\lambda_2\lambda_3(3\lambda_2 - 1), \quad \psi_7 = \frac{9}{2}\lambda_2\lambda_3(3\lambda_3 - 1), \qquad (5.25)$$

$$\psi_8 = \frac{9}{2}\lambda_1\lambda_3(3\lambda_3 - 1), \quad \psi_9 = \frac{9}{2}\lambda_1\lambda_3(3\lambda_1 - 1),$$

$$\psi_{10} = 27\lambda_1\lambda_2\lambda_3.$$

As shown in the rectangular case, the Lagrangian triangular elements possess continuity of the trial function across interelement boundaries and thus throughout the entire region of the underlying problem.

For every $r \geq 3$, there are $I_r = (r - 1)(r - 2)/2$ internal nodes within each triangular element K. That is, the number of these internal nodes increases at a rate of r^2 as r increases. These internal nodes can be eliminated in a similar fashion as for the serendipity elements carried out in the previous section. This elimination process cannot guarantee that the resulting trial function is a complete polynomial of order r. To preserve the polynomial completeness, one or more equations of constraints are required to relate the coefficients of the trial function. As $r = 3$, only nine nodal parameters exist after the internal node (centroid) is deleted from element K. To be sufficient to allow determination of the 10 polynomial coefficients, an equation of the following constraint can be imposed, for example:

$$\sum_{i=1}^{3} \frac{1}{6} p_i - \sum_{i=1}^{9} \frac{1}{4} p_i + p_{10} = 0, \tag{5.26}$$

where p_i is the value of the trial function p_h at the ith node, $i = 1, 2, \ldots, 10$. The 10th node is the centroid (Fig. 5.13). With this constraint, the constrained cubic basis functions at the nine nodes on the edges of the element K become

$$\psi_i = \frac{1}{2} \lambda_i (3\lambda_i - 1)(3\lambda_i - 2) - \frac{9}{2} \lambda_1 \lambda_2 \lambda_3, \quad i = 1, 2, 3,$$

$$\psi_4 = \frac{9}{2} \lambda_1 \lambda_2 (3\lambda_1 - 1) + \frac{27}{4} \lambda_1 \lambda_2 \lambda_3, \quad \psi_5 = \frac{9}{2} \lambda_1 \lambda_2 (3\lambda_2 - 1) + \frac{27}{4} \lambda_1 \lambda_2 \lambda_3,$$

$$\psi_6 = \frac{9}{2} \lambda_2 \lambda_3 (3\lambda_2 - 1) + \frac{27}{4} \lambda_1 \lambda_2 \lambda_3, \quad \psi_7 = \frac{9}{2} \lambda_2 \lambda_3 (3\lambda_3 - 1) + \frac{27}{4} \lambda_1 \lambda_2 \lambda_3,$$

$$\psi_8 = \frac{9}{2} \lambda_1 \lambda_3 (3\lambda_3 - 1) + \frac{27}{4} \lambda_1 \lambda_2 \lambda_3, \quad \psi_9 = \frac{9}{2} \lambda_1 \lambda_3 (3\lambda_1 - 1) + \frac{27}{4} \lambda_1 \lambda_2 \lambda_3.$$

5.2.3. Hermitian triangular elements

Hermitian triangular elements involve the nodal values of a function and its partial derivatives as degrees of freedom as in the Hermitian rectangular case. In general, one uses polynomials of degree $r = 2\gamma + 1$ ($\gamma = 1, 2, \ldots$) to construct basis functions $\{\psi\}$ for this family of elements. The polynomials of this type have T_r coefficients to determine, where

$$T_r = \frac{1}{2}(r + 1)(r + 2) = \frac{1}{2}(2\gamma + 2)(2\gamma + 3) = (\gamma + 1)(2\gamma + 3), \quad \gamma = 1, 2, \ldots.$$

On each triangle K, these T_r coefficients can be determined by the function and its partial derivatives (i.e., the degrees of freedom):

$$D_{\boldsymbol{\alpha}} \psi(\lambda_{i,1}, \lambda_{i,2}, \lambda_{i,3}), \quad |\boldsymbol{\alpha}| = \alpha_1 + \alpha_2 \leq \gamma, \ i = 1, 2, 3,$$

$$D_{\boldsymbol{\alpha}} \psi \left(\frac{1}{3}, \frac{1}{3}, \frac{1}{3} \right), \qquad |\boldsymbol{\alpha}| \leq \gamma - 1, \tag{5.27}$$

where $(\lambda_{i,1}, \lambda_{i,2}, \lambda_{i,3})$ are the area coordinates of three vertices of triangle K, $i = 1, 2, 3$, $(1/3, 1/3, 1/3)$ are the coordinates of the centroid of K, and the 2-tuple $\boldsymbol{\alpha} = (\alpha_1, \alpha_2) \in Z_+^2$ (two copies of the set of nonnegative integers). The notation $D_{\boldsymbol{\alpha}}\psi$ stands for a partial derivative of ψ. For example, for $\boldsymbol{\alpha} = (1, 0)$, $D_{\boldsymbol{\alpha}}\psi$ indicates the first partial derivative of ψ with respect to x_1, i.e., $D_{\boldsymbol{\alpha}}\psi = \partial\psi/\partial x_1$, and for $\boldsymbol{\alpha} = (1, 1)$, $D_{\boldsymbol{\alpha}}\psi = \partial^2\psi/\partial x_1 \partial x_2$.

Case $\gamma = 1$. For $\gamma = 1$ and thus $r = 3$, the Hermitian element on each triangle K is a complete polynomial of degree three, which has 10 degrees of freedom. These degrees of freedom can be determined by the nodal values of the function and the first partial derivatives at three vertices and the function value at the centroid of the triangle K, i.e., the 10 nodal values. The 10 basis (shape) functions on K are denoted by

$$\psi_{\boldsymbol{\alpha},i}, \quad i = 1, 2, 3, \quad \boldsymbol{\alpha} = (\alpha_1, \alpha_2) \in Z_+^2, \quad \psi_{(0,0),4}.$$

For $\boldsymbol{\alpha} = (0, 0)$, $\psi_{(0,0),i}$ indicates the basis function associated with the function value at each vertex; if $\boldsymbol{\alpha} = (1, 0)$, the functions $\psi_{(1,0),i}$ correspond to the first partial derivative in x_1; as $\boldsymbol{\alpha} = (0, 1)$, $\psi_{(0,1),i}$ are the basis functions corresponding to the first partial derivative in x_2 at three vertices, $i = 1, 2, 3$. Finally, $\psi_{(0,0),4}$ is the basis function associated with the function value at the centroid of triangle K (see Fig. 5.14, where the little circles indicate the first derivatives as the degrees of freedom). By the definition of basis functions, they satisfy

$$D_{\boldsymbol{\alpha}}\psi_{\boldsymbol{\beta},i}(\lambda_{j,1}, \lambda_{j,2}, \lambda_{j,3}) = \delta_{ij}\delta_{\boldsymbol{\alpha}\boldsymbol{\beta}}, \qquad |\boldsymbol{\alpha}| \leq 1, \ |\boldsymbol{\beta}| \leq 1,$$
$$i = 1, 2, 3, \ j = 1, 2, 3, 4, \quad (5.28)$$
$$D_{\boldsymbol{\alpha}}\psi_{(0,0),4}(\lambda_{j,1}, \lambda_{j,2}, \lambda_{j,3}) = \delta_{4j}\delta_{\boldsymbol{\alpha}(0,0)}, \quad |\boldsymbol{\alpha}| \leq 1, \ j = 1, 2, 3, 4,$$

where δ_{ij} and $\delta_{\boldsymbol{\alpha}\boldsymbol{\beta}}$ are the Kronecker symbols

$$\delta_{ij} = \begin{cases} 1 & \text{if } i = j, \\ 0 & \text{if } i \neq j, \end{cases} \quad \delta_{\boldsymbol{\alpha}\boldsymbol{\beta}} = \begin{cases} 1 & \text{if } \boldsymbol{\alpha} = \boldsymbol{\beta}, \\ 0 & \text{if } \boldsymbol{\alpha} \neq \boldsymbol{\beta}. \end{cases}$$

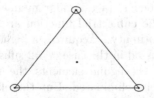

Figure 5.14 A cubic Hermitian triangular element.

Obviously, each basis function has 10 independent relations given in system (5.28), and it is uniquely determined. These basis functions are

$$\psi_{(0,0),i} = \lambda_i^3 + 3\lambda_i^2(\lambda_j + \lambda_m) - 7\lambda_1\lambda_2\lambda_3,$$

$$\psi_{(1,0),i} = \left(\frac{\partial x_1}{\partial \lambda_2}\right)_i (\lambda_i^2\lambda_j - \lambda_1\lambda_2\lambda_3) + \left(\frac{\partial x_1}{\partial \lambda_3}\right)_i (\lambda_i^2\lambda_m - \lambda_1\lambda_2\lambda_3),$$

$$\psi_{(0,1),i} = \left(\frac{\partial x_2}{\partial \lambda_2}\right)_i (\lambda_i^2\lambda_j - \lambda_1\lambda_2\lambda_3) + \left(\frac{\partial x_2}{\partial \lambda_3}\right)_i (\lambda_i^2\lambda_m - \lambda_1\lambda_2\lambda_3),$$

$$\psi_{(0,0),4} = 27\lambda_1\lambda_2\lambda_3,$$

(5.29)

where $\{i, j, m\}$ are cyclic permutations of the indices $\{1, 2, 3\}$. It follows from Eq. (5.15) that

$$x_1 = x_{1,1} + (x_{2,1} - x_{1,1})\lambda_2 + (x_{3,1} - x_{1,1})\lambda_3,$$
$$x_2 = x_{1,2} + (x_{2,2} - x_{1,2})\lambda_2 + (x_{3,2} - x_{1,2})\lambda_3.$$

Consequently, the first partial derivatives in (5.29) are given as

$$\frac{\partial x_1}{\partial \lambda_2} = x_{2,1} - x_{1,1}, \quad \frac{\partial x_1}{\partial \lambda_3} = x_{3,1} - x_{1,1},$$

$$\frac{\partial x_2}{\partial \lambda_2} = x_{2,2} - x_{1,2}, \quad \frac{\partial x_2}{\partial \lambda_3} = x_{3,2} - x_{1,2}.$$

(5.30)

The interelement compatibility of this element is of interest. Along each edge of a triangle, the trial function p_h is represented by a cubic polynomial in ℓ where ℓ is measured along the edge. At each vertex node, p_h, $\partial p_h/\partial x_1$, and $\partial p_h/\partial x_2$ are given as nodal values. The derivative $\partial p_h/\partial \ell$ in the ℓ direction can be obtained as a linear combination of $\partial p_h/\partial x_1$ and $\partial p_h/\partial x_2$, and is thus known at each vertex node. Along each edge, therefore, the values of p_h and $\partial p_h/\partial \ell$ at its end points are known, providing sufficient conditions for solving for the four coefficients of a cubic polynomial. These cubic representations along the common edge of two adjacent triangles are uniquely pegged on the same nodal values, causing interelement continuity of the trial function to occur. In a similar manner, it can be seen that the first derivatives of the cubic trial function are not continuous across interelements. If such continuity is required, a higher-order representation is needed such as is discussed in the case $\gamma = 2$ subsequently.

As in the Hermitian rectangular elements, the internal node (i.e., centroid) of a triangle can be eliminated. But the elimination process will reduce the polynomial order of the trial function. For the resulting trial function to be a complete cubic polynomial, an independent relation, such as the following one, can be imposed as one of 10 degrees of freedom on each triangle K:

$$p_h(\mathbf{x}_0) = \frac{1}{3} \sum_{l=i}^{m} p_h(\mathbf{x}_l) - \frac{1}{6} \sum_{l=i}^{m} \nabla p_h(\mathbf{x}_l) \cdot (\mathbf{x}_l - \mathbf{x}_0), \qquad (5.31)$$

where $\mathbf{x}_i$, $\mathbf{x}_j$, and $\mathbf{x}_m$ are the coordinates of the three vertices of K, $\mathbf{x}_0 = (\mathbf{x}_i + \mathbf{x}_j + \mathbf{x}_m)/3$ (the centroid of K), and ∇p_h is the gradient of the trial function p_h. With this and the degrees of freedom given in the first equation of Eq. (5.28), the basis functions for the Hermitian cubic element become

$$\psi_{(0,0),i} = \lambda_i^3 + 3\lambda_i^2(\lambda_j + \lambda_m) + 2\lambda_1\lambda_2\lambda_3,$$

$$\psi_{(1,0),i} = \left(\frac{\partial x_1}{\partial \lambda_2}\right)_i \left(\lambda_i^2 \lambda_j + \frac{1}{2}\lambda_1\lambda_2\lambda_3\right) + \left(\frac{\partial x_2}{\partial \lambda_3}\right)_i \left(\lambda_i^2 \lambda_m + \frac{1}{2}\lambda_1\lambda_2\lambda_3\right),$$

$$\psi_{(0,1),i} = \left(\frac{\partial x_2}{\partial \lambda_2}\right)_i \left(\lambda_i^2 \lambda_j + \frac{1}{2}\lambda_1\lambda_2\lambda_3\right) + \left(\frac{\partial x_1}{\partial \lambda_3}\right)_i \left(\lambda_i^2 \lambda_m + \frac{1}{2}\lambda_1\lambda_2\lambda_3\right).$$

$$(5.32)$$

This Hermitian cubic element is called the *constrained cubic Zienkiewicz triangle*.

Case $\gamma = 2$. We now consider the case where $\gamma = 2$ and so $r = 5$; the Hermitian element on each triangle K is a complete polynomial of degree five, which has 21 degrees of freedom. Using condition (5.27), these degrees of freedom can be the nodal values of the function and all the first and second partial derivatives at three vertices and the function value and the first partial derivatives at the centroid of the triangle K, altogether 21 nodal values (see Fig. 5.15, where the little and large circles indicate the first and second derivatives as the degrees of freedom). This element was used in Example 3.5.

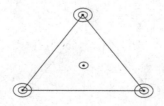

Figure 5.15 A quintic Hermitian element.

The basis functions on each triangle K are uniquely determined by

$$D_{\boldsymbol{\alpha}}\psi_{\boldsymbol{\beta},i}(\lambda_{j,1},\lambda_{j,2},\lambda_{j,3}) = \delta_{ij}\delta_{\boldsymbol{\alpha}\boldsymbol{\beta}}, \quad |\boldsymbol{\alpha}| \leq 2, \ |\boldsymbol{\beta}| \leq 2,$$
$$i = 1,2,3, \ j = 1,2,3,4,$$
$$D_{\boldsymbol{\alpha}}\psi_{\boldsymbol{\beta},4}(\lambda_{j,1},\lambda_{j,2},\lambda_{j,3}) = \delta_{4j}\delta_{\boldsymbol{\alpha}\boldsymbol{\beta}}, \quad |\boldsymbol{\alpha}| \leq 2, \ |\boldsymbol{\beta}| \leq 1,$$
$$j = 1,2,3,4. \tag{5.33}$$

It can be shown that interelement continuity of the trial function occurs for both the function and its first partial derivatives.

The degrees of freedom for the quintic Hermitian element can be replaced with the following relations:

$$D_{\boldsymbol{\alpha}}\psi(\lambda_{i,1},\lambda_{i,2},\lambda_{i,3}), \quad |\boldsymbol{\alpha}| = \alpha_1 + \alpha_2 \leq 2, \ i = 1,2,3,$$
$$\frac{\partial \psi}{\partial \boldsymbol{\nu}}(\mathbf{x}_{ij}), \qquad i < j, \ i,j = 1,2,3, \tag{5.34}$$

where we recall that $\partial\psi/\partial\boldsymbol{\nu}$ is the normal derivative of ψ outward to element K and $\mathbf{x}_{12}$, $\mathbf{x}_{23}$, and $\mathbf{x}_{13}$ are the midpoints of the edges of K (Fig. 5.16). That is, the three degrees of freedom associated with the centroid node are replaced by the three normal derivatives. This Hermitian triangular element is often known as *Argyris' triangle*, which again has the continuity of the trial function and its first partial derivatives throughout the entire region.

One can reduce the number of degrees of freedom by restricting to the class of polynomials of degree five whose normal derivatives on each edge of triangle K are polynomials of degree three rather than four. For this class of polynomials, the normal derivative along an edge is uniquely determined by the derivatives at its endpoints (vertices). The number of degrees of freedom on each $K \in K_h$ for this *reduced Argyris' triangle* (preferably, *Bell's triangle*) is 18 (Fig. 5.17):

$$D_{\boldsymbol{\alpha}}\psi(\lambda_{i,1},\lambda_{i,2},\lambda_{i,3}), \quad |\boldsymbol{\alpha}| = \alpha_1 + \alpha_2 \leq 2, \ i = 1,2,3. \tag{5.35}$$

Interelement continuity in the trial function and its first partial derivatives is retained in this process.

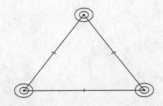

Figure 5.16   Argyris' triangle.

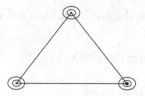

Figure 5.17 Bell's triangle.

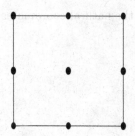

Figure 5.18 Bi-quadratic element.

5.3. Exercises

5.1. Referring to Fig. 5.1 (also see Fig. 5.18), find the concrete expressions of the basis functions of the Lagrangian rectangular element in the case $m = n = 2$.

5.2. Use systems (5.6) and (5.8) to find $\psi_{00,2}(\xi_1, \xi_2)$, $\psi_{10,2}(\xi_1, \xi_2)$, $\psi_{01,2}(\xi_1, \xi_2)$, and $\psi_{11,2}(\xi_1, \xi_2)$.

5.3. Derive the cubic Lagrangian basis functions given in system (5.25) in detail.

5.4. Let K be a triangle with vertices $\mathbf{x}_i$, $i = 1, 2, 3$. Show that if $v \in P_r(K)$ (the set of polynomials of degree r defined on K) vanishes on the edge $\mathbf{x}_2\mathbf{x}_3$, then v is of the form

$$v(\mathbf{x}) = \lambda_1(\mathbf{x})w(\mathbf{x}), \quad \mathbf{x} \in K,$$

where $w \in P_{r-1}(K)$ and λ_1 is the first natural coordinate.

5.5. Let K be a triangle with vertices $\mathbf{x}_i$, $i = 1, 2, 3$, and edge midpoints $\mathbf{x}_{ij}$, $i < j$, $i, j = 1, 2, 3$. Show that a function $v \in P_2(K)$ (the set of quadratic polynomials defined on K) has the representation

$$v(\mathbf{x}) = \sum_{i=1}^{3} v(\mathbf{x}_i)\lambda_i(\mathbf{x})(2\lambda_i(\mathbf{x}) - 1)$$
$$+ \sum_{i,j=1;\ i<j}^{3} 4v(\mathbf{x}_{ij})\lambda_i(\mathbf{x})\lambda_j(\mathbf{x}), \quad \mathbf{x} \in K, \tag{5.36}$$

where $(\lambda_1, \lambda_2, \lambda_3)$ are the natural coordinates of the triangle K.

Chapter 6

Three-Dimensional Elements and their Properties

The one- and two-dimensional finite elements introduced in the previous two chapters are now extended to three dimensions. Three-dimensional elements inherently involve a large number of degrees of freedom. To preserve interelement continuity for the trial function, elements with vertex nodes are emphasized in this chapter. Hexahedral, tetrahedral, and pentahedral elements are described in Secs. 6.1–6.3, respectively. A new concept related to the generation of elements, *isoparametric elements*, is discussed in Sec. 6.4. The selection of elements for a particular problem is considered in Sec. 6.5. The generation of elements has been so far for polygonal domains. In Sec. 6.6, general domains with curved boundaries are treated. Finally, quadrature rules for one-, two-, and three-dimensional elements are discussed in Sec. 6.7.

6.1. Hexahedral elements

The hexahedral family of elements are also referred to as brick elements or rectangular parallelepiped elements for the obvious reason.

Let Ω be a rectangular domain in space $\mathbb{R}^3$, and K_h be a partition of Ω into non overlapping *hexahedra* such that their faces are parallel to the x_1-, x_2-, and x_3-coordinate axes, respectively. We require that no vertex of any hexahedron lie in the interior of an edge or face of another hexahedron.

The tensor products of one-dimensional polynomials of degree r in the x_1-, x_2-, and x_3-directions result in the three-dimensional polynomials:

$$v(x_1, x_2, x_3) = \sum_{m=0}^{r} \sum_{j=0}^{r} \sum_{i=0}^{r} v_{ijm} x_1^i x_2^j x_2^m, \quad r \geq 0, \qquad (6.1)$$

where x_1, x_2, and x_3 are variables and the coefficients v_{ijm} are real numbers. The total number of terms in such a function v is $(r+1)^3$. For $r = 1$, the function v is a *trilinear polynomial*:

$$v(x_1, x_2, x_3) = v_{000} + v_{100}x_1 + v_{010}x_2 + v_{001}x_3 + v_{110}x_1x_2$$
$$+ v_{011}x_2x_3 + v_{101}x_1x_3 + v_{111}x_1x_2x_3.$$

The most commonly used hexahedral elements are the Lagrangian and serendipity types, which are described in this section.

6.1.1. Lagrangian hexahedral elements

Similar to the two-dimensional case, three-dimensional Lagrangian elements have basis (shape) functions that are tensor products of the Lagrangian interpolation functions. The first element of this family has eight vertex nodes (Fig. 6.1). At each node the value of a function is specified, giving a total of eight nodal degrees of freedom. Using the local coordinates (ξ_1, ξ_2, ξ_3) and choosing the origin at the centroid of the element, the basis functions on the three-dimensional reference element $[-1, 1] \times [-1, 1] \times [-1, 1]$ are

$$\psi_i = \frac{1}{8}(1 + \xi_1\xi_{i,1})(1 + \xi_2\xi_{i,2})(1 + \xi_3\xi_{i,3}), \quad i = 1, 2, \ldots, 8, \qquad (6.2)$$

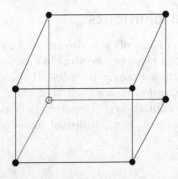

Figure 6.1   Trilinear element.

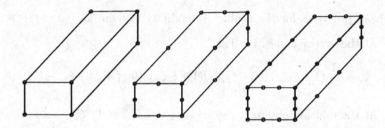

Figure 6.2 The first three members of three-dimensional serendipity elements.

where $(\xi_{i,1}, \xi_{i,2}, \xi_{i,3})$ are the coordinates of the ith vertex of the reference element in the (ξ_1, ξ_2, ξ_3)-coordinate system, $i = 1, 2, \ldots, 8$.

In addition to the corner nodes, higher order Lagrangian hexahedral elements may have edge, face, and interior nodes (see Exercise 6.1). These elements, however, are not widely used.

6.1.2. Serendipity elements

As was the case for two-dimensional serendipity elements, the three-dimensional serendipity elements do not involve internal nodes. The first three members of this family of elements are given in Fig. 6.2, and they have 8, 20, and 32 degrees of freedom, respectively. The "linear" element is given in Eq. (6.2). Using the same local (ξ_1, ξ_2, ξ_3)-coordinate system as in Eq. (6.2), the basis functions for the "quadratic" serendipity element are

- At the vertex nodes, $i = 1, 2, \ldots, 8$,

$$\psi_i = \frac{1}{8}(1 + \xi_1 \xi_{i,1})(1 + \xi_2 \xi_{i,2})(1 + \xi_3 \xi_{i,3})(\xi_1 \xi_{i,1} + \xi_2 \xi_{i,2} + \xi_3 \xi_{i,3} - 2);$$
(6.3)

- At the midside nodes, $\xi_{i,1} = 0$, $\xi_{i,2} = \pm 1$, and $\xi_{i,3} = \pm 1$,

$$\psi_i = \frac{1}{4}(1 - \xi_1^2)(1 + \xi_2 \xi_{i,2})(1 + \xi_3 \xi_{i,3});$$
(6.4)

- At the midside nodes, $\xi_{i,1} = \pm 1$, $\xi_{i,2} = \pm 1$, and $\xi_{i,3} = 0$,

$$\psi_i = \frac{1}{4}(1 + \xi_1 \xi_{i,1})(1 + \xi_2 \xi_{i,2})(1 - \xi_3^2);$$
(6.5)

- At the midside nodes, $\xi_{i,1} = \pm 1$, $\xi_{i,2} = 0$, and $\xi_{i,3} = \pm 1$,

$$\psi_i = \frac{1}{4}(1 + \xi_1 \xi_{i,1})(1 - \xi_2^2)(1 + \xi_3 \xi_{i,3}).$$
(6.6)

The basis functions for the "cubic" serendipity element are

- At the vertex nodes, $i = 1, 2, \ldots, 8$,

$$\psi_i = \frac{1}{64}(1 + \xi_1 \xi_{i,1})(1 + \xi_2 \xi_{i,2})(1 + \xi_3 \xi_{i,3})(9[\xi_1^2 + \xi_2^2 + \xi_3^2] - 19); \quad (6.7)$$

- At the midside nodes, $\xi_{i,1} = \pm \frac{1}{3}$, $\xi_{i,2} = \pm 1$,   and $\xi_{i,3} = \pm 1$,

$$\psi_i = \frac{9}{64}(1 - \xi_1^2)(1 + 9\xi_1 \xi_{i,1})(1 + \xi_2 \xi_{i,2})(1 + \xi_3 \xi_{i,3}); \quad (6.8)$$

- At the midside nodes, $\xi_{i,1} = \pm 1$, $\xi_{i,2} = \pm \frac{1}{3}$,   and $\xi_{i,3} = \pm 1$,

$$\psi_i = \frac{9}{64}(1 - \xi_2^2)(1 + \xi_1 \xi_{i,1})(1 + 9\xi_2 \xi_{i,2})(1 + \xi_3 \xi_{i,3}); \quad (6.9)$$

- At the midside nodes, $\xi_{i,1} = \pm 1$, $\xi_{i,2} = \pm 1$,   and $\xi_{i,3} = \pm \frac{1}{3}$,

$$\psi_i = \frac{9}{64}(1 - \xi_3^2)(1 + \xi_1 \xi_{i,1})(1 + \xi_2 \xi_{i,2})(1 + 9\xi_3 \xi_{i,3}). \quad (6.10)$$

These three families of serendipity elements have considerable practical applications.

6.2. Tetrahedral elements

The most popular three-dimensional elements are the tetrahedral elements. However, it is occasionally difficult to partition a domain into only elements of this type. For this reason, tetrahedral elements are often combined with hexahedral and pentahedral elements introduced in the previous and next sections, respectively.

Let K_h be a partition of a polygonal domain $\Omega \subset \mathbb{R}^3$ into non overlapping *tetrahedra* such that no vertex of any tetrahedron lies in the interior of an edge or face of another tetrahedron. A complete polynomial of degree r in variables x_1, x_2, and x_3 is

$$v(\mathbf{x}) = \sum_{0 \leq i+j+m \leq r} v_{ijm} x_1^i x_2^j x_2^m, \quad r \geq 0,$$

where the coefficients v_{ijm} are real numbers. The number of terms in such a polynomial of degree r is $(r+1)(r+2)(r+3)/6$. For $r = 1$, v is a linear function in three variables:

$$v(\mathbf{x}) = v_{000} + v_{100}x_1 + v_{010}x_2 + v_{001}x_3.$$

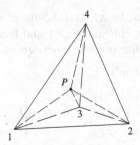

Figure 6.3 The volume coordinates.

For $r = 2$, it is quadratic:

$$v(\mathbf{x}) = v_{000} + v_{100}x_1 + v_{010}x_2 + v_{001}x_3 + v_{200}x_1^2 + v_{110}x_1x_2$$
$$+ v_{101}x_1x_3 + v_{002}x_2^2 + v_{011}x_2x_3 + v_{002}x_3^2.$$

6.2.1. Natural coordinates in three dimensions

In the three-dimensional case, the natural coordinates are the volume coordinates, which are the counterparts of the area coordinates in two dimensions. For any point P on a tetrahedron K with the vertex coordinates $(x_{1,1}, x_{1,2}, x_{1,3})$, $(x_{2,1}, x_{2,2}, x_{2,3})$, $(x_{3,1}, x_{3,2}, x_{3,3})$, and $(x_{4,1}, x_{4,2}, x_{4,3})$ (Fig. 6.3), the volume coordinates $(\lambda_1, \lambda_2, \lambda_3, \lambda_4)$ are defined as

$$\lambda_1 = \frac{\Delta_1}{\Delta}, \quad \lambda_2 = \frac{\Delta_2}{\Delta}, \quad \lambda_3 = \frac{\Delta_3}{\Delta}, \quad \lambda_4 = \frac{\Delta_4}{\Delta}, \qquad (6.11)$$

where Δ is the volume of the entire tetrahedron K and Δ_i is the volume subtended by the point P and the face opposite the ith node.

It is clear that

$$0 \leq \lambda_1, \lambda_2, \lambda_3, \lambda_4 \leq 1, \quad \lambda_1 + \lambda_2 + \lambda_3 + \lambda_4 = 1. \qquad (6.12)$$

The tetrahedron faces correspond to $\lambda_1 = 0$, $\lambda_2 = 0$, $\lambda_3 = 0$, and $\lambda_4 = 0$, and the opposite vertices to $\lambda_1 = 1$, $\lambda_2 = 1$, $\lambda_3 = 1$, and $\lambda_4 = 1$, respectively.

Similar to the two-dimensional case, the relationship between the Cartesian coordinates (x_1, x_2, x_3) and the volume coordinates $(\lambda_1, \lambda_2, \lambda_3, \lambda_4)$ is

$$\begin{pmatrix} x_1 \\ x_2 \\ x_3 \\ 1 \end{pmatrix} = \begin{pmatrix} x_{1,1} & x_{2,1} & x_{3,1} & x_{4,1} \\ x_{1,2} & x_{2,2} & x_{3,2} & x_{4,2} \\ x_{1,3} & x_{2,3} & x_{3,3} & x_{4,3} \\ 1 & 1 & 1 & 1 \end{pmatrix} \begin{pmatrix} \lambda_1 \\ \lambda_2 \\ \lambda_3 \\ \lambda_4 \end{pmatrix}. \qquad (6.13)$$

The volume coordinates $\lambda_1, \lambda_2, \lambda_3$, and λ_4 are analogous to basis (shape) functions in that they have the values of 1 and 0 at the four vertex nodes. In fact, they are the local linear basis functions on tetrahedron K.

We define the determinants

$$a_1 = \begin{vmatrix} x_{2,1} & x_{2,2} & x_{2,3} \\ x_{3,1} & x_{3,2} & x_{3,3} \\ x_{4,1} & x_{4,2} & x_{4,3} \end{vmatrix}, \quad b_1 = - \begin{vmatrix} 1 & x_{2,2} & x_{2,3} \\ 1 & x_{3,2} & x_{3,3} \\ 1 & x_{4,2} & x_{4,3} \end{vmatrix},$$

$$c_1 = - \begin{vmatrix} x_{2,1} & 1 & x_{2,3} \\ x_{3,1} & 1 & x_{3,3} \\ x_{4,1} & 1 & x_{4,3} \end{vmatrix}, \quad d_1 = - \begin{vmatrix} x_{2,1} & x_{2,2} & 1 \\ x_{3,1} & x_{3,2} & 1 \\ x_{4,1} & x_{4,2} & 1 \end{vmatrix}.$$

The other quantities a_i, b_i, and c_i can be obtained by cyclic permutation of the subscripts $\{1, 2, 3, 4\}$. Note that the signs in these equations depend on the sense in which the nodes are identified. The above relations apply to a right-handed Cartesian coordinate system when the nodes 1, 2, and 3 are numbered counterclockwise as viewed from node 4.

System (6.13) can be solved for $\lambda_1, \lambda_2, \lambda_3$, and λ_4 in terms of x_1, x_2, and x_3:

$$\begin{pmatrix} \lambda_1 \\ \lambda_2 \\ \lambda_3 \\ \lambda_4 \end{pmatrix} = \frac{1}{6\Delta} \begin{pmatrix} a_1 & b_1 & c_1 & d_1 \\ a_2 & b_2 & c_2 & d_2 \\ a_3 & b_3 & c_3 & d_3 \\ a_4 & b_4 & c_4 & d_4 \end{pmatrix} \begin{pmatrix} 1 \\ x_1 \\ x_2 \\ x_3 \end{pmatrix}, \tag{6.14}$$

where

$$6\Delta = \begin{vmatrix} 1 & x_{1,1} & x_{1,2} & x_{1,3} \\ 1 & x_{2,1} & x_{2,2} & x_{2,3} \\ 1 & x_{3,1} & x_{3,2} & x_{3,3} \\ 1 & x_{4,1} & x_{4,2} & x_{4,3} \end{vmatrix}.$$

A local volume matrix originally obtained through reference to the global system can be transferred to the natural system of volume coordinates by transformation (6.13). Generally, the element contributions are then integrals of the form

$$\int_K \lambda_1^l \lambda_2^m \lambda_3^n \lambda_4^q \, dx_1 dx_2 dx_3,$$

which can be analytically computed as

$$\int_K \lambda_1^l \lambda_2^m \lambda_3^n \lambda_4^q \, dx_1 dx_2 dx_3 = 6\Delta \frac{l!m!n!q!}{(l + m + n + q + 3)!}, \tag{6.15}$$

where l, m, n, and q are nonnegative integers.

6.2.2. Natural coordinates in d-dimensions

The concept of natural coordinates can be generalized to *simplices* in any dimensional Euclidean space $\mathbb{R}^d$, $d > 3$. In $\mathbb{R}^d$, a (nondegenerate) d-simplex is the convex hull K of $d + 1$ points $\mathbf{x}_i = (x_{i,1}, x_{i,2}, \ldots, x_{i,d}) \in \mathbb{R}^d$, which are termed the vertices of the d-simplex, $i = 1, 2, \ldots, d + 1$, and which are such that the matrix

$$\mathbf{A} = \begin{pmatrix} x_{1,1} & x_{2,1} & \cdots & x_{d+1,1} \\ x_{1,2} & x_{2,2} & \cdots & x_{d+1,2} \\ \cdot & \cdot & \cdots & \cdot \\ \cdot & \cdot & \cdots & \cdot \\ \cdot & \cdot & \cdots & \cdot \\ x_{1,d} & x_{2,d} & \cdots & x_{d+1,d} \\ 1 & 1 & \cdots & 1 \end{pmatrix} \tag{6.16}$$

is nonsingular (i.e., $|\mathbf{A}| \neq 0$); equivalently, the $d + 1$ points $\mathbf{x}_i$ are not contained in a hyperplane. This convex hull K is

$$K = \left\{ \mathbf{x} = \sum_{i=1}^{d+1} \lambda_i \mathbf{x}_i : 0 \leq \lambda_i \leq 1, i = 1, 2, \ldots, d + 1, \ \sum_{i=1}^{d+1} \lambda_i = 1 \right\}. \tag{6.17}$$

Note that a 2-simplex is a triangle and a 3-simplex is a tetrahedron.

The natural coordinates (or barycentric coordinates) $\lambda_i = \lambda_i(\mathbf{x})$, $i = 1, 2, \ldots, d + 1$, of any point $\mathbf{x} = (x_1, x_2, \ldots, x_d) \in \mathbb{R}^d$ in K with respect to the $d + 1$ points $\mathbf{x}_i$ are the unique solutions of the linear system

$$\sum_{i=1}^{d+1} x_{i,j} \lambda_i = x_j, \quad j = 1, 2, \ldots, d,$$

$$\sum_{i=1}^{d+1} \lambda_i = 1, \tag{6.18}$$

whose coefficient matrix is exactly the matrix $\mathbf{A}$ given in Eq. (6.16). It follows from inspection of linear system (6.18) that the natural coordinates can be expressed as

$$\lambda_i = \sum_{j=1}^{d} b_{i,j} x_j + b_{i,d+1}, \quad i = 1, 2, \ldots, d + 1, \tag{6.19}$$

where the matrix $(b_{i,j})_{(d+1) \times (d+1)}$ is the inverse of matrix $\mathbf{A}$. The barycenter (or the center of gravity) of a d-simplex K is the point in K whose all natural coordinates equal $1/(d + 1)$.

6.2.3. Lagrangian tetrahedral elements

As in Lagrangian triangular elements, tetrahedral elements can be constructed by selecting a sufficient number of nodes to allow a unique determination for the coefficients in the chosen polynomial trial function. A complete polynomial of order r in three dimensions contains $T_r = (r + 1)(r + 2)(r + 3)/6$ coefficients, and a tetrahedral element must accordingly contain the same number T_r of nodes.

The first three elements of this family, i.e., the 4-, 10-, and 20-node elements, are illustrated in Fig. 6.4 and correspond to the complete linear, quadratic, and cubic trial functions, respectively. Each node has solely one degree of freedom, the function value at this node. The basis (shape) functions on each tetrahedron K can be easily described.

For the 4-node element, the basis functions are linear:

$$\psi_i = \lambda_i, \quad i = 1, 2, 3, 4, \tag{6.20}$$

which correspond to the four vertices $\mathbf{x}_i$ of tetrahedron $K, i = 1, 2, 3, 4$.

We define the midpoints of the edges (Fig. 6.4) by

$$\mathbf{x}_{ij} = \frac{1}{2}(\mathbf{x}_i + \mathbf{x}_j), \quad 1 \le i < j \le 4.$$

For the 10-node element, the quadratic basis functions associated with the vertices are

$$\psi_i = \lambda_i(2\lambda_i - 1), \quad i = 1, 2, 3, 4, \tag{6.21}$$

and the basis functions associated with the midpoints $\mathbf{x}_{ij}$ are

$$\psi_{ij} = 4\lambda_i\lambda_j, \quad 1 \le i < j \le 4. \tag{6.22}$$

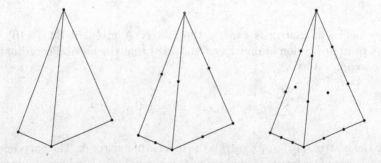

Figure 6.4   The first three Lagrangian tetrahedral elements.

We define the points on the edges

$$\mathbf{x}_{iij} = \frac{1}{3}(2\mathbf{x}_i + \mathbf{x}_j), \quad 1 \le i,j \le 4, \ i \ne j,$$

and the points on the faces

$$\mathbf{x}_{ijm} = \frac{1}{3}(\mathbf{x}_i + \mathbf{x}_j + \mathbf{x}_m), \quad 1 \le i < j < m \le 4.$$

Now, for the 20-node element, the cubic basis functions are

$$\psi_i = \frac{1}{2}\lambda_i(3\lambda_i - 1)(3\lambda_i - 2), \quad i = 1,2,3,4,$$

$$\psi_{iij} = \frac{9}{2}\lambda_i\lambda_j(3\lambda_i - 1), \quad 1 \le i,j \le 4, \ i \ne j, \qquad (6.23)$$

$$\psi_{ijm} = 27\lambda_i\lambda_j\lambda_k, \quad 1 \le i < j < m \le 4,$$

which are associated with the nodes $\mathbf{x}_i$, $\mathbf{x}_{iij}$, and $\mathbf{x}_{ijm}$, respectively. Similar Lagrangian tetrahedral elements with polynomials of higher degree can be defined, but they are not often used.

6.2.4. Hermitian tetrahedral elements

The first example of the Hermitian tetrahedral elements is based on complete cubic polynomials, which have 20 coefficients to determine. The same number of degrees of freedom consist of the values of the function and its three first partial derivatives at the vertex nodes $\mathbf{x}_i$ of the tetrahedron $K, i = 1,2,3,4$, and the values of the function at the centroids of the four faces (Fig. 6.5).

The basis (shape) functions corresponding to the values of the function at the vertices are

$$\psi_{(0,0,0),i} = 3\lambda_i^2 - 2\lambda_i^3 - 7\lambda_i \sum_{1 \le j < m \le 4, j \ne i, m \ne i} \lambda_j\lambda_m, \quad i = 1,2,3,4, \quad (6.24)$$

and these functions at the face centroids are (Fig. 6.5)

$$\psi_{(0,0,0),4+i} = 27\lambda_1\lambda_2\lambda_3\lambda_4/\lambda_i, \quad i = 1,2,3,4. \qquad (6.25)$$

To introduce the basis functions corresponding to the values of the first partial derivatives at the vertex nodes, let E_i be the cofactors of the four entries in the first row of the matrix, $i = 1,2,3,4$:

$$A = \begin{vmatrix} 1 & 1 & 1 & 1 \\ x_{1,1} & x_{2,1} & x_{3,1} & x_{4,1} \\ x_{1,2} & x_{2,2} & x_{3,2} & x_{4,2} \\ x_{1,3} & x_{2,3} & x_{3,3} & x_{4,3} \end{vmatrix}. \qquad (6.26)$$

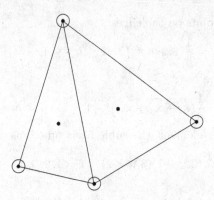

Figure 6.5   The cubic Hermitian tetrahedral elements.

Recall that a cofactor associated with an element in a square matrix is equal to the determinant of the matrix formed by removing the row and column in which the element appears from the given determinant. The entries of E_i are denoted by e_{lm}^i $(l, m = 1, 2, 3)$, and the cofactor of e_{lm}^i is indicated by A_{lm}^i, $i = 1, 2, 3, 4$. Finally, we define the 3-tuples $\boldsymbol{\alpha}_1 = (1, 0, 0)$, $\boldsymbol{\alpha}_2 = (0, 1, 0)$, and $\boldsymbol{\alpha}_3 = (0, 0, 1)$. Now, for $l = 1, 2, 3$, the rest of the basis functions are

$$
\begin{aligned}
\psi_{\boldsymbol{\alpha}_l, i} = A\{\lambda_i^2 (A_{lj}^i \lambda_j &+ A_{lm}^i \lambda_m + A_{ln}^i \lambda_n) \\
&- \lambda_i ([A_{lj}^i + A_{lm}^i]\lambda_j \lambda_m + [A_{lj}^i + A_{ln}^i]\lambda_j \lambda_n \qquad (6.27) \\
&+ [A_{lm}^i + A_{ln}^i]\lambda_m \lambda_n)\}/E_i, \quad i = 1, 2, 3, 4,
\end{aligned}
$$

where (i, j, m, n) are cyclic permutations of $(1, 2, 3, 4)$.

As in the Hermitian triangular elements, the internal nodes (i.e., the centroid nodes of the faces) of a tetrahedron can be eliminated. But the elimination process will reduce the polynomial order of the trial function; namely, the element without the degrees of freedom on these nodes will become an incomplete cubic polynomial. Finally, we mention that the Lagrangian and Hermitian tetrahedral elements satisfy the interelement compatibility of continuity throughout the entire region.

6.3. Pentahedral elements

Pentahedral elements, in the shape of triangular prisms and thus also known as *prismatic elements*, are quite often used in conjunction with hexahedral elements. The basis functions are constructed by forming products of triangular interpolation functions with Lagrange or serendipity functions in the remaining direction.

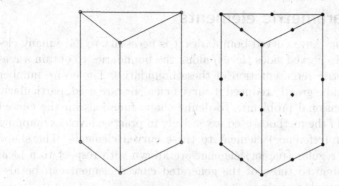

Figure 6.6 Pentahedral elements.

Let $\Omega \subset \mathbb{R}^3$ (the space) be a domain of the form $\Omega = G \times [l_1, l_2]$, where $G \subset \mathbb{R}^2$ is a planar domain and l_1 and l_2 are real numbers. Let K_h be a partition of Ω into *prisms* such that their base is a triangle in the (x_1, x_2)-plane with three vertical edges parallel to the x_3-axis (Fig. 6.6). The tensor products of complete polynomials of degree r in variables x_1 and x_2 and of degree l in variable x_3 are

$$v(\mathbf{x}) = \sum_{0 \leq i+j \leq r} \sum_{m=0}^{l} v_{ijm} x_1^i x_2^j x_3^m, \quad \mathbf{x} = (x_1, x_2, x_3),$$

where the coefficients v_{ijm} are real numbers. The number of terms in each function of this type is $(r+1)(r+2)(l+1)/2$. For $l = 1$ and $r = 1$, the function v is of the form

$$v(\mathbf{x}) = v_{000} + v_{100}x_1 + v_{010}x_2 + v_{001}x_3 + v_{101}x_1x_3 + v_{011}x_2x_3.$$

The first two elements of this family, i.e., the 6-node and 18-node elements, are illustrated in Fig. 6.6. The shape functions can be easily determined using the natural coordinates $(\lambda_1, \lambda_2, \lambda_3, \xi)$, where $(\lambda_1, \lambda_2, \lambda_3)$ are the barycentric coordinates in the x_1x_2-plane and $-1 \leq \xi \leq 1$ is the one-dimensional natural coordinate in the vertical direction. The simplest pentahedral element has the following shape functions:

$$\psi_i = \frac{1}{2}\lambda_i(1 + \xi), \qquad i = 1, 2, 3,$$

$$\psi_i = \frac{1}{2}\lambda_{i-3}(1 - \xi), \quad i = 4, 5, 6,$$

(6.28)

which correspond to the function values at the six nodes, respectively.

6.4. Isoparametric elements

In problems that have curved boundaries it is necessary to have many elements that have curved sides (faces) along the boundaries to obtain a reasonable geometric representation of these boundaries. The entire number of elements can be greatly reduced if curved elements are used, particularly for three-dimensional problems. To define shape functions on the curved elements, one of the methods used extensively in practice involves mapping from a regular (reference) element to these curved elements. The shape functions of a regular (parent) element are known with respect to a local coordinate system so those of the generated curved element can be also determined through mapping.

The coordinates of the reference element are indicated by (ξ_1, ξ_2, ξ_3), and the Lagrange shape functions on this element are denoted by $\psi_j(\xi_1, \xi_2, \xi_3)$, $j = 1, 2, \ldots, l$. Then the mapping from the local coordinates (ξ_1, ξ_2, ξ_3) to the global coordinates (x_1, x_2, x_3) is through the shape function relations:

$$x_i = F_i(\xi_1, \xi_2, \xi_3) = \sum_{j=1}^{l} \psi_j(\xi_1, \xi_2, \xi_3) x_{j,i}, \quad i = 1, 2, 3, \qquad (6.29)$$

where $(x_{j,1}, x_{j,2}, x_{j,3})$ are the coordinates of the jth node on the original element with respect to the global system. The right-hand side of Eq. (6.29) is a polynomial in the local coordinates (ξ_1, ξ_2, ξ_3).

From Fig. 6.7 it can be seen that any point in the reference element with the local coordinates (ξ_1, ξ_2, ξ_3) has a corresponding point with the global coordinates (x_1, x_2, x_3) in the generated element through the transformation $\mathbf{F} = (F_1, F_2, F_3)$ in Eq. (6.29). The value of the trial function at point (x_1, x_2, x_3) is the same as its value at the corresponding (ξ_1, ξ_2, ξ_3). If the transformation $\mathbf{F}$ is one-to-one (i.e., different points on the reference element correspond to different points on the generated element), the generated element is termed an *isoparametric finite element*, and it is said to be *isoparametrically equivalent* to the parent element.

For it to be one-to-one, the Jacobian of the transformation $\mathbf{F}$ must not equal zero:

$$\det \mathbf{G} = \begin{vmatrix} \dfrac{\partial x_1}{\partial \xi_1} & \dfrac{\partial x_1}{\partial \xi_2} & \dfrac{\partial x_1}{\partial \xi_3} \\[2mm] \dfrac{\partial x_2}{\partial \xi_1} & \dfrac{\partial x_2}{\partial \xi_2} & \dfrac{\partial x_2}{\partial \xi_3} \\[2mm] \dfrac{\partial x_3}{\partial \xi_1} & \dfrac{\partial x_3}{\partial \xi_2} & \dfrac{\partial x_3}{\partial \xi_3} \end{vmatrix} \neq 0.$$

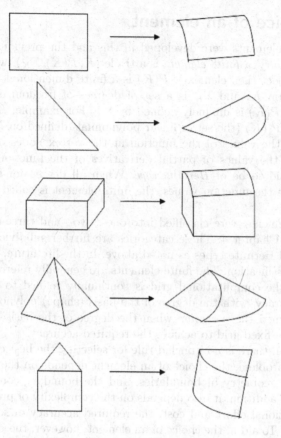

Figure 6.7 Isoparametric elements.

In this case, the integral on the generated element K can be changed to the reference element $\hat{K}$:

$$\int_K g(x_1, x_2, x_3)dx_1dx_2dx_3 = \int_{\hat{K}} \hat{g}(\xi_1, \xi_2, \xi_3)|\det \mathbf{G}|d\xi_1 d\xi_2 d\xi_3,$$

where $\hat{g}(\xi_1, \xi_2, \xi_3) = g(\mathbf{F}(\xi_1, \xi_2, \xi_3))$.

In the case where the transformation $\mathbf{F}$ is a lower representation than the trial function, the generated element is termed *subparametic*. For example, when the transformation $\mathbf{F}$ is linear and the trial function is represented using polynomials of order two, the subparametic case occurs. On the other hand, if the transformation $\mathbf{F}$ is a higher representation than the trial function, the generated element is termed *superparametic*. Examples of the transformation $\mathbf{F}$ are given in Sec. 6.6 (also see Exercise 6.2).

6.5. Choice of an element

A variety of elements were developed in this and the previous two chapters. In summary, a *finite element* is a triple $(K, P(K), \Sigma_K)$, where K is a geometric object (i.e., element), $P(K)$ is a finite dimensional linear space of functions on K, and Σ_K is a set of degrees of freedom, such that a function $v \in P(K)$ is uniquely defined by Σ_K. For example, K is a triangle, $P(K) = P_1(K)$ (the set of linear polynomials defined on K), and Σ_K is the set of the values of the function at the vertex nodes of K. When Σ_K includes the values of partial derivatives of the function, the finite element is said to be of *Hermite type*. When all degrees of freedom are given only by the function values, the finite element is called a *Lagrange* element.

Finite elements were classified into one-, two-, and three-dimensional categories in Chapter 4. These categories are further subdivided into the Lagrange and Hermite types as noted above. In the literature, there exists a broader classification. The finite elements are generally referred to as the *h-version* if the computational grid is continually refined to achieve the required accuracy, with the degree of the basis (shape) polynomials fixed. They are termed the *p-version* when the degree of the basis functions is increased on a fixed grid to achieve the required accuracy.

In general, there is no ironclad rule for selecting the best element. For a particular problem, the choice of an element depends on the type of the problem, the geometry of boundaries, and the boundary conditions and their type. In addition, it also depends on the complexity of programming, the computational effort and cost, the required accuracy of solution, and other factors. To aid in the choice of an element, however, there are certain guidelines to follow. Trial and test functions must be *admissible* according to the admissible function space used for the underlying problem. If the boundaries of the problem are regular, elements of simple geometry are usually chosen. To match irregular boundaries, the choice is between many regular elements or few, more complex, isoparametric elements. Finally, there exists considerable advantage in selecting elements that have their nodal parameters concentrated at vertices. Derivative elements can be valuable where the solution involves derivatives.

6.6. General domains

In the construction of finite elements so far, except briefly in Sec. 6.4, we have assumed that the domain Ω is polygonal. In this section, we consider the case where Ω is curved. For simplicity of the subsequent discussion, we focus on two space dimensions.

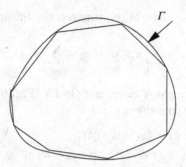

Figure 6.8 Linear approximation to a curved boundary Γ.

For a two-dimensional domain Ω, the simplest approximation for its curved boundary Γ is a polygonal line (Fig. 6.8). The resulting error is of order $\mathcal{O}(h^2)$, where h is the mesh size as usual (see Exercise 6.3). To obtain a more accurate approximation, we can approximate the boundary Γ with piecewise polynomials of degree $r \geq 2$. The error in this approximation becomes $\mathcal{O}(h^{r+1})$. In the partition of such an approximated domain, the elements close to Γ then have one curved edge as described in Sec. 6.4.

Isoparametrically equivalent elements were introduced to handle curved elements in Sec. 6.4. An alternative approach is adopted here. Referring to Fig. 6.9, consider a curved triangle K along the boundary of a domain that is sufficiently smooth. The straight edges P_1P_2 and P_2P_3 lie inside the domain, and the curved edge $\overline{P_1P_3}$ is on the boundary. The coordinates of these three vertices in a global system are indicated by $(x_{1,1}, x_{1,2})$, $(x_{2,1}, x_{2,2})$, and $(x_{3,1}, x_{3,2})$, respectively. When the curved edge $\overline{P_1P_3}$ is approximated by a line segment between vertices P_1 and P_3, the resulting regular triangle

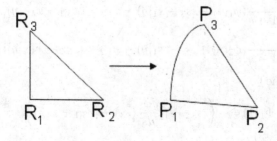

Figure 6.9 Mapping from a regular triangle to a curved triangle.

is denoted by K_1. Using natural coordinates, the linear transformation

$$x_1 = \sum_{i=1}^{3} x_{i,1} \lambda_i, \quad x_2 = \sum_{i=1}^{3} x_{i,2} \lambda_i \tag{6.30}$$

maps the reference triangle $\hat{K}$ onto triangle K_1 (Fig. 6.9).

For the notational convenience, set

$$\lambda_1 = 1 - \xi_1 - \xi_2, \quad \lambda_2 = \xi_1, \qquad \lambda_3 = \xi_2,$$
$$\bar{x}_{j,1} = x_{j,1} - x_{1,1}, \quad \bar{x}_{j,2} = x_{j,2} - x_{1,2}, \quad j = 2, 3.$$

Then, the transformation (6.30) can be rewritten as

$$\begin{aligned}
x_1 &= F_1^0(\xi_1, \xi_2) = x_{1,1} + \bar{x}_{2,1}\xi_1 + \bar{x}_{3,1}\xi_2, \\
x_2 &= F_2^0(\xi_1, \xi_2) = x_{1,2} + \bar{x}_{2,2}\xi_1 + \bar{x}_{3,2}\xi_2,
\end{aligned} \tag{6.31}$$

through which the vertices P_1, P_2, and P_3 of K_1 correspond to the vertices of $\hat{K}$. That is, there is a one-to-one correspondence between the regular triangles $\hat{K}$ and K_1. Note that if linear polynomial interpolations are used for the trial function, the transformation (6.31) is isoparametric according to the definition given in Sec. 6.4. If higher-order interpolations are used, it will become subparametric.

Now, the question is how to modify this transformation so that the line segment on the reference triangle $\hat{K}$, which was mapped onto the line $P_1 P_3$, corresponds to the curve $\overline{P_1 P_3}$. Assume that the parametric equations of this curve are

$$x_1 = \phi_1(s), \quad x_2 = \phi_2(s), \quad s_1 \leq s \leq s_3, \tag{6.32}$$

such that

$$(x_{1,1}, x_{1,2}) = \big(\phi_1(s_1), \phi_2(s_1)\big), \quad (x_{3,1}, x_{3,2}) = \big(\phi_1(s_3), \phi_2(s_3)\big). \tag{6.33}$$

For $0 \leq \eta < 1$, introduce the functions

$$\begin{aligned}
\Phi_1(\eta) &= \frac{1}{1-\eta}\big(\phi_1(s_1 + [s_3 - s_1]\eta) - x_{1,1} - [x_{3,1} - x_{1,1}]\eta\big), \\
\Phi_2(\eta) &= \frac{1}{1-\eta}\big(\phi_2(s_1 + [s_3 - s_1]\eta) - x_{1,2} - [x_{3,2} - x_{1,2}]\eta\big);
\end{aligned} \tag{6.34}$$

for $\eta = 1$, define

$$\begin{aligned}
\Phi_1(1) &= -\left([s_3 - s_1]\frac{d\phi_1}{ds}(s_3) - [x_{3,1} - x_{1,1}]\right), \\
\Phi_2(1) &= -\left([s_3 - s_1]\frac{d\phi_2}{ds}(s_3) - [x_{3,2} - x_{1,2}]\right).
\end{aligned} \tag{6.35}$$

Note that

$$\Phi_1(0) = \Phi_2(0) = 0. \tag{6.36}$$

A new transformation is now defined as

$$\begin{aligned}
x_1 &= F_1^0(\xi_1, \xi_2) + (1 - \xi_1 - \xi_2)\Phi_1(\xi_2), \\
x_2 &= F_2^0(\xi_1, \xi_2) + (1 - \xi_1 - \xi_2)\Phi_2(\xi_2),
\end{aligned} \tag{6.37}$$

which modifies the transformation in Eq. (6.31). We now check that this transformation maps the reference triangle $\hat{K}$ onto the curved triangle K.

First, as $\xi_2 = 0$, $\lambda_3 = 0$, which corresponds to the line segment P_1P_2. Applying relation (6.36), the transformation (6.37) becomes

$$\begin{aligned}
x_1 &= F_1^0(\xi_1, 0) = x_{1,1} + \bar{x}_{2,1}\xi_1, \quad 0 \le \xi_1 \le 1, \\
x_2 &= F_2^0(\xi_1, 0) = x_{1,2} + \bar{x}_{2,2}\xi_1, \quad 0 \le \xi_1 \le 1,
\end{aligned}$$

which maps the edge R_1R_2 onto P_1P_2.

Second, the equation $1 - \xi_1 - \xi_2 = 0$ corresponds to the line segment R_2R_3. In this case, the transformation (6.37) reduces to

$$x_1 = F_1^0(\xi_1, \xi_2), \quad x_2 = F_2^0(\xi_1, \xi_2), \quad 0 \le \xi_1, \quad \xi_2 \le 1,$$

which maps R_2R_3 onto the line segment P_2P_3.

Finally, on the line segment R_1R_3, $\xi_1 = 0$ and $\le \xi_2 \le 1$. Then, the transformation (6.37) becomes

$$\begin{aligned}
x_1 &= x_{1,1} + [x_{3,1} - x_{1,1}]\xi_2 + (1 - \xi_2)\Phi_1(\xi_2), \quad 0 \le \xi_2 \le 1, \\
x_2 &= x_{1,2} + [x_{3,2} - x_{1,2}]\xi_2 + (1 - \xi_2)\Phi_2(\xi_2), \quad 0 \le \xi_2 \le 1.
\end{aligned}$$

Applying the definition of Φ_1 and Φ_2 in Eq. (6.34), we see that

$$x_1 = \phi_1(s_1 + (s_3 - s_1)\xi_2), \quad x_2 = \phi_2(s_1 + (s_3 - s_1)\xi_2), \quad 0 \le \xi_2 \le 1. \tag{6.38}$$

Set $s = s_1 + (s_3 - s_1)\xi_2$, so the relations in (6.38) become

$$x_1 = \phi_1(s), \quad x_2 = \phi_2(s), \quad s_1 \le s \le s_3,$$

which are exactly the parametric equations of the curve $\overline{P_1P_3}$. Hence, we see that the line segment R_1R_3 is mapped onto the curve $\overline{P_1P_3}$ through the transformation (6.37). Therefore, this transformation maps the reference element $\hat{K}$ onto the curved element K, and the shape functions on the latter can be defined through those on this reference element using this transformation.

6.7. Quadrature rules

Some of the integrals encountered earlier cannot be evaluated analytically, and must be computed numerically. There exist a variety of numerical integration formulas (*quadrature rules*), such as the Newton–Cotes, Gauss, and Gauss–Legendre quadratures for evaluating these integrals (Hoffman, 1992). Some principles of the widely used Gauss quadrature rule, together with tables of its convenient numerical coefficients, are summarized below.

6.7.1. One dimension

In the one-dimensional case, through a local coordinate transformation, it suffices to evaluate integrals on the reference interval $[-1, 1]$. The Gauss quadrature utilizes certain integration points ξ_i on $[-1, 1]$ and weights $w_i > 0$ such that the approximate integral

$$\int_{-1}^{1} f(\xi) \, d\xi \approx \sum_{i=1}^{r} w_i f(\xi_i) \tag{6.39}$$

is exact for any polynomial of degree $2r - 1, r \geq 2$. Because a polynomial of degree $2r - 1$ has $2r$ degrees of freedom, with r integration points, Eq. (6.39) can be used to determine uniquely the $2r$ unknowns ξ_i and $w_i, i = 1, 2, \ldots, r$. The resulting approximation error in using this quadrature is of order $O(h^{2r})$, where h is the integration spacing.

For example, in the case $r = 2$, Eq. (6.39) is required to be exact for polynomials of degree $2 \times 2 - 1 = 3$. Assume that the integration points are symmetric with respect to the origin: $\xi_1 = -\xi_2 = -a$ and $w_1 = w_2 = w$. Then, for $f(\xi) = a_0 + a_1\xi + a_2\xi^2 + a_3\xi^3$, we see that

$$\int_{-1}^{1} f(\xi) \, d\xi = \int_{-1}^{1} (a_0 + a_1\xi + a_2\xi^2 + a_3\xi^3) \, d\xi = 2a_0 + \frac{2}{3}a_2.$$

On the other hand, the right-hand side of Eq. (6.39) gives

$$\sum_{i=1}^{2} w_i f(\xi_i) = w_1 f(-a) + w_2 f(a) = 2w(a_0 + a_2 a^2).$$

The error between these two equations is

$$\epsilon = \int_{-1}^{1} f(\xi) \, d\xi - \sum_{i=1}^{2} w_i f(\xi_i) = 2a_0 + \frac{2}{3}a_2 - 2w(a_0 + a_2 a^2)$$

$$= 2a_0(w - 1) + 2a_2 \left(\frac{1}{3} - wa^2 \right),$$

Table 6.1 Integration points and weights in the one-dimensional Gauss quadrature.

r	ξ_i	w_i
2	$\pm$0.577 350 269 189 626	1.000 000 000 000 000
3	$\pm$0.774 596 669 214 483	0.555 555 555 555 555
	0.000 000 000 000 000	0.888 888 888 888 889
4	$\pm$0.861 136 311 594 053	0.347 854 845 137 454
	$\pm$0.339 981 043 584 856	0.652 145 154 862 546
5	$\pm$0.906 179 845 938 664	0.236 926 885 056 189
	$\pm$0.538 469 310 105 683	0.478 628 670 499 366
	0.000 000 000 000 000	0.568 885 888 888 889
6	$\pm$0.932 469 514 203 152	0.171 324 492 379 170
	$\pm$0.661 209 386 466 265	0.360 761 573 948 139
	$\pm$0.238 619 186 083 179	0.467 913 934 572 691

which is required to be zero for any a_0 and a_2. That is

$$w = 1, \quad \frac{1}{3} - wa^2 = 0.$$

Consequently, $w = 1$ and $a = \pm 1/\sqrt{3}$. Therefore, for $r = 2$, the Gauss quadrature is

$$\int_{-1}^{1} f(\xi)\, d\xi \approx f\left(-\frac{1}{\sqrt{3}}\right) + \left(\frac{1}{\sqrt{3}}\right). \tag{6.40}$$

This approach can be used to determine integration points ξ_i on $[-1, 1]$ and weights w_i for any $r \geq 2$ (see Exercise 6.4), which is given in Table 6.1 for $r \leq 6$.

6.7.2. Rectangles and bricks

The most obvious approach of evaluating the integral

$$\int_{-1}^{1} \int_{-1}^{1} f(\xi_1, \xi_2)\, d\xi_1 d\xi_2$$

is to first fix the second variable ξ_2

$$\int_{-1}^{1} f(\xi_1, \xi_2)\, d\xi_1 = \sum_{i=1}^{r} w_i f(\xi_{i,1}, \xi_2) = g(\xi_2).$$

Then, we calculate the outer integral in a similar way:

$$\int_{-1}^{1} g(\xi_2)\, d\xi_2 = \sum_{j=1}^{r} w_j g(\xi_{j,2}) = \sum_{j=1}^{r} w_j \sum_{i=1}^{r} w_i f(\xi_{i,1}, \xi_{j,2}).$$

Consequently, we obtain

$$\int_{-1}^{1} \int_{-1}^{1} f(\xi_1, \xi_2) \, d\xi_1 d\xi_2 \approx \sum_{i=1}^{r} \sum_{j=1}^{r} w_i w_j f(\xi_{i,1}, \xi_{j,2}). \qquad (6.41)$$

In the three-dimensional case, a similar approach yields

$$\int_{-1}^{1} \int_{-1}^{1} \int_{-1}^{1} f(\xi_1, \xi_2, \xi_3) \, d\xi_1 d\xi_2 d\xi_3 \approx \sum_{i,j,m=1}^{r} w_i w_j w_m f(\xi_{i,1}, \xi_{j,2}, \xi_{m,3}).$$
$$(6.42)$$

In the quadratures (6.41) and (6.42) the numbers of integration points in each integration direction are the same. Obviously, this is not necessary; on occasion it may be of advantage to use different numbers in each direction.

6.7.3. Triangles and tetrahedra

When the element K is a triangle or tetrahedron, one can also use the *quadrature rule* of Gauss type:

$$\int_K f(\mathbf{x}) \, d\mathbf{x} \approx \sum_{i=1}^{m} w_i f(\mathbf{x}_i), \qquad (6.43)$$

where $w_i > 0$ and $\mathbf{x}_i$ are certain weights and integration points on the element K, respectively. If the quadrature rule (6.43) is exact for polynomials of degree r, i.e.,

$$\int_K f(\mathbf{x}) \, d\mathbf{x} = \sum_{i=1}^{m} w_i f(\mathbf{x}_i), \quad f \in P_r(K), \qquad (6.44)$$

then, the error in using (6.43) can be bounded by (Ciarlet and Raviart, 1972)

$$\left| \int_K f(\mathbf{x}) \, d\mathbf{x} - \sum_{i=1}^{m} w_i f(\mathbf{x}_i) \right| \le C h_K^{r+1} \sum_{|\boldsymbol{\alpha}|=r+1} \int_K |D_{\boldsymbol{\alpha}} f(\mathbf{x})| \, d\mathbf{x},$$

where $r > 0$, h_K is the diameter of K, and $D_{\boldsymbol{\alpha}} f$ indicates a partial derivative of f, with $\boldsymbol{\alpha} = (\alpha_1, \alpha_2, \ldots, \alpha_d)$ and $|\boldsymbol{\alpha}| = \alpha_1 + \alpha_2 + \cdots + \alpha_d$, $d = 2$ or 3.

Table 6.2 Integration points and weights over the reference triangle.

r	$(\lambda_{i,1}, \lambda_{i,2}, \lambda_{i,3})$	$2w_i$
1	$\left(\frac{1}{3}, \frac{1}{3}, \frac{1}{3}\right)$	1.0
2	$\left(0, \frac{1}{2}, \frac{1}{2}\right)$	$\frac{1}{3}$
	$\left(\frac{1}{2}, 0, \frac{1}{2}\right)$	$\frac{1}{3}$
	$\left(\frac{1}{2}, \frac{1}{2}, 0\right)$	$\frac{1}{3}$
3	$(1, 0, 0)$	$\frac{1}{20}$
	$(0, 1, 0)$	$\frac{1}{20}$
	$(0, 0, 1)$	$\frac{1}{20}$
	$\left(0, \frac{1}{2}, \frac{1}{2}\right)$	$\frac{2}{15}$
	$\left(\frac{1}{2}, 0, \frac{1}{2}\right)$	$\frac{2}{15}$
	$\left(\frac{1}{2}, \frac{1}{2}, 0\right)$	$\frac{2}{15}$
	$\left(\frac{1}{3}, \frac{1}{3}, \frac{1}{3}\right)$	$\frac{9}{20}$

Table 6.3 Integration points and weights over the reference tetrahedron.

r	$(\lambda_{i,1}, \lambda_{i,2}, \lambda_{i,3}, \lambda_{i,4})$	$6w_i$
1	$\left(\frac{1}{4}, \frac{1}{4}, \frac{1}{4}, \frac{1}{4}\right)$	1.0
2	$(0.5854102, 0.1381966, 0.1381966, 0.1381966)$	$\frac{1}{4}$
	$(0.1381966, 0.5854102, 0.1381966, 0.1381966)$	$\frac{1}{4}$
	$(0.1381966, 0.1381966, 0.5854102, 0.1381966)$	$\frac{1}{4}$
	$(0.1381966, 0.1381966, 0.1381966, 0.5854102)$	$\frac{1}{4}$
3	$(1, 0, 0, 0)$	$\frac{1}{40}$
	$(0, 1, 0, 0)$	$\frac{1}{40}$
	$(0, 0, 1, 0)$	$\frac{1}{40}$
	$(0, 0, 0, 1)$	$\frac{1}{40}$
	$\left(0, \frac{1}{3}, \frac{1}{3}, \frac{1}{3}\right)$	$\frac{9}{40}$
	$\left(\frac{1}{3}, 0, \frac{1}{3}, \frac{1}{3}\right)$	$\frac{9}{40}$
	$\left(\frac{1}{3}, \frac{1}{3}, 0, \frac{1}{3}\right)$	$\frac{9}{40}$
	$\left(\frac{1}{3}, \frac{1}{3}, \frac{1}{3}, 0\right)$	$\frac{9}{40}$

Triangular case

In the triangular case, it is sufficient to evaluate integrals on the reference element with vertices $(0,0)$, $(1,0)$, and $(0,1)$. In terms of the area coordinates an integral has the form

$$\int_0^1 \int_0^{1-\lambda_1} f(\lambda_1, \lambda_2, \lambda_3) \, d\lambda_2 d\lambda_1 \approx \sum_{i=1}^m w_i f(\lambda_{i,1}, \lambda_{i,2}, \lambda_{i,3}). \qquad (6.45)$$

Three examples are presented in Table 6.2, where r indicates the maximum degree of polynomials for which Eq. (6.45) holds.

Tetrahedral case

Similarly, on the reference tetrahedron, in terms of the volume coordinates an integral is

$$\int_0^1 \int_0^{1-\lambda_1} \int_0^{1-\lambda_1-\lambda_2} f(\lambda_1, \lambda_2, \lambda_3, \lambda_4) \, d\lambda_3 d\lambda_2 d\lambda_1$$

$$\approx \sum_{i=1}^m w_i f(\lambda_{i,1}, \lambda_{i,2}, \lambda_{i,3}, \lambda_{i,4}). \qquad (6.46)$$

Some examples are given in Table 6.3.

6.8. Exercises

6.1. State the tri-quadratic basis (shape) functions of Lagrangian type on the three-dimensional reference element $[-1,1] \times [-1,1] \times [-1,1]$.

6.2. Let $\hat{K} = (0,1) \times (0,1)$ be the unit square with vertices $\hat{\mathbf{x}}_i, i = 1,2,3,4$, $P(\hat{K}) = Q_1(\hat{K})$, and $\Sigma_{\hat{K}}$ be the degrees of freedom corresponding to the values at $\hat{\mathbf{x}}_i$. If K is a convex quadrilateral, define an appropriate mapping $\mathbf{F}: \hat{K} \to K$ so that an isoparametric finite element $(K, P(K), \Sigma_K)$ can be defined in the form

$$P(K) = \{v : v(\mathbf{x}) = \hat{v}(\mathbf{F}^{-1}(\mathbf{x})), \ \mathbf{x} \in K, \ \hat{v} \in P(\hat{K})\},$$

Σ_K consists of function values at $\mathbf{x}_i = \mathbf{F}(\hat{\mathbf{x}}_i), \ i = 1,2,3,4$.

6.3. Suppose that Γ is a circle with diameter L and that Γ_h is a polygonal approximation of Γ with vertices on Γ and maximal edge length equal to h. Show that the maximal distance from Γ to Γ_h is of the order $h^2/4L$ (see Sec. 6.6).

6.4. Find the integration points ξ_i on $[-1,1]$ and weights $w_i > 0$ for the integral (6.39), $i = 1,2,\ldots,r$, as $r = 3$.

6.5. Verify that the quadrature formula (6.45) holds exactly for the function $f(\lambda_1, \lambda_2, \lambda_3) = \lambda_1^3$, with the integration weights and points given in Table 6.2 in the case $r = 3$.

6.6. Check that the quadrature formula (6.46) holds exactly for the function $f(\lambda_1, \lambda_2, \lambda_3, \lambda_4) = \lambda_1^3$, with the integration weights and points given in Table 6.3 in the case $r = 3$.

Chapter 7

Finite Elements for Transient and Nonlinear Problems

The finite element method has been developed and studied for stationary differential equations, i.e., time-independent equations. This chapter is devoted to a brief extension of this method to transient problems, i.e., time-dependent problems. An extension to nonlinear transient problems is also briefly touched upon. After an introduction of one-dimensional transient problems, semi discrete (i.e., continuous in time) and fully discrete finite element schemes for multidimensional transient problems are described in Sec. 7.1. Several time approximation approaches are discussed for nonlinear transient problems in Sec. 7.2. Below, $\partial/\partial t$ denotes the derivative with respect to time t.

7.1. Finite elements for transient problems

In this section, we briefly study the finite element method for a *transient* (parabolic) problem in a bounded domain $\Omega \subset \mathbb{R}^d$, $d \geq 1$, with boundary Γ:

$$
\begin{aligned}
\phi \frac{\partial p}{\partial t} - \nabla \cdot (\mathbf{a} \nabla p) &= f && \text{in } \Omega \times J, \\
p &= 0 && \text{on } \Gamma \times J, \\
p(\cdot, 0) &= p_0 && \text{in } \Omega,
\end{aligned}
\tag{7.1}
$$

where $J = (0, T]$ ($T > 0$) is the time interval of interest, ϕ, f, $\mathbf{a}$ (tensor), and p_0 are given functions, and d is the dimension number. A typical problem is heat conduction in an inhomogeneous body Ω with heat capacity ϕ and conductivity tensor $\mathbf{a}$. The third equation in system (7.1) is an initial condition with an initial datum p_0. We first present a *semi discrete* approximation scheme where problem (7.1) is discretized only in space using the finite element method. Then, we consider *fully discrete* approximation schemes where the time discretization is based on the *backward* (or *forward*) *Euler method* or the *Crank-Nicholson method*. For more details on the finite element method for transient problems, refer to Thomée (1984).

7.1.1. A one-dimensional model problem

To see some of the major properties of the solution to problem (7.1), we consider the following one-dimensional version that models heat conduction in a bar with endpoints 0 and π:

$$
\begin{aligned}
\frac{\partial p}{\partial t} - \frac{\partial^2 p}{\partial x^2} &= 0, & 0 < x < \pi, \, t \in J, \\
p(0, t) = p(\pi, t) &= 0, & t \in J, \\
p(x, 0) &= p_0(x), & 0 < x < \pi.
\end{aligned}
\tag{7.2}
$$

An application of separation of variables yields

$$
p(x, t) = \sum_{j=1}^{\infty} p_0^j e^{-j^2 t} \sin(jx),
\tag{7.3}
$$

where the *Fourier coefficients* p_0^j of the initial datum p_0 are given by

$$
p_0^j = \sqrt{\frac{2}{\pi}} \int_0^\pi p_0(x) \sin(jx) \, dx, \qquad j = 1, 2, \ldots.
$$

Note that $\left\{ \sqrt{\frac{2}{\pi}} \sin(jx) \right\}_{j=1}^{\infty}$ forms an *orthonormal system* in the sense

$$
\frac{2}{\pi} \int_0^\pi \sin(jx) \sin(mx) \, dx = \begin{cases} 1 & \text{if } j = m, \\ 0 & \text{if } j \neq m. \end{cases}
\tag{7.4}
$$

It follows from Eq. (7.3) that the solution p is a linear combination of sine waves $\sin(jx)$ with amplitudes $p_0^j e^{-j^2 t}$ and frequencies j. Because $e^{-j^2 t}$ is very small for $j^2 t$ moderately large, each component $\sin(jx)$ lives on a time scale of order $\mathcal{O}(j^{-2})$. Consequently, high-frequency components are quickly damped, and the solution p becomes smoother as t increases. This property can be also seen from the following stability estimates:

$$\left(\int_0^\pi p^2(x,t)\, dx \right)^{1/2} \leq \left(\int_0^\pi p_0^2(x)\, dx \right)^{1/2}, \qquad t \in J,$$

$$\left(\int_0^\pi \left| \frac{\partial p}{\partial t}(x,t) \right|^2 dx \right)^{1/2} \leq \frac{C}{t} \left(\int_0^\pi p_0^2(x)\, dx \right)^{1/2}, \quad t \in J,$$

(7.5)

where C is a positive constant. We show these two estimates formally (a proof that is not concerned with any of the convergence questions). From Eqs. (7.3) and (7.4), it follows that

$$\int_0^\pi p^2(x,t)\, dx = \frac{\pi}{2} \sum_{j=1}^\infty \left(p_0^j \right)^2 e^{-2j^2 t} \leq \frac{\pi}{2} \sum_{j=1}^\infty \left(p_0^j \right)^2 = \int_0^\pi p_0^2(x)\, dx.$$

Also, note that

$$\frac{\partial p}{\partial t} = \sum_{j=1}^\infty p_0^j \left(-j^2 \right) e^{-j^2 t} \sin(jx),$$

so that

$$\int_0^\pi \left| \frac{\partial p}{\partial t}(x,t) \right|^2 dx = \frac{\pi}{2} \sum_{j=1}^\infty \left(p_0^j \right)^2 \left(-j^2 \right)^2 e^{-2j^2 t}.$$

Using the fact $0 \leq \gamma^2 e^{-\gamma} \leq C$ for any $\gamma \geq 0$, we see that

$$\int_0^\pi \left| \frac{\partial p}{\partial t}(x,t) \right|^2 dx \leq \frac{C}{t^2} \int_0^\pi p_0^2(x)\, dx, \quad t > 0.$$

Thus, estimates (7.5) were proven.

It follows from the second estimate in Eq. (7.5) that if $(\int_0^\pi p_0^2(x)\, dx)^{1/2} < \infty$, then $\left(\int_0^\pi \left| \frac{\partial p}{\partial t}(x,t) \right|^2 dx \right)^{1/2} = \mathcal{O}(t^{-1})$ as $t \to 0$. An initial phase (for t small) where certain derivatives of p are large is referred to as an *initial transient*. In general, the solution p of a parabolic problem has an initial transient. It will become smoother as t increases. This observation is very important when the parabolic problem is numerically solved. It is desirable to vary the grid size (in space and time) according to the smoothness of p. For a region where p is nonsmooth, a fine grid is used; for a region where p becomes smoother, the grid size is increased. That is, an *adaptive finite element method* should be employed, which is discussed in Chapter 11. We mention that transients may also occur for $t > 0$ if the boundary data or a source term (the right-hand side function in Eq. (7.1)) changes abruptly in time.

7.1.2. A semi discrete scheme in space

We now go back to problem (7.1). For simplicity, we study a special case of this problem where $\phi = 1$ and $\mathbf{a} = \mathbf{I}$ (the identity tensor). We use the admissible function space

$$V = \left\{ \text{Functions } v\colon v \text{ is a continuous function on } \Omega, \nabla v \text{ is} \right.$$

$$\left. \text{piecewise continuous and bounded on } \Omega, \text{ and } v = 0 \text{ on } \Gamma \right\}.$$

As in Sec. 2.2, using Green's formula (2.9), problem (7.1) is written in the variational form: find $p : J \to V$ such that

$$\int_\Omega \frac{\partial p}{\partial t} v \, d\mathbf{x} + \int_\Omega \nabla p \cdot \nabla v \, d\mathbf{x} = \int_\Omega fv \, d\mathbf{x} \qquad \forall v \in V, t \in J, \tag{7.6}$$

$$p(\mathbf{x}, 0) = p_0(\mathbf{x}) \qquad\qquad\qquad \forall \mathbf{x} \in \Omega.$$

Let V_h be a finite element subspace of V. Replacing V in system (7.6) by V_h, we have the finite element method: find $p_h : J \to V_h$ such that

$$\int_\Omega \frac{\partial p_h}{\partial t} v \, d\mathbf{x} + \int_\Omega \nabla p_h \cdot \nabla v \, d\mathbf{x} = \int_\Omega fv \, d\mathbf{x} \qquad \forall v \in V_h, t \in J,$$

$$\int_\Omega p_h(\mathbf{x}, 0) v(\mathbf{x}) \, d\mathbf{x} = \int_\Omega p_0(\mathbf{x}) v(\mathbf{x}) \, d\mathbf{x} \qquad \forall v \in V_h. \tag{7.7}$$

This system is discretized in space, but continuous in time. For this reason, system (7.7) is called a *semi discrete scheme*. Let the basis functions in V_h be denoted by φ_i, $i = 1, 2, \ldots, M$, and express p_h as

$$p_h(\mathbf{x}, t) = \sum_{i=1}^M p_i(t) \varphi_i(\mathbf{x}), \qquad (\mathbf{x}, t) \in \Omega \times J. \tag{7.8}$$

For $j = 1, 2, \ldots, M$, we take $v = \varphi_j$ in (7.7) and utilize (7.8) to see that, for $t \in J$,

$$\sum_{i=1}^M \int_\Omega \varphi_i \varphi_j \, d\mathbf{x} \frac{dp_i}{dt} + \sum_{i=1}^M \int_\Omega \nabla \varphi_i \cdot \nabla \varphi_j \, d\mathbf{x} \, p_i = \int_\Omega f\varphi_j \, d\mathbf{x},$$

$$j = 1, 2, \ldots, M,$$

$$\sum_{i=1}^M \int_\Omega \varphi_i \varphi_j \, d\mathbf{x} \, p_i(0) = \int_\Omega p_0 \varphi_j \, d\mathbf{x}, \qquad j = 1, 2, \ldots, M,$$

which, in matrix form, is given by

$$\mathbf{B}\frac{d\mathbf{p}(t)}{dt} + \mathbf{A}\mathbf{p}(t) = \mathbf{f}(t), \quad t \in J,$$

$$\mathbf{B}\mathbf{p}(0) = \mathbf{p}_0,$$

(7.9)

where the $M \times M$ matrices $\mathbf{A}$ and $\mathbf{B}$ and the vectors $\mathbf{p}$, $\mathbf{f}$, and $\mathbf{p}_0$ are

$$\mathbf{A} = (a_{i,j}), \qquad a_{i,j} = \int_\Omega \nabla\varphi_i \cdot \nabla\varphi_j \, d\mathbf{x},$$

$$\mathbf{B} = (b_{i,j}), \qquad b_{i,j} = \int_\Omega \varphi_i\varphi_j \, d\mathbf{x},$$

$$\mathbf{p} = (p_j), \qquad \mathbf{f} = (f_j), \quad f_j = \int_\Omega f\varphi_j \, d\mathbf{x},$$

$$\mathbf{p}_0 = ((p_0)_j), \qquad (p_0)_j = \int_\Omega p_0\varphi_j \, d\mathbf{x}.$$

Both $\mathbf{A}$ and $\mathbf{B}$ are symmetric and positive definite, as shown in the stationary case. Their *condition numbers* are of the order $\mathcal{O}(h^{-2})$ and $\mathcal{O}(1)$ as $h \to 0$ (Chen, 2005), respectively. Recall that for a symmetric matrix, its condition number is defined as the ratio of its largest eigenvalue to its smallest eigenvalue (see Sec. 2.5.1). For this reason, the matrices $\mathbf{A}$ and $\mathbf{B}$ are referred to as the *stiffness* and *mass* matrices, respectively. Thus system (7.9) is a *stiff system* of ordinary differential equations (ODEs). The usual way to solve an ODE system is to discretize the time derivative as well. One approach is to exploit the numerical methods developed already for ODEs. Because of the large number of simultaneous equations, special numerical methods for transient partial differential problems have been developed independent of the methods for ODEs. This is discussed in the next section.

We show a stability result for system (7.7) with $f = 0$. We choose $v = p_h(\mathbf{x}, t)$ in the first equation of Eq. (7.7) to obtain

$$\int_\Omega \frac{\partial p_h}{\partial t} p_h \, d\mathbf{x} + \int_\Omega \nabla p_h \cdot \nabla p_h \, d\mathbf{x} = 0,$$

which gives

$$\frac{1}{2}\frac{d}{dt}\int_\Omega p_h^2(\mathbf{x}, t) \, d\mathbf{x} + \int_\Omega \nabla p_h \cdot \nabla p_h \, d\mathbf{x} = 0.$$

Also, take $v = p_h(\mathbf{x}, 0)$ in the second equation of system (7.7) and use Cauchy's or Cauchy–Schwartz's inequality (1.30) to see that

$$\int_\Omega p_h^2(\mathbf{x}, 0) \, d\mathbf{x} \le \int_\Omega p_0^2(\mathbf{x}) \, d\mathbf{x}.$$

Then, it follows that

$$\int_\Omega p_h^2(\mathbf{x},t)\ d\mathbf{x} + 2\int_0^t \int_\Omega |\nabla p_h(\mathbf{x},\ell)|^2\ d\mathbf{x}d\ell$$

$$= \int_\Omega p_h^2(\mathbf{x},0)\ d\mathbf{x} \le \int_\Omega p_0^2(\mathbf{x})\ d\mathbf{x}. \tag{7.10}$$

Consequently, we obtain

$$\left(\int_\Omega p_h^2(\mathbf{x},t)\ d\mathbf{x}\right)^{1/2} \le \left(\int_\Omega p_0^2(\mathbf{x})\ d\mathbf{x}\right)^{1/2}, \qquad t \in J, \tag{7.11}$$

because the second term in the left-hand side of inequality (7.10) is non-negative. This inequality is similar to the first inequality in system (7.5). In fact, the latter inequality can be shown in the same manner as for inequality (7.11).

The derivation of an error estimate for system (7.7) is much more elaborate than that for a stationary problem. We just state an estimate for the case where V_h is the space of piecewise linear functions on a *quasiuniform* triangulation of Ω in the sense that there is a positive constant β_1, independent of h, such that

$$h_K \ge \beta_1 h, \tag{7.12}$$

where we recall that $h_K = \text{diam}(K)$, $K \in K_h$, and $h = \max\{h_K : K \in K_h\}$. Condition (7.12) says that all elements $K \in K_h$ are roughly of the same size. Under this condition, the error estimate reads as follows (Thomée, 1984; Johnson, 1994):

$$\max_{t \in J} \left(\int_\Omega (p - p_h)^2(\mathbf{x},t)\ d\mathbf{x}\right)^{1/2} \le Ch^2 \left(1 + \left|\ln \frac{T}{h^2}\right|\right), \tag{7.13}$$

where the constant C depends on the second partial derivatives in space of the solution p on Ω. Due to the presence of $\ln h^{-2}$, this estimate is *almost optimal*.

7.1.3. Fully discrete schemes

We consider three fully discrete schemes: the backward and forward Euler methods and the Crank-Nicholson method.

The backward Euler method

Let $0 = t^0 < t^1 < \cdots < t^N = T$ be a partition of J into subintervals $J^n = (t^{n-1}, t^n)$, with length $\Delta t^n = t^n - t^{n-1}$. For a generic function v

of time, set $v^n = v(t^n)$. The *backward Euler method* for the semi-discrete version (7.7) is: find $p_h^n \in V_h$, $n = 1, 2, \ldots, N$, such that

$$\int_\Omega \frac{p_h^n - p_h^{n-1}}{\Delta t^n} v \, d\mathbf{x} + \int_\Omega \nabla p_h^n \cdot \nabla v \, d\mathbf{x} = \int_\Omega f^n v \, d\mathbf{x} \qquad \forall v \in V_h,$$

$$\int_\Omega p_h^0 v \, d\mathbf{x} = \int_\Omega p_0 v \, d\mathbf{x} \qquad \forall v \in V_h. \tag{7.14}$$

Note that Eq. (7.14) comes from replacing the time derivative in system (7.7) by the first-order difference quotient $(p_h^n - p_h^{n-1})/\Delta t^n$. This replacement results in a discretization error of order $\mathcal{O}(\Delta t^n)$. As in system (7.9), system (7.14) can be expressed in matrix form

$$(\mathbf{B} + \mathbf{A}\Delta t^n) \, \mathbf{p}^n = \mathbf{B}\mathbf{p}^{n-1} + \mathbf{f}^n \Delta t^n,$$

$$\mathbf{B}\mathbf{p}(0) = \mathbf{p}_0, \tag{7.15}$$

where

$$p_h^n = \sum_{i=1}^M p_i^n \varphi_i, \quad n = 0, 1, \ldots, N,$$

and

$$\mathbf{p}^n = (p_1^n, p_2^n, \ldots, p_M^n)^T.$$

Clearly, system (7.15) is an *implicit* scheme; i.e., we need to solve a system of linear equations at each time step.

We can also show a basic *stability* estimate for Eq. (7.14) in the case $f = 0$. Choosing $v = p_h^n$ in Eq. (7.14), we see that

$$\int_\Omega |p_h^n|^2 \, d\mathbf{x} - \int_\Omega p_h^{n-1} p_h^n \, d\mathbf{x} + \int_\Omega |\nabla p_h^n|^2 \, d\mathbf{x} \, \Delta t^n = 0.$$

It follows from Cauchy's inequality (1.30) that

$$\int_\Omega p_h^{n-1} p_h^n \, d\mathbf{x} \leq \left(\int_\Omega (p_h^{n-1})^2 \, d\mathbf{x} \right)^{1/2} \left(\int_\Omega (p_h^n)^2 \, d\mathbf{x} \right)^{1/2}$$

$$\leq \frac{1}{2} \left(\int_\Omega (p_h^{n-1})^2 \, d\mathbf{x} + \int_\Omega (p_h^n)^2 \, d\mathbf{x} \right).$$

Consequently, we get

$$\frac{1}{2} \left(\int_\Omega (p_h^{n-1})^2 \, d\mathbf{x} - \int_\Omega (p_h^n)^2 \, d\mathbf{x} \right) + \int_\Omega |\nabla p_h^n|^2 \, d\mathbf{x} \, \Delta t^n \leq 0.$$

We sum over n and use the second equation in system (7.14) to give

$$\int_\Omega \left(p_h^j\right)^2 \, d\mathbf{x} + 2\sum_{n=1}^j \int_\Omega |\nabla p_h^n|^2 \, d\mathbf{x} \, \Delta t^n \leq \int_\Omega \left(p_h^0\right)^2 \, d\mathbf{x} \leq \int_\Omega \left(p^0\right)^2 \, d\mathbf{x}.$$

The second term in the left-hand side of this equation is nonnegative; so, we obtain the stability result

$$\left(\int_\Omega \left(p_h^j\right)^2 \, d\mathbf{x}\right)^{1/2} \leq \left(\int_\Omega (p^0)^2 \, d\mathbf{x}\right)^{1/2}, \qquad j = 0, 1, \ldots, N. \qquad (7.16)$$

Note that Eq. (7.16) holds regardless of the size of the time steps Δt^j. In other words, the backward Euler method (7.14) is *unconditionally stable*. This is a very desirable feature of a time discretization scheme for a parabolic problem.

We remark that an estimate for the error $p - p_h$ can be derived. The error stems from a combination of the space and time discretizations. When V_h is the finite element space of piecewise linear functions, the error $p^n - p_h^n$ ($0 \leq n \leq N$) in the norm $(\int_\Omega (p^n - p_h^n)^2 \, d\mathbf{x})^{1/2}$ is of order $\mathcal{O}\left(\Delta t + h^2\right)$ (Thomée, 1984) under appropriate smoothness assumptions on the solution p, where $\Delta t = \max\{\Delta t^j, 1 \leq j \leq N\}$.

The Crank-Nicholson method

The *Crank-Nicholson method* for system (7.7) is defined as follows: find $p_h^n \in V_h$, $n = 1, 2, \ldots, N$, such that

$$\int_\Omega \frac{p_h^n - p_h^{n-1}}{\Delta t^n} v \, d\mathbf{x} + \int_\Omega \nabla\frac{p_h^n + p_h^{n-1}}{2} \cdot \nabla v \, d\mathbf{x} = \int_\Omega \frac{f^n + f^{n-1}}{2} v \, d\mathbf{x}$$

$$\forall v \in V_h,$$

$$\int_\Omega p_h^0 v \, d\mathbf{x} = \int_\Omega p_0 v \, d\mathbf{x} \qquad \forall v \in V_h.$$

$$(7.17)$$

In the present case, the difference quotient $(p_h^n - p_h^{n-1})/\Delta t^n$ now replaces the time derivative $\partial p(t^{n-1/2})/\partial t$ at the midpoint $t^{n-1/2} = (t^{n-1} + t^n)/2$, and $(\nabla p_h^n + \nabla p_h^{n-1})/2$ and $(f^n + f^{n-1})/2$ replace $\nabla p_h^{n-1/2}$ and $f^{n-1/2}$, respectively. The resulting discretization error is $\mathcal{O}\left((\Delta t^n)^2\right)$. Similarly to system (7.15), the linear system from (7.17) is

$$\left(\mathbf{B} + \frac{\Delta t^n}{2}\mathbf{A}\right)\mathbf{p}^n = \left(\mathbf{B} - \frac{\Delta t^n}{2}\mathbf{A}\right)\mathbf{p}^{n-1} + \frac{\mathbf{f}^n + \mathbf{f}^{n-1}}{2}\Delta t^n, \qquad (7.18)$$

$$\mathbf{B}\mathbf{p}(0) = \mathbf{p}_0,$$

for $n = 1, 2, \ldots, N$. Again, this is an implicit method. When $f = 0$, by taking $v = (p_h^n + p_h^{n-1})/2$ in Eq. (7.17) one can show that the stability result (7.16) unconditionally holds for the Crank-Nicholson method, too (see Exercise 7.1). For the piecewise linear finite element space V_h, for each n the error $p^n - p_h^n$ in the norm $\left(\int_\Omega (p^n - p_h^n)^2 \, d\mathbf{x} \right)^{1/2}$ is of order $\mathcal{O}\left((\Delta t)^2 + h^2 \right)$ this time. Note that the Crank-Nicholson method is more accurate in time than the backward Euler method and is slightly more expensive from the computational point of view.

The forward Euler method

We conclude with the *forward Euler method* for system (7.7): find $p_h^n \in V_h$, $n = 1, 2, \ldots, N$, such that

$$\int_\Omega \frac{p_h^n - p_h^{n-1}}{\Delta t^n} v \, d\mathbf{x} + \int_\Omega \nabla p_h^{n-1} \cdot \nabla v \, d\mathbf{x} = \int_\Omega f^{n-1} v \, d\mathbf{x} \qquad \forall v \in V_h,$$

$$\int_\Omega p_h^0 v \, d\mathbf{x} = \int_\Omega p_0 v \, d\mathbf{x} \qquad\qquad\qquad \forall v \in V_h.$$

$$(7.19)$$

Its corresponding matrix form is

$$\mathbf{B}\mathbf{p}^n = (\mathbf{B} - \mathbf{A}\Delta t^n)\,\mathbf{p}^{n-1} + \mathbf{f}^{n-1}\Delta t^n,$$
$$\mathbf{B}\mathbf{p}(0) = \mathbf{p}_0.$$

$$(7.20)$$

We perform a stability analysis for this system by introducing the Cholesky decomposition $\mathbf{B} = \mathbf{D}\mathbf{D}^T$ (Axelsson, 1994; Chen, 2005) and using the new variable $\mathbf{q} = \mathbf{D}^T\mathbf{p}$, where $\mathbf{D}^T$ is the transpose of $\mathbf{D}$. Then, system (7.20) is of the simpler form

$$\mathbf{q}^n = (\mathbf{I} - \tilde{\mathbf{A}}\Delta t^n)\mathbf{q}^{n-1} + \mathbf{D}^{-1}\mathbf{f}^{n-1}\Delta t^n,$$
$$\mathbf{q}(0) = \mathbf{D}^{-1}\mathbf{p}_0,$$

$$(7.21)$$

where $\tilde{\mathbf{A}} = \mathbf{D}^{-1}\mathbf{A}\mathbf{D}^{-T}$. Clearly, system (7.21) is an *explicit scheme* in $\mathbf{q}$. A stability result similar to system (7.16) can be proven only under the *stability condition*

$$\Delta t^n \leq Ch^2, \qquad n = 1, 2, \ldots, N,$$

$$(7.22)$$

where C is a constant independent of Δt and h. This can be seen as follows: with $f = 0$, the first equation of system (7.21) becomes

$$\mathbf{q}^n = \left(\mathbf{I} - \tilde{\mathbf{A}}\Delta t^n \right) \mathbf{q}^{n-1}.$$

$$(7.23)$$

We define the matrix norm

$$\|\tilde{\mathbf{A}}\| = \max_{\boldsymbol{\eta} \in \mathbb{R}^M, \boldsymbol{\eta} \neq 0} \frac{\|\tilde{\mathbf{A}}\boldsymbol{\eta}\|}{\|\boldsymbol{\eta}\|},$$

where $\|\boldsymbol{\eta}\|$ is the Euclidean norm of $\boldsymbol{\eta} = (\eta_1, \eta_2, \ldots, \eta_M)$: $\|\boldsymbol{\eta}\|^2 = \eta_1^2 + \eta_2^2 + \ldots + \eta_M^2$. Assume that the symmetric matrix $\tilde{\mathbf{A}}$ has eigenvalues μ_i, $i = 1, 2, \ldots, M$. Then, we see that (Axelsson, 1994)

$$\|\tilde{\mathbf{A}}\| = \max_{i=1,2,\ldots,M} |\mu_i|.$$

Thus, it follows that

$$\|\mathbf{I} - \tilde{\mathbf{A}}\Delta t^n\| = \max_{i=1,2,\ldots,M} |1 - \mu_i \Delta t^n|.$$

Let the maximum occur as $i = M$, e.g. Then

$$\|\mathbf{I} - \tilde{\mathbf{A}}\Delta t^n\| \leq 1,$$

only if $\mu_M \Delta t^n \leq 2$. Since $\mu_M = \mathcal{O}(h^{-2})$ (Chen, 2005), $\Delta t^n \leq 2/\mu_M = \mathcal{O}(h^2)$, which is the stability condition (7.22).

The stability condition (7.22) requires that the time step be sufficiently small. In other words, the forward Euler method (7.19) is *conditionally stable*. This condition is very restrictive, particularly for long-time integration. In contrast, the backward Euler and Crank-Nicholson methods are unconditionally stable, but require more work per time step. These two methods are more efficient for parabolic problems since the extra cost involved at each step for an implicit method is more than compensated for by the fact that bigger time steps can be utilized.

7.2. Finite elements for nonlinear problems

In this section, we briefly consider an application of the finite element method to the *nonlinear transient problem*

$$
\begin{aligned}
c(p)\frac{\partial p}{\partial t} - \nabla \cdot \big(a(p)\nabla p\big) &= f(p) \quad &&\text{in } \Omega \times J, \\
p &= 0 \quad &&\text{on } \Gamma \times J, \qquad (7.24) \\
p(\cdot, 0) &= p_0 \quad &&\text{in } \Omega,
\end{aligned}
$$

where $c(p) = c(\mathbf{x}, t, p)$, $a(p) = a(\mathbf{x}, t, p)$, and $f(p) = f(\mathbf{x}, t, p)$ depend on the unknown p itself. In problem (7.24) and below, for notational convenience, we drop the dependence of these coefficients on $\mathbf{x}$ and t. We assume that Eq. (7.24) admits a unique solution. Furthermore, we assume that the coefficients $c(p)$, $a(p)$, and $f(p)$ are *globally Lipschitz continuous* in the variable p; i.e., for some constants C_c, C_a, and C_f, they satisfy

$$\begin{aligned}
|c(p_1) - c(p_2)| &\leq C_c|p_1 - p_2|, \quad p_1, p_2 \in \mathbb{R},\\
|a(p_1) - a(p_2)| &\leq C_a|p_1 - p_2|, \quad p_1, p_2 \in \mathbb{R},\\
|f(p_1) - f(p_2)| &\leq C_f|p_1 - p_2|, \quad p_1, p_2 \in \mathbb{R}.
\end{aligned} \tag{7.25}$$

With the admissible function space V defined as in the previous section, problem (7.24) can be written in the variational form: find $p : J \to V$ such that

$$\begin{aligned}
\int_\Omega c(p)\frac{\partial p}{\partial t}v \, d\mathbf{x} + \int_\Omega a(p)\nabla p \cdot \nabla v \, d\mathbf{x} &= \int_\Omega f(p)v \, d\mathbf{x} \quad \forall v \in V, t \in J,\\
p(\mathbf{x}, 0) &= p_0(\mathbf{x}) \qquad\qquad\qquad\quad \forall \mathbf{x} \in \Omega.
\end{aligned} \tag{7.26}$$

Let V_h be a finite element subspace of V. The finite element version of system (7.26) is: find $p_h : J \to V_h$ such that

$$\begin{aligned}
\int_\Omega c(p_h)\frac{\partial p_h}{\partial t}v \, d\mathbf{x} + \int_\Omega a(p_h)\nabla p_h \cdot \nabla v \, d\mathbf{x} &= \int_\Omega f(p_h)v \, d\mathbf{x} \quad \forall v \in V_h,\\
\int_\Omega p_h(\mathbf{x}, 0)v \, d\mathbf{x} &= \int_\Omega p_0 v \, d\mathbf{x} \qquad\qquad \forall v \in V_h.
\end{aligned} \tag{7.27}$$

As for system (7.9), after the introduction of basis functions in V_h, system (7.27) can be stated in matrix form:

$$\begin{aligned}
\mathbf{C}(\mathbf{p})\frac{d\mathbf{p}}{dt} + \mathbf{A}(\mathbf{p})\mathbf{p} &= \mathbf{f}(\mathbf{p}), \quad t \in J,\\
\mathbf{B}\mathbf{p}(0) &= \mathbf{p}_0.
\end{aligned} \tag{7.28}$$

Under the assumption that the coefficient $c(p)$ is bounded below by a positive constant, this nonlinear system of ODEs locally has a unique solution. In fact, because of assumption (7.25) on c, a, and f, the solution $\mathbf{p}(t)$ exists for all t. Several approaches for solving Eq. (7.28) are discussed in this section.

7.2.1. Linearization approach

The nonlinear system (7.28) can be linearized by allowing the nonlinearities to lag one time step behind. Thus, the backward Euler method for problem

(7.24) takes the form: find $p_h^n \in V_h$, $n = 1, 2, \ldots, N$, such that

$$\int_\Omega c\left(p_h^{n-1}\right) \frac{p_h^n - p_h^{n-1}}{\Delta t^n} v \, d\mathbf{x} + \int_\Omega a\left(p_h^{n-1}\right) \nabla p_h^n \cdot \nabla v \, d\mathbf{x}$$

$$= \int_\Omega f\left(p_h^{n-1}\right) v \, d\mathbf{x} \qquad \forall v \in V_h, \qquad (7.29)$$

$$\int_\Omega p_h^0 v \, d\mathbf{x} = \int_\Omega p_0 v \, d\mathbf{x} \qquad \forall v \in V_h.$$

In matrix form, it is given by

$$\mathbf{C}\left(\mathbf{p}^{n-1}\right) \frac{\mathbf{p}^n - \mathbf{p}^{n-1}}{\Delta t^n} + \mathbf{A}(\mathbf{p}^{n-1})\mathbf{p}^n = \mathbf{f}(\mathbf{p}^{n-1}),$$

$$\mathbf{Bp}(0) = \mathbf{p}_0. \qquad (7.30)$$

Note that system (7.30) is a system of linear equations for the unknown vector $\mathbf{p}^n$, which can be solved using *iterative algorithms*. When the space V_h is the finite element space of piecewise linear functions, the error $p^n - p_h^n$ ($0 \leq n \leq N$) in the norm $(\int_\Omega (p^n - p_h^n)^2 \, d\mathbf{x})^{1/2}$ is of order $\mathcal{O}\left(\Delta t + h^2\right)$ as for problem (7.1) under appropriate smoothness assumptions on p and for Δt small enough (Thomée, 1984; Chen-Douglas, 1991). We may use the Crank-Nicholson discretization method in system (7.29). However, the linearization decreases the order of the time discretization error to $\mathcal{O}(\Delta t)$, giving $\mathcal{O}\left(\Delta t + h^2\right)$ overall. This is true for any higher-order time discretization method with the present linearization technique. This drawback can be overcome by using *extrapolation techniques* in the linearization of the coefficients c, a, and f; see the next section. Combined with an appropriate extrapolation, the Crank-Nicholson method can be shown to produce an error of order $\mathcal{O}\left((\Delta t)^2\right)$ in time (Douglas, 1961; Thomée, 1984). In general, higher-order extrapolations increase data storage.

7.2.2. Extrapolation time approach

In the *extrapolation approach*, the pressure at the new time level is obtained from the previous two time levels:

$$p_h^{(n+1)^*} = p_h^n + \frac{\Delta t^{n+1}}{\Delta t^n}(p_h^n - p_h^{n-1}), \quad n = 0, 1, \ldots, N-1, \qquad (7.31)$$

and the corresponding method for problem (7.24) is: find $p_h^{n+1} \in V_h$, $n = 0, 1, \ldots, N - 1$, such that

$$\int_\Omega c\left(p_h^{(n+1)^*}\right) \frac{p_h^{n+1} - p_h^n}{\Delta t^{n+1}} v\, d\mathbf{x} + \int_\Omega a\left(p_h^{(n+1)^*}\right) \nabla p_h^{n+1} \cdot \nabla v\, d\mathbf{x}$$

$$= \int_\Omega f\left(p_h^{(n+1)^*}\right) v\, d\mathbf{x} \qquad \forall v \in V_h, \quad (7.32)$$

$$\int_\Omega p_h^0 v\, d\mathbf{x} = \int_\Omega p_0 v\, d\mathbf{x} \qquad \forall v \in V_h.$$

Its matrix form is

$$\mathbf{C}\left(\mathbf{p}^{(n+1)^*}\right) \frac{\mathbf{p}^{n+1} - \mathbf{p}^n}{\Delta t^{n+1}} + \mathbf{A}\left(\mathbf{p}^{(n+1)^*}\right)\mathbf{p}^{n+1} = \mathbf{f}\left(\mathbf{p}^{(n+1)^*}\right),$$

$$\mathbf{B}\mathbf{p}(0) = \mathbf{p}_0.$$

$$(7.33)$$

Note that system (7.33) is again a system of linear equations for the unknown vector $\mathbf{p}^{n+1}$. Figure 7.1 indicates the pressure values used to compute the nonlinear coefficients in this approach. It shows that the closer the pressure function in time to a linear function, the more accurate the approximation $p^{(n+1)^*}$ to $p^{(n+1)}$. Note that p^{-1} is required to compute p^1 at the first time level t^1, which is generally not given and needs special care in the first iteration.

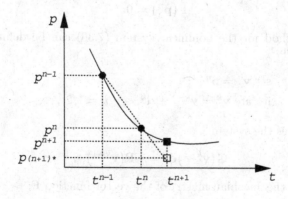

Figure 7.1 The extrapolation approach.

7.2.3. Implicit time approximation

We now consider a fully *implicit time approximation* scheme for problem (7.24): find $p_h^n \in V_h$, $n = 1, 2, \ldots, N$, such that

$$\int_\Omega c\,(p_h^n)\, \frac{p_h^n - p_h^{n-1}}{\Delta t^n}\, v\; dx + \int_\Omega a\,(p_h^n)\, \nabla p_h^n \cdot \nabla v\; dx$$

$$= \int_\Omega f\,(p_h^n)\, v\; dx \qquad \forall v \in V_h, \qquad (7.34)$$

$$\int_\Omega p_h^0 v\; dx = \int_\Omega p_0 v\; dx \qquad \forall v \in V_h.$$

Its matrix form is

$$\mathbf{C}\,(\mathbf{p}^n)\, \frac{\mathbf{p}^n - \mathbf{p}^{n-1}}{\Delta t^n} + \mathbf{A}(\mathbf{p}^n)\mathbf{p}^n = \mathbf{f}(\mathbf{p}^n),$$
$$\mathbf{B}\mathbf{p}(0) = \mathbf{p}_0. \qquad (7.35)$$

Now, system (7.35) is a system of nonlinear equations in the unknown vector $\mathbf{p}^n$, which must be solved at each time step via an iteration method. Let us consider *Newton's method* (or Newton–Raphson's method). Note that the first equation of system (7.35) can be rewritten as

$$\left(\mathbf{A}(\mathbf{p}^n) + \frac{1}{\Delta t^n} \mathbf{C}\,(\mathbf{p}^n) \right) \mathbf{p}^n - \frac{1}{\Delta t^n} \mathbf{C}\,(\mathbf{p}^n)\, \mathbf{p}^{n-1} - \mathbf{f}(\mathbf{p}^n) = 0.$$

We express this equation as

$$\mathbf{F}(\mathbf{p}^n) = \mathbf{0}. \qquad (7.36)$$

Newton's method for the nonlinear system (7.36) can be defined in the form: given $\mathbf{p}^{n-1}$,

$$\text{set } \mathbf{v}^0 = \mathbf{p}^{n-1};$$
$$\text{iterate } \mathbf{v}^k = \mathbf{v}^{k-1} + \mathbf{d}^k, \qquad k = 1, 2, \ldots,$$

where $\mathbf{d}^k$ solves the system

$$\mathbf{G}(\mathbf{v}^{k-1})\mathbf{d}^k = -\mathbf{F}(\mathbf{v}^{k-1}),$$

with $\mathbf{G}$ being the Jacobian matrix of the vector function $\mathbf{F}$:

$$\mathbf{G} = \left(\frac{\partial F_i}{\partial p_j} \right)_{i,j=1,2,\ldots,M}.$$

If the matrix $\mathbf{G}(\mathbf{p}^n)$ is nonsingular and the second partial derivatives of $\mathbf{F}$ are bounded, Newton's method converges quadratically in a neighborhood of the solution $\mathbf{p}^n$; i.e., there are constants $\epsilon > 0$ and C such that if $\left| \mathbf{v}^{k-1} - \mathbf{p}^n \right| \leq \epsilon$, then

$$\left| \mathbf{v}^k - \mathbf{p}^n \right| \leq C \left| \mathbf{v}^{k-1} - \mathbf{p}^n \right|^2.$$

The main difficulty with Newton's method is to get a sufficiently good initial guess. Once it is obtained, Newton's method converges with very few iterations. This method is a very powerful iteration method for strongly nonlinear problems. There are many variants of Newton's method available in the literature (Ostrowski, 1973; Rheinboldt, 1998). We remark that the Crank-Nicholson discretization procedure can be used in system (7.34) as well. In the present implicit case, this procedure generates second-order accuracy in time. Numerical experience has indicated that the Crank-Nicholson procedure may not be a good choice for nonlinear parabolic equations because it can be unstable for such stiff equations.

7.2.4. Explicit time approximation

We end this chapter with the application of a forward, explicit time approximation method to problem (7.24): find $p_h^n \in V_h$, $n = 1, 2, \dots, N$, such that

$$\left(c\left(p_h^n\right) \frac{p_h^n - p_h^{n-1}}{\Delta t^n}, v \right) + \left(a\left(p_h^{n-1}\right) \nabla p_h^{n-1}, \nabla v \right)$$

$$= \left(f\left(p_h^{n-1}\right), v \right) \qquad \forall v \in V_h, \qquad (7.37)$$

$$\left(p_h^0, v\right) = (p_0, v) \quad \forall v \in V_h.$$

In matrix form, it is written as follows

$$\mathbf{C}\left(\mathbf{p}^n\right) \frac{\mathbf{p}^n - \mathbf{p}^{n-1}}{\Delta t^n} + \mathbf{A}\left(\mathbf{p}^{n-1}\right)\mathbf{p}^{n-1} = \mathbf{f}\left(\mathbf{p}^{n-1}\right),$$

$$\mathbf{B}\mathbf{p}(0) = \mathbf{p}_0. \qquad (7.38)$$

Note that the only nonlinearity is in matrix $\mathbf{C}$. This system can be solved via any standard nonlinear solution method (Ostrowski, 1973; Rheinboldt, 1998).

For the explicit method (7.37) to be stable in the sense discussed in Sec. 7.1.3, a *stability condition* of the following type must be satisfied:

$$\Delta t^n \leq C h^2, \qquad n = 1, 2, \dots, N, \qquad (7.39)$$

where the constant C now depends on the coefficients c and a (inequality (7.22)). Unfortunately, this condition on the time steps is very restrictive for long-time integration, as noted earlier.

In summary, we have developed linearization, extrapolation, implicit, and explicit time approximation approaches for numerically solving problem (7.24). In terms of computational effort, the explicit approach is the simplest at each time step; however, it requires an impracticable stability restriction. The linearization approach is more practical, but it reduces the order of accuracy in time for high-order time discretization methods (unless the extrapolation approach is exploited). An efficient and accurate method is the fully implicit approach; the extra cost involved at each time step for this implicit method is more than compensated for by the fact that larger time steps may be taken, particularly when Newton's method with a good initial guess is employed. Modified implicit methods such as *semi implicit* (quasi-implicit) *methods* (Aziz-Settari, 1979) can be applied; for a given physical problem, the linearization approach should be applied to weak nonlinearity, while the implicit one should be used for strong nonlinearity (Chen *et al.*, 2000).

7.3. Exercises

7.1. Show the stability result (7.16) for Crank-Nicholson's method (7.17) with $f = 0$. What can be said if $f \neq 0$?

7.2. Consider the time-dependent problem

$$\frac{\partial p}{\partial t} - \nabla \cdot (\mathbf{a}\nabla p) + \boldsymbol{\beta} \cdot \nabla p = f \quad \text{in } \Omega \times J,$$
$$p = 0 \qquad\qquad\qquad \text{on } \Gamma \times J,$$
$$p(\cdot, 0) = p_0 \qquad\qquad \text{in } \Omega,$$

where $\mathbf{a}$ is a $d \times d$ matrix ($d = 2$ or 3), $\boldsymbol{\beta}$ is a constant vector, and f and p_0 are given functions. Extend the methods (7.7), (7.14), (7.17), and (7.19) to this problem and show a stability inequality similar to inequality (7.16) for the method (7.14) in the case $f = 0$.

Chapter 8

Application to Solid Mechanics

The finite element method is the most widely used tool for evaluating deformations and stresses of elastic and inelastic bodies subject to loads. An *elastic body* is a solid for which the additional deformation generated by an increment of stress completely disappears when the increment is removed. Examples include two-dimensional plane stress or plane strain distributions, axisymmetrical solids, plate bending, shells, and fully three-dimensional elastic bodies. In this chapter, we briefly discuss the application of the finite element method to some of these elastic bodies: plan stress and plane strain, axisymmetrical stress analysis, and three-dimensional stress analysis. In Sec. 8.1, we develop equilibrium relations, kinematics, and material laws for plan stress and plane strain, and obtain their finite element solution. Then, these studies are extended to three-dimensional stress and axisymmetrical stress analysis, respectively, in Secs. 8.2 and 8.3.

8.1. Plane stress and plane strain

In both problems of plan stress and plane strain, the displacement field is uniquely given in a horizontal x_1x_2-plane. The only stresses and strains that must be considered are the three components in this plane. In the case of plane stress, all other components of stress are zero and thus give no contribution to internal work. In plane strain, the stress in the direction orthogonal to the x_1x_2-plane is not zero. However, by definition, the strain in that direction equals zero, and thus this stress does not make any contribution to internal work, either.

8.1.1. Kinematics

In general, for a three-dimensional body Ω, it is initially in a natural state, which can be described by a mapping

$$\mathbf{R} : \Omega \to \mathbb{R}^3.$$

We write this mapping in the form

$$\mathbf{R}(\mathbf{x}) = \mathbf{ID}(\mathbf{x}) + \mathbf{u}(\mathbf{x}), \qquad \mathbf{x} \in \Omega, \tag{8.1}$$

where $\mathbf{ID}$ is the identity function and $\mathbf{u}$ is the *displacement*. We often deal with the case where the displacement is small.

The gradient tensor is defined

$$\nabla \mathbf{R} = \begin{pmatrix} \dfrac{\partial R_1}{\partial x_1} & \dfrac{\partial R_1}{\partial x_2} & \dfrac{\partial R_1}{\partial x_3} \\ \dfrac{\partial R_2}{\partial x_1} & \dfrac{\partial R_2}{\partial x_2} & \dfrac{\partial R_2}{\partial x_3} \\ \dfrac{\partial R_3}{\partial x_1} & \dfrac{\partial R_3}{\partial x_2} & \dfrac{\partial R_3}{\partial x_3} \end{pmatrix},$$

where $\mathbf{R} = (R_1, R_2, R_3)$. If the determinant of $\nabla \mathbf{R}$ is positive, i.e.,

$$\det(\nabla \mathbf{R}) > 0,$$

the mapping $\mathbf{R}$ represents a *deformation*. The matrix

$$\mathbf{C} = \nabla \mathbf{R}^T \nabla \mathbf{R} \tag{8.2}$$

represents a transformation of the body and is termed the *Cauchy–Green strain tensor*. Its deviation from the identity is referred to as the *strain*:

$$\mathbf{E} = \frac{1}{2}(\mathbf{C} - \mathbf{I}). \tag{8.3}$$

It follows from Eqs. (8.1)–(8.3) that

$$E_{ij} = \frac{1}{2}\left(\frac{\partial u_i}{\partial x_j} + \frac{\partial u_j}{\partial x_i}\right) + \frac{1}{2}\sum_{k=1}^{3} \frac{\partial u_i}{\partial x_k}\frac{\partial u_j}{\partial x_k}, \qquad i, j = 1, 2, 3.$$

In *linear elasticity* the quadratic terms are assumed small and neglected, and the components of the resulting strain are denoted by

$$\epsilon_{ij} = \frac{1}{2}\left(\frac{\partial u_i}{\partial x_j} + \frac{\partial u_j}{\partial x_i}\right), \qquad i, j = 1, 2, 3. \tag{8.4}$$

This quantity (*strain*) is one of the most important quantities in elasticity theory.

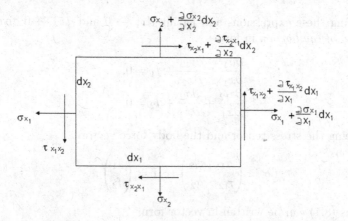

Figure 8.1 The free body diagram of an infinitesimal element.

For the two-dimensional stress and strain problems, the primary strains of interest are $\epsilon_1 = \epsilon_{11}$, $\epsilon_2 = \epsilon_{22}$, and $\gamma_{12} = 2\epsilon_{12}$. That is

$$\epsilon_1 = \frac{\partial u_1}{\partial x_1}, \quad \epsilon_2 = \frac{\partial u_2}{\partial x_2}, \quad \gamma_{12} = \frac{\partial u_1}{\partial x_2} + \frac{\partial u_2}{\partial x_1}. \tag{8.5}$$

8.1.2. Equilibrium

Let σ_1 and σ_2 be the normal stresses in the x_1- and x_2-directions, respectively, and τ_{12} be the shear stress. As in the previous section, the corresponding strains are ϵ_1, ϵ_2, and γ_{12}. Furthermore, let f_1 and f_2 be the body forces per unit area (or per unit volume assuming unit thickness perpendicular to the horizontal plane) in the x_1- and x_2-directions, respectively, which are supposed to be positive when acted along the positive axes.

Consider a free body diagram of an infinitesimal element, as illustrated in Fig. 8.1, with dimensions dx_1 and dx_2. All stress components shown in this figure are positive. Summation of the forces in the x_1- and x_2-directions gives

$$\left(\sigma_1 + \frac{\partial \sigma_1}{\partial x_1}dx_1\right)dx_2 - \sigma_1 dx_2 + \left(\tau_{12} + \frac{\partial \tau_{12}}{\partial x_2}dx_2\right)dx_1$$
$$-\tau_{12}dx_1 + f_1 dx_1 dx_2 = 0,$$
$$\left(\tau_{12} + \frac{\partial \tau_{12}}{\partial x_1}dx_1\right)dx_2 - \tau_{12}dx_2 + \left(\sigma_2 + \frac{\partial \sigma_2}{\partial x_2}dx_2\right)dx_1$$
$$-\sigma_2 dx_1 + f_2 dx_1 dx_2 = 0. \tag{8.6}$$

Simplifying these expressions and letting $dx_1 \to 0$ and $dx_2 \to 0$ give the *equations of equilibrium* in form

$$\frac{\partial \sigma_1}{\partial x_1} + \frac{\partial \tau_{12}}{\partial x_2} + f_1 = 0,$$
$$\frac{\partial \tau_{12}}{\partial x_1} + \frac{\partial \sigma_2}{\partial x_2} + f_2 = 0. \tag{8.7}$$

Introducing the stress tensor and the body force vector

$$\boldsymbol{\sigma} = \begin{pmatrix} \sigma_1 & \tau_{12} \\ \tau_{12} & \sigma_2 \end{pmatrix}, \quad \mathbf{f} = \begin{pmatrix} f_1 \\ f_2 \end{pmatrix},$$

equations (8.7) can be written in vector form

$$\nabla \cdot \boldsymbol{\sigma} + \mathbf{f} = \mathbf{0}, \qquad \mathbf{x} \in \Omega, \tag{8.8}$$

where Ω is the reference configuration of the body under consideration.

8.1.3. Material laws

The equilibrium relation (8.8) and the strain relation (8.5) comprise five equations, and they are not sufficient to determine the eight unknowns

$$\boldsymbol{\sigma} = \begin{pmatrix} \sigma_1 \\ \sigma_2 \\ \tau_{12} \end{pmatrix}, \quad \boldsymbol{\epsilon} = \begin{pmatrix} \epsilon_1 \\ \epsilon_2 \\ \gamma_{12} \end{pmatrix}, \quad \mathbf{u} = \begin{pmatrix} u_1 \\ u_2 \end{pmatrix},$$

where the notation is abused: the stress and strain tensor notation is used to represent their respective column vectors as well; their concrete meanings will become clear in the subsequent context. The other three necessary equations arise from constitutive relationships, i.e., *material laws*. An important task in solid mechanics is to find these laws, which show how the deformation of a body depends on material properties and applied forces.

In general, the relationship between the stress vector $\boldsymbol{\sigma}$ and the strain vector $\boldsymbol{\epsilon}$ is

$$\boldsymbol{\sigma} = \mathbf{D}\boldsymbol{\epsilon}, \tag{8.9}$$

where $\mathbf{D}$ is an *elasticity matrix* that describes appropriate material properties.

Plane stress–isotropic material

A state of deformation in which all the strain components are constant throughout the body is called a *homogeneous deformation*. On the other hand, if the properties of the body are identical in all directions, the material is termed *isotropic*. In the case where the displacement is small and the material is isotropic, the relationship between the stress and strain is the *linear Hooke's law*:

$$\epsilon_1 = \frac{1}{E}\sigma_1 - \frac{\nu}{E}\sigma_2,$$

$$\epsilon_2 = -\frac{\nu}{E}\sigma_1 + \frac{1}{E}\sigma_2, \tag{8.10}$$

$$\gamma_{12} = \frac{2(1+\nu)}{E}\tau_{12},$$

where E and ν are the *modulus of elasticity* (*Young modulus*) and the *Poisson ratio* (*contraction ratio*), respectively. Solving Eq. (8.10) for the stresses produces the material matrix $\mathbf{D}$:

$$\mathbf{D} = \frac{E}{1-\nu^2}\begin{pmatrix} 1 & \nu & 0 \\ \nu & 1 & 0 \\ 0 & 0 & \dfrac{1-\nu}{2} \end{pmatrix}. \tag{8.11}$$

Plane strain–isotropic material

In the case of plane strain, a normal stress σ_3 exists in addition to the three other stress components in the x_1x_2-plane. In the case of isotropic thermal expansion, the relationship between strain and stress is

$$\epsilon_1 = \frac{1}{E}\sigma_1 - \frac{\nu}{E}\sigma_2 - \frac{\nu}{E}\sigma_3,$$

$$\epsilon_2 = -\frac{\nu}{E}\sigma_1 + \frac{1}{E}\sigma_2 - \frac{\nu}{E}\sigma_3, \tag{8.12}$$

$$\gamma_{12} = \frac{2(1+\nu)}{E}\tau_{12},$$

and

$$\epsilon_3 = -\frac{\nu}{E}\sigma_1 - \frac{\nu}{E}\sigma_2 + \frac{1}{E}\sigma_3 = 0. \tag{8.13}$$

Eliminating σ_3 and solving for the remaining three stress components give

$$\mathbf{D} = \frac{E(1-\nu)}{(1+\nu)(1-2\nu)}\begin{pmatrix} 1 & \dfrac{\nu}{1-\nu} & 0 \\ \dfrac{\nu}{1-\nu} & 1 & 0 \\ 0 & 0 & \dfrac{1-2\nu}{2(1-\nu)} \end{pmatrix}. \tag{8.14}$$

Anisotropic material

For a completely anisotropic material, 21 independent elastic constants are necessary to define completely the three-dimensional stress and strain relationship. If a two-dimensional analysis is applied, a symmetry of properties must exist, implying at most six independent constants in the property matrix $\mathbf{D}$. That is, it is possible to define

$$\mathbf{D} = \begin{pmatrix} d_{11} & d_{12} & d_{13} \\ d_{12} & d_{22} & d_{23} \\ d_{13} & d_{23} & d_{33} \end{pmatrix}$$

for describing the most general two-dimensional behavior.

A case of particular interest is a "stratified" case or transversely isotropic material where a rotational symmetry of properties exists within the plane of the strata. Such a material has only five independent elastic constants. In this special case, the general stress and strain relationship (Lekhnitskii, 1963) is, taking now the x_2-axis as perpendicular to the strata (Fig. 8.2),

$$
\begin{aligned}
\epsilon_1 &= \frac{1}{E_1}\sigma_1 - \frac{\nu_2}{E_2}\sigma_2 - \frac{\nu_1}{E_1}\sigma_3, \\
\epsilon_2 &= -\frac{\nu_2}{E_2}\sigma_1 + \frac{1}{E_2}\sigma_2 - \frac{\nu_2}{E_2}\sigma_3, \\
\epsilon_3 &= -\frac{\nu_1}{E_1}\sigma_1 - \frac{\nu_2}{E_2}\sigma_2 + \frac{1}{E_1}\sigma_3 = 0, \\
\gamma_{12} &= \frac{1}{G_2}\tau_{12}, \\
\gamma_{23} &= \frac{1}{G_2}\tau_{23}, \\
\gamma_{13} &= \frac{2(1+\nu_1)}{E_1}\tau_{13},
\end{aligned}
\tag{8.15}
$$

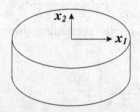

Figure 8.2 Stratified material.

where the constants E_1 and ν_1 are associated with the behavior in the strata plane and E_2, ν_2, and G_2 with the direction normal to this plane.

In the case of plane stress, with $E_1/E_2 = l$ and $G_2/E_2 = m$, the matrix $\mathbf{D}$ becomes

$$\mathbf{D} = \frac{E_2}{1 - l\nu_2^2} \begin{pmatrix} l & l\nu_2 & 0 \\ l\nu_2 & 1 & 0 \\ 0 & 0 & m(1 - l\nu_2^2) \end{pmatrix}. \tag{8.16}$$

For plane strain, it is

$$\mathbf{D} = \frac{E_2}{(1 + \nu_1)(1 - \nu_1 - 2l\nu_2^2)}$$
$$\times \begin{pmatrix} l(1 - l\nu_2^2) & l\nu_2(1 + \nu_1) & 0 \\ l\nu_2(1 + \nu_1) & 1 - \nu_1^2 & 0 \\ 0 & 0 & m(1 + \nu_1)(1 - \nu_1 - 2l\nu_2^2) \end{pmatrix}. \tag{8.17}$$

Equations (8.5), (8.8), and (8.9) constitute the basic laws for a homogeneous, isotropic (or anisotropic), elastic body, with matrix $\mathbf{D}$ given by Eqs. (8.11), (8.14), (8.16), or (8.17). With appropriate boundary conditions, they are used to determine the displacement $\mathbf{u}$, the stress σ, and the strain ϵ.

8.1.4. Boundary conditions

Let the boundary Γ be decomposed into two parts Γ_D and Γ_N, where Γ_D is fixed and a surface force $\mathbf{g}$ is applied on Γ_N (Fig. 8.3). That is

$$\begin{aligned} \mathbf{u} &= \mathbf{0} && \text{on } \Gamma_D, \\ \sigma \cdot \nu &= \mathbf{g} && \text{on } \Gamma_N, \end{aligned} \tag{8.18}$$

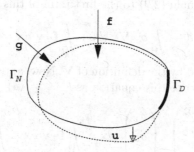

Figure 8.3   An elastic body Ω.

where $\boldsymbol{\nu}$ is the outward unit normal to Γ_N. If $\Gamma_D = \emptyset$ (respectively, $\Gamma_N = \emptyset$), the boundary value problem is called a *pure traction* (respectively, *pure displacement*) problem.

8.1.5. The finite element method

In this section, we consider the standard finite element method for solving Eqs. (8.5), (8.8), (8.9), and (8.18). They can be solved in different approaches, such as in displacement form or in mixed stress-displacement form (Ciarlet, 1978; Chen, 2005). Only the displacement approach is discussed. Furthermore, only the simplest triangular elements are addressed.

For notational convenience, we introduce a differential operator tensor **B** by

$$
\boldsymbol{\epsilon} = \begin{pmatrix} \epsilon_1 \\ \epsilon_2 \\ \gamma_{12} \end{pmatrix} = \begin{pmatrix} \dfrac{\partial}{\partial x_1} & 0 \\ 0 & \dfrac{\partial}{\partial x_2} \\ \dfrac{\partial}{\partial x_2} & \dfrac{\partial}{\partial x_1} \end{pmatrix} \begin{pmatrix} u_1 \\ u_2 \end{pmatrix} \equiv \mathbf{Bu}. \tag{8.19}
$$

We use the admissible real function space

$$
\mathbf{V} = \Big\{ \text{Vector functions } \mathbf{v}\colon \mathbf{v} \text{ is continuous on } \Omega, \frac{\partial \mathbf{v}}{\partial x_1} \text{ and } \frac{\partial \mathbf{v}}{\partial x_2} \text{ are}
$$

$$
\text{piecewise continuous and bounded on } \Omega, \text{ and } \mathbf{v} = \mathbf{0} \text{ on } \Gamma_D \Big\}.
$$

Multiplying both sides of Eq. (8.8) by a test function $\mathbf{v} \in \mathbf{V}$ and integrating the resulting equation on Ω, we see that

$$
\int_\Omega (\nabla \cdot \boldsymbol{\sigma}) \cdot \mathbf{v} \, d\mathbf{x} + \int_\Omega \mathbf{f} \cdot \mathbf{v} \, d\mathbf{x} = 0, \quad \mathbf{v} \in \mathbf{V}.
$$

Applying Green's formula (2.9) to the first term of this equation implies

$$
\int_{\Gamma_N} (\boldsymbol{\sigma} \cdot \boldsymbol{\nu}) \cdot \mathbf{v} \, d\ell - \int_\Omega \boldsymbol{\sigma} \cdot \nabla \mathbf{v} \, d\mathbf{x} + \int_\Omega \mathbf{f} \cdot \mathbf{v} \, d\mathbf{x} = 0, \quad \mathbf{v} \in \mathbf{V},
$$

where $\mathbf{v}|_{\Gamma_D} = \mathbf{0}$ is used by the definition of $\mathbf{V}$. Now, applying the boundary condition (8.18), we write this equation as

$$
\int_\Omega \begin{pmatrix} \dfrac{\partial v_1}{\partial x_1} & 0 & \dfrac{\partial v_1}{\partial x_2} \\ 0 & \dfrac{\partial v_2}{\partial x_2} & \dfrac{\partial v_2}{\partial x_1} \end{pmatrix} \cdot \begin{pmatrix} \sigma_1 \\ \sigma_2 \\ \tau_{12} \end{pmatrix} d\mathbf{x} = \int_\Omega \mathbf{f} \cdot \mathbf{v} \, d\mathbf{x} + \int_{\Gamma_N} \mathbf{g} \cdot \mathbf{v} \, d\ell, \tag{8.20}
$$

where $\mathbf{v} = (v_1, v_2)^T$.

We introduce the notation

$$\mathbf{B}^T\mathbf{v} = \mathbf{B}^T \begin{pmatrix} v_1 \\ v_2 \end{pmatrix} = \begin{pmatrix} \dfrac{\partial v_1}{\partial x_1} & 0 & \dfrac{\partial v_1}{\partial x_2} \\ 0 & \dfrac{\partial v_2}{\partial x_2} & \dfrac{\partial v_2}{\partial x_1} \end{pmatrix}. \tag{8.21}$$

Then, by Eqs. (8.9) and (8.19), system (8.20) becomes: find $\mathbf{u} \in \mathbf{V}$ such that

$$\int_{\Omega} \mathbf{B}^T\mathbf{v}\mathbf{D}\mathbf{B}\mathbf{u}\, dx = \int_{\Omega} \mathbf{f} \cdot \mathbf{v}\, dx + \int_{\Gamma_N} \mathbf{g} \cdot \mathbf{v}\, d\ell, \quad \mathbf{v} \in \mathbf{V}. \tag{8.22}$$

This is a Galerkin variational formulation of the two-dimensional plane stress or plane strain problem. If the boundary part Γ_D has a positive length, system (8.22) can be shown to have a unique solution (see Exercise 8.1). In the case of the pure traction problem, system (8.22) is solvable under a compatibility condition between $\mathbf{f}$ and $\mathbf{g}$ (Brenner-Scott, 1994). In the present two-dimensional case, for system (8.22) to have a solution in the pure traction case, the necessary and sufficient condition is

$$\int_{\Omega} \mathbf{f} \cdot \mathbf{v}\, dx + \int_{\Gamma} \mathbf{g} \cdot \mathbf{v}\, d\ell = 0,$$

for any function $\mathbf{v}$ of the form: $\mathbf{v}^T = \mathbf{b}^T + c(x_2, -x_1)$, where $\mathbf{b}$ is any constant vector and c is any constant.

For simplicity, let Ω be a convex polygonal domain, and let K_h be a regular triangulation of Ω into triangles as in Chapter 2. With the admissible space $\mathbf{V}$ defined above, we introduce the finite element space

$$\mathbf{V}_h = \{\text{Functions } \mathbf{v}: \mathbf{v} \text{ is a continuous function on } \Omega, v_i \text{ is linear}$$
$$\text{on each triangle } K \in K_h, \ i = 1, 2, \text{ and } \mathbf{v} = \mathbf{0} \text{ on } \Gamma_D\},$$

where $\mathbf{v}^T = (v_1, v_2)$. Now, the finite element method in the displacement formulation can be stated: find $\mathbf{u}_h \in \mathbf{V}_h$ such that

$$\int_{\Omega} \mathbf{B}^T\mathbf{v}\mathbf{D}\mathbf{B}\mathbf{u}_h\, dx = \int_{\Omega} \mathbf{f} \cdot \mathbf{v}\, dx + \int_{\Gamma_N} \mathbf{g} \cdot \mathbf{v}\, d\ell, \quad \mathbf{v} \in \mathbf{V}_h. \tag{8.23}$$

It can be shown that the discrete problem (8.23) has a unique solution if the boundary part Γ_D has a positive length (see Exercise 8.2). For the pure traction problem, $\mathbf{f}$ and $\mathbf{g}$ must satisfy a compatibility condition, as noted above. It can be proven that the following error estimate between the exact solution $\mathbf{u}$ and the finite element solution $\mathbf{u}_h$ holds

(Chen, 2005):

$$\left(\int_\Omega |\nabla(\mathbf{u} - \mathbf{u}_h)|^2 \, d\mathbf{x} \right)^{1/2} \le C(E, \nu)h, \tag{8.24}$$

where the constant C also depends on the second partial derivatives of the solution $\mathbf{u}$.

For a fixed pair of (E, ν), estimate (8.24) gives a satisfactory convergence result. But the convergence of the finite element solution to the exact solution is not uniform in ν as $h \to 0$. In particular, the performance of the finite element method deteriorates as $\nu \to \infty$. This phenomenon is known as *locking* (*Poisson locking* or *volume locking*). There are several approaches to reducing the effects of locking such as the *mixed* and *nonconforming finite element methods* (Ciarlet, 1978; Chen, 2005).

On each triangle $K \in K_h$, the area coordinates $(\lambda_1, \lambda_2, \lambda_3)$ (see Chapter 5) can be used to represent the approximate solution $\mathbf{u}_h^T = (u_{h1}, u_{h2})$:

$$u_{h1} = u_{11}\lambda_1 + u_{21}\lambda_2 + u_{31}\lambda_3, \quad u_{h2} = u_{12}\lambda_1 + u_{22}\lambda_2 + u_{32}\lambda_3.$$

Then, it follows from the definition (8.19) of the differential operator $\mathbf{B}$ that

$$
\mathbf{B}\mathbf{u}_h = \begin{pmatrix} \dfrac{\partial u_{h1}}{\partial x_1} \\ \dfrac{\partial u_{h2}}{\partial x_2} \\ \dfrac{\partial u_{h1}}{\partial x_2} + \dfrac{\partial u_{h2}}{\partial x_1} \end{pmatrix}
$$

$$
= \begin{pmatrix} \dfrac{\partial \lambda_1}{\partial x_1} & 0 & \dfrac{\partial \lambda_2}{\partial x_1} & 0 & \dfrac{\partial \lambda_3}{\partial x_1} & 0 \\ 0 & \dfrac{\partial \lambda_1}{\partial x_2} & 0 & \dfrac{\partial \lambda_2}{\partial x_2} & 0 & \dfrac{\partial \lambda_3}{\partial x_2} \\ \dfrac{\partial \lambda_1}{\partial x_2} & \dfrac{\partial \lambda_1}{\partial x_1} & \dfrac{\partial \lambda_2}{\partial x_2} & \dfrac{\partial \lambda_2}{\partial x_1} & \dfrac{\partial \lambda_3}{\partial x_2} & \dfrac{\partial \lambda_3}{\partial x_1} \end{pmatrix} \begin{pmatrix} u_{11} \\ u_{12} \\ u_{21} \\ u_{22} \\ u_{31} \\ u_{32} \end{pmatrix}, \tag{8.25}
$$

which we write, with the obvious definitions of $\mathbf{B}_h$ and $\mathbf{U}$ from the right-hand side of the second equality of Eq. (8.25), as

$$\mathbf{B}\mathbf{u}_h = \mathbf{B}_h \mathbf{U}.$$

Note that the partial derivatives $\partial \lambda_i / \partial x_j$ can be expressed in terms of the coordinates of the vertices of triangle K (see Sec. 5.2). Taking $\mathbf{v} = (\lambda_1, 0)$,

$(0, \lambda_1)$, $(\lambda_2, 0)$, $(0, \lambda_2)$, $(\lambda_3, 0)$, and $(0, \lambda_3)$ consecutively in Eq. (8.23) results in the linear system on each triangle K:

$$\int_K \mathbf{B}_h^T \mathbf{D} \mathbf{B}_h \ d\mathbf{x} \ \mathbf{U} = \mathbf{F} + \mathbf{G}, \tag{8.26}$$

where

$$\mathbf{F} = \begin{pmatrix} \int_K f_1 \lambda_1 \ d\mathbf{x} \\ \int_K f_2 \lambda_1 \ d\mathbf{x} \\ \int_K f_1 \lambda_2 \ d\mathbf{x} \\ \int_K f_2 \lambda_2 \ d\mathbf{x} \\ \int_K f_1 \lambda_3 \ d\mathbf{x} \\ \int_K f_2 \lambda_3 \ d\mathbf{x} \end{pmatrix}, \quad \mathbf{G} = \begin{pmatrix} \int_{\Gamma_N} g_1 \lambda_1 \ d\ell \\ \int_{\Gamma_N} g_2 \lambda_1 \ d\ell \\ \int_{\Gamma_N} g_1 \lambda_2 \ d\ell \\ \int_{\Gamma_N} g_2 \lambda_2 \ d\ell \\ \int_{\Gamma_N} g_1 \lambda_3 \ d\ell \\ \int_{\Gamma_N} g_2 \lambda_3 \ d\ell \end{pmatrix}.$$

Both of the matrices $\mathbf{B}_h$ and $\mathbf{D}$ are constant (independent of variables x_1 and x_2); so, the element matrix contribution becomes

$$\mathbf{B}_h^T \mathbf{D} \mathbf{B}_h |K|, \tag{8.27}$$

where $|K|$ is the area of the triangle K. This expression holds for both plane stress and plane strain. The global system matrix and right-hand vector can be assembled as in Sec. 2.4.

When shape functions of other types are used, we first calculate matrix $\mathbf{B}_h$ as illustrated in Eq. (8.25) and then insert it into Eq. (8.26). The row size of $\mathbf{B}_h$ is always three while the column size equals twice the number of nodes per element since the solution has two components per node for the present plane stress and plane strain problems. The body and boundary forces at the nodes can be similarly evaluated.

8.2. Three-dimensional solids

In this section, we generalize the finite element method in the previous section to three-dimensional solids. Such problems embrace clearly all the practical cases, although for some, various two-dimensional approximations give an adequate and more economical model.

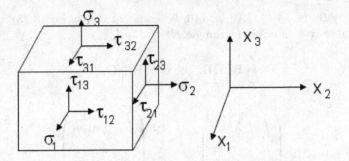

Figure 8.4 A three-dimensional solid element.

The equations of equilibrium can be derived in a similar fashion as in the previous section:

$$\frac{\partial \sigma_1}{\partial x_1} + \frac{\partial \tau_{12}}{\partial x_2} + \frac{\partial \tau_{13}}{\partial x_3} + f_1 = 0,$$

$$\frac{\partial \tau_{12}}{\partial x_1} + \frac{\partial \sigma_2}{\partial x_2} + \frac{\partial \tau_{23}}{\partial x_3} + f_2 = 0, \qquad (8.28)$$

$$\frac{\partial \tau_{13}}{\partial x_1} + \frac{\partial \tau_{23}}{\partial x_2} + \frac{\partial \sigma_3}{\partial x_3} + f_3 = 0,$$

where the stresses are shown in Fig. 8.4 in the positive direction and f_1, f_2, and f_3 are body forces per unit volume. Again, after introducing the stress tensor and the body force vector

$$\boldsymbol{\sigma} = \begin{pmatrix} \sigma_1 & \tau_{12} & \tau_{13} \\ \tau_{12} & \sigma_2 & \tau_{23} \\ \tau_{13} & \tau_{23} & \sigma_3 \end{pmatrix}, \quad \mathbf{f} = \begin{pmatrix} f_1 \\ f_2 \\ f_3 \end{pmatrix},$$

Equations (8.28) can be written in vector form

$$\nabla \cdot \boldsymbol{\sigma} + \mathbf{f} = \mathbf{0}, \qquad \mathbf{x} \in \Omega, \qquad (8.29)$$

where Ω is the reference configuration of a three-dimensional body.

By abusing the notation, we introduce the stress and strain column vectors:

$$\boldsymbol{\sigma} = \begin{pmatrix} \sigma_1 \\ \sigma_2 \\ \sigma_3 \\ \tau_{12} \\ \tau_{23} \\ \tau_{13} \end{pmatrix}, \quad \boldsymbol{\epsilon} = \begin{pmatrix} \epsilon_1 \\ \epsilon_2 \\ \epsilon_3 \\ \gamma_{12} \\ \gamma_{23} \\ \gamma_{13} \end{pmatrix}.$$

Then, a relationship between these two vectors is

$$\sigma = \mathbf{D}\epsilon. \tag{8.30}$$

With complete anisotropy the material property matrix $\mathbf{D}$ can contain 21 independent constants. While no difficulty presents itself in computation when dealing with such materials, because the multiplication will never be performed explicitly, we state an example for isotropic material:

$$\mathbf{D} = \frac{E}{(1+\nu)(1-2\nu)} \begin{pmatrix} 1-\nu & \nu & \nu & 0 & 0 & 0 \\ \nu & 1-\nu & \nu & 0 & 0 & 0 \\ \nu & \nu & 1-\nu & 0 & 0 & 0 \\ 0 & 0 & 0 & \dfrac{1-2\nu}{2} & 0 & 0 \\ 0 & 0 & 0 & 0 & \dfrac{1-2\nu}{2} & 0 \\ 0 & 0 & 0 & 0 & 0 & \dfrac{1-2\nu}{2} \end{pmatrix}. \tag{8.31}$$

Finally, the kinematic equation is

$$\epsilon = \begin{pmatrix} \epsilon_1 \\ \epsilon_2 \\ \epsilon_3 \\ \gamma_{12} \\ \gamma_{23} \\ \gamma_{13} \end{pmatrix} = \begin{pmatrix} \dfrac{\partial u_1}{\partial x_1} \\ \dfrac{\partial u_2}{\partial x_2} \\ \dfrac{\partial u_3}{\partial x_3} \\ \dfrac{\partial u_1}{\partial x_2} + \dfrac{\partial u_2}{\partial x_1} \\ \dfrac{\partial u_2}{\partial x_3} + \dfrac{\partial u_3}{\partial x_2} \\ \dfrac{\partial u_1}{\partial x_3} + \dfrac{\partial u_3}{\partial x_1} \end{pmatrix} = \begin{pmatrix} \dfrac{\partial}{\partial x_1} & 0 & 0 \\ 0 & \dfrac{\partial}{\partial x_2} & 0 \\ 0 & 0 & \dfrac{\partial}{\partial x_3} \\ \dfrac{\partial}{\partial x_2} & \dfrac{\partial}{\partial x_1} & 0 \\ 0 & \dfrac{\partial}{\partial x_3} & \dfrac{\partial}{\partial x_2} \\ \dfrac{\partial}{\partial x_3} & 0 & \dfrac{\partial}{\partial x_1} \end{pmatrix} \begin{pmatrix} u_1 \\ u_2 \\ u_3 \end{pmatrix},$$

which we write as

$$\epsilon = \mathbf{B}\mathbf{u}. \tag{8.32}$$

The boundary conditions can be imposed as in Eq. (8.18). Now, Eqs. (8.29), (8.30), and (8.32) determine the variables σ, ϵ, and $\mathbf{u}$.

The admissible real function space is

$$\mathbf{V} = \left\{ \text{Vector functions } \mathbf{v} \colon \mathbf{v} \text{ is continuous on } \Omega, \frac{\partial \mathbf{v}}{\partial x_1}, \frac{\partial \mathbf{v}}{\partial x_2}, \text{ and } \frac{\partial \mathbf{v}}{\partial x_3} \right.$$

$$\left. \text{are piecewise continuous and bounded on } \Omega, \text{ and } \mathbf{v} = \mathbf{0} \text{ on } \Gamma_D \right\}.$$

Then, the three-dimensional variational formulation can be derived in a similar manner as in two dimensions: find $\mathbf{u} \in \mathbf{V}$ such that

$$\int_\Omega \mathbf{B}^T \mathbf{v} \mathbf{D} \mathbf{B} \mathbf{u} \, dx = \int_\Omega \mathbf{f} \cdot \mathbf{v} \, dx + \int_{\Gamma_N} \mathbf{g} \cdot \mathbf{v} \, d\ell, \quad \mathbf{v} \in \mathbf{V}. \tag{8.33}$$

The simplest three-dimensional continuum element is a tetrahedron. Let K_h be a partition of a polygonal domain $\Omega \subset \mathbb{R}^3$ into non overlapping *tetrahedra* such that no vertex of any tetrahedron lies in the interior of an edge or face of another tetrahedron. The finite element space is

$\mathbf{V}_h = \{$Functions $\mathbf{v}$: $\mathbf{v}$ is a continuous function on Ω, v_i is linear

on each tetrahedron $K \in K_h$, $i = 1, 2, 3$, and $\mathbf{v} = \mathbf{0}$ on $\Gamma_D\}$,

where $\mathbf{v}^T = (v_1, v_2, v_3)$. Now, the finite element solution $\mathbf{u}_h \in \mathbf{V}_h$ satisfies the system

$$\int_\Omega \mathbf{B}^T \mathbf{v} \mathbf{D} \mathbf{B} \mathbf{u}_h \, dx = \int_\Omega \mathbf{f} \cdot \mathbf{v} \, dx + \int_{\Gamma_N} \mathbf{g} \cdot \mathbf{v} \, d\ell, \quad \mathbf{v} \in \mathbf{V}_h. \tag{8.34}$$

On each tetrahedron $K \in K_h$, the solution $\mathbf{u}_h$ is represented using the volume coordinates λ_i (see Sec. 6.2):

$$\mathbf{u}_h = \sum_{i=1}^4 \begin{pmatrix} u_1 \\ u_2 \\ u_3 \end{pmatrix}_i \lambda_i.$$

As a result, it follows from the definition of the differential operator $\mathbf{B}$ that Eq. (8.32) can be replaced at the discrete level:

$$\mathbf{B} \mathbf{u}_h = \mathbf{B}_h \mathbf{U},$$

where the matrix $\mathbf{B}_h$ equals

$$\begin{pmatrix}
\frac{\partial \lambda_1}{\partial x_1} & 0 & 0 & \frac{\partial \lambda_2}{\partial x_1} & 0 & 0 & \frac{\partial \lambda_3}{\partial x_1} & 0 & 0 & \frac{\partial \lambda_4}{\partial x_1} & 0 & 0 \\
0 & \frac{\partial \lambda_1}{\partial x_2} & 0 & 0 & \frac{\partial \lambda_2}{\partial x_2} & 0 & 0 & \frac{\partial \lambda_3}{\partial x_2} & 0 & 0 & \frac{\partial \lambda_4}{\partial x_2} & 0 \\
0 & 0 & \frac{\partial \lambda_1}{\partial x_3} & 0 & 0 & \frac{\partial \lambda_2}{\partial x_3} & 0 & 0 & \frac{\partial \lambda_3}{\partial x_3} & 0 & 0 & \frac{\partial \lambda_4}{\partial x_3} \\
\frac{\partial \lambda_1}{\partial x_2} & \frac{\partial \lambda_1}{\partial x_1} & 0 & \frac{\partial \lambda_2}{\partial x_2} & \frac{\partial \lambda_2}{\partial x_1} & 0 & \frac{\partial \lambda_3}{\partial x_2} & \frac{\partial \lambda_3}{\partial x_1} & 0 & \frac{\partial \lambda_4}{\partial x_2} & \frac{\partial \lambda_4}{\partial x_1} & 0 \\
0 & \frac{\partial \lambda_1}{\partial x_3} & \frac{\partial \lambda_1}{\partial x_2} & 0 & \frac{\partial \lambda_2}{\partial x_3} & \frac{\partial \lambda_2}{\partial x_2} & 0 & \frac{\partial \lambda_3}{\partial x_3} & \frac{\partial \lambda_3}{\partial x_2} & 0 & \frac{\partial \lambda_4}{\partial x_3} & \frac{\partial \lambda_4}{\partial x_2} \\
\frac{\partial \lambda_1}{\partial x_3} & 0 & \frac{\partial \lambda_1}{\partial x_1} & \frac{\partial \lambda_2}{\partial x_3} & 0 & \frac{\partial \lambda_2}{\partial x_1} & \frac{\partial \lambda_3}{\partial x_3} & 0 & \frac{\partial \lambda_3}{\partial x_1} & \frac{\partial \lambda_4}{\partial x_3} & 0 & \frac{\partial \lambda_4}{\partial x_1}
\end{pmatrix},$$

and $\mathbf{U} = (u_{11}, u_{12}, u_{13}, u_{21}, u_{22}, u_{23}, u_{31}, u_{32}, u_{33}, u_{41}, u_{42}, u_{43})^T$. Therefore, in a similar argument as for Eq. (8.26), we obtain the local linear system on each tetrahedron K:

$$\mathbf{B}_h^T \mathbf{D} \mathbf{B}_h |K| \mathbf{U} = \mathbf{F} + \mathbf{G}, \qquad (8.35)$$

where $|K|$ is the volume of K and

$$\mathbf{F} = \left(\int_K f_1 \lambda_i \, d\mathbf{x}, \int_K f_2 \lambda_i \, d\mathbf{x}, \int_K f_3 \lambda_i \, d\mathbf{x} \right)_{i=1-4}^T,$$

$$\mathbf{G} = \left(\int_{\Gamma_N} g_1 \lambda_i \, d\ell, \int_{\Gamma_N} g_2 \lambda_i \, d\ell; \int_{\Gamma_N} g_3 \lambda_i \, d\ell \right)_{i=1-4}^T.$$

8.3. Axisymmetric solids

The problem of determining stress distribution in solids of revolution (axisymmetric bodies) under axisymmetric forces is of important interest (Fig. 8.5). The mathematical consideration is very analogous to that of plane stress and plane strain presented in Sec. 8.1, since both cases are two-dimensional. By symmetry, the two components of displacement in any plane section of the solid along its axis of symmetry completely determine the state of strain and thus the state of stress. In particular, two shear stress components $\tau_{r\theta}$ and $\tau_{z\theta}$ vanish in a cylindrical (r, θ, z)-coordinate system, where r, θ, and z represent the radial, circumferential, and axial directions,

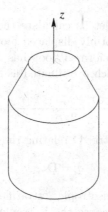

Figure 8.5 An axisymmetric solid.

respectively. Consequently, the stress and strain vectors reduce to

$$\boldsymbol{\sigma} = \begin{pmatrix} \sigma_r \\ \sigma_\theta \\ \sigma_z \\ \tau_{rz} \end{pmatrix}, \quad \boldsymbol{\epsilon} = \begin{pmatrix} \epsilon_r \\ \epsilon_\theta \\ \epsilon_z \\ \gamma_{rz} \end{pmatrix}.$$

In the present case, the equations of equilibrium in the r- and z-directions are

$$\frac{\partial \sigma_r}{\partial r} + \frac{\partial \tau_{rz}}{\partial z} + \frac{\sigma_r - \sigma_\theta}{r} + f_r = 0,$$

$$\frac{\partial \sigma_z}{\partial z} + \frac{\partial \tau_{rz}}{\partial r} + \frac{\tau_{rz}}{r} + f_z = 0,$$

$$(8.36)$$

where f_r and f_z are the body forces per unit volume in the r- and z-directions, respectively. Because the stress components are independent of θ, the equilibrium equation in the θ-direction is identically satisfied. In addition, the kinematic equation is

$$\begin{pmatrix} \epsilon_r \\ \epsilon_\theta \\ \epsilon_z \\ \gamma_{rz} \end{pmatrix} = \begin{pmatrix} \dfrac{\partial u_r}{\partial r} \\ \dfrac{u_r}{r} \\ \dfrac{\partial u_z}{\partial z} \\ \dfrac{\partial u_r}{\partial z} + \dfrac{\partial u_z}{\partial r} \end{pmatrix}, \quad (8.37)$$

where u_r and u_z indicate the displacements in the radial and axial directions, respectively.

To see that $\epsilon_\theta = u_r/r$, let us calculate the circumferential strain ϵ_θ if a hoop of radius r is uniformly displaced along the radial direction by displacement u_r. Then the deformed hoop has a uniform strain along the circumferential direction, which is evaluated as follows:

$$\epsilon_\theta = \frac{2\pi(r + u_r) - 2\pi r}{2\pi r} = \frac{u_r}{r}.$$

Finally, the elasticity matrix $\mathbf{D}$ relating the stress to the strain

$$\boldsymbol{\sigma} = \mathbf{D}\boldsymbol{\epsilon} \qquad (8.38)$$

must be determined. The anisotropic, "stratified" material considered in Sec. 8.1 is first discussed because the isotropic case can then be simply deduced as a special case.

8.3.1. Anisotropic material

With the z-axis representing the normal to the planes of stratification (Fig. 8.5), Eqs. (8.15) can be rewritten as

$$
\begin{aligned}
\epsilon_r &= -\frac{\nu_2}{E_2}\sigma_z + \frac{1}{E_1}\sigma_r - \frac{\nu_1}{E_1}\sigma_\theta, \\
\epsilon_\theta &= -\frac{\nu_2}{E_2}\sigma_z - \frac{\nu_1}{E_1}\sigma_r + \frac{1}{E_1}\sigma_\theta, \\
\epsilon_z &= \frac{1}{E_2}\sigma_z - \frac{\nu_2}{E_2}\sigma_r - \frac{\nu_2}{E_2}\sigma_\theta, \\
\gamma_{rz} &= \frac{1}{G_2}\tau_{rz}.
\end{aligned}
\tag{8.39}
$$

With $E_1/E_2 = l$ and $G_2/E_2 = m$, we see that

$$
\mathbf{D} = \frac{E_2}{(1+\nu_1)(1-\nu_1-2l\nu_2^2)}
$$

$$
\times \begin{pmatrix}
l(1-l\nu_2^2) & l(\nu_1+l\nu_2^2) & l\nu_2(1+\nu_1) & 0 \\
l(\nu_1+l\nu_2^2) & l(1-l\nu_2^2) & l\nu_2(1+\nu_1) & 0 \\
l\nu_2(1+\nu_1) & l\nu_2(1+\nu_1) & 1-\nu_1^2 & 0 \\
0 & 0 & 0 & m(1+\nu_1)(1-\nu_1-2l\nu_2^2)
\end{pmatrix}.
\tag{8.40}
$$

8.3.2. Isotropic material

For isotropic material, matrix $\mathbf{D}$ is obtained by taking

$$
E_1 = E_2 = E, \text{ i.e., } l = 1, \quad \nu_1 = \nu_2 = \nu,
$$

and using the well-known relationship between the elastic constants

$$
\frac{G_2}{E_2} = \frac{G}{E} = m = \frac{1}{2(1+\nu)}.
$$

With these simplifications, Eq. (8.40) reduces to

$$
\mathbf{D} = \frac{E}{(1+\nu)(1-2\nu)}
\begin{pmatrix}
1-\nu & \nu & \nu & 0 \\
\nu & 1-\nu & \nu & 0 \\
\nu & \nu & 1-\nu & 0 \\
0 & 0 & 0 & \dfrac{1-2\nu}{2}
\end{pmatrix}.
\tag{8.41}
$$

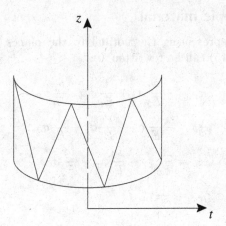

Figure 8.6 A ring element.

To develop an element stiffness matrix, we use the linear triangular element again. We replace variables x_1 and x_2 in the shape functions with variables r and z, respectively, for the axisymmetric problem. In addition, the axisymmetric element is a ring element as illustrated in Fig. 8.6. Substitution of the shape functions into the kinematic equation (8.37) yields

$$
\epsilon = \begin{pmatrix}
\dfrac{\partial \lambda_1}{\partial r} & 0 & \dfrac{\partial \lambda_2}{\partial r} & 0 & \dfrac{\partial \lambda_3}{\partial r} & 0 \\[2mm]
\dfrac{\lambda_1}{r} & 0 & \dfrac{\lambda_2}{r} & 0 & \dfrac{\lambda_3}{r} & 0 \\[2mm]
0 & \dfrac{\partial \lambda_1}{\partial z} & 0 & \dfrac{\partial \lambda_2}{\partial z} & 0 & \dfrac{\partial \lambda_3}{\partial z} \\[2mm]
\dfrac{\partial \lambda_1}{\partial z} & \dfrac{\partial \lambda_1}{\partial r} & \dfrac{\partial \lambda_2}{\partial z} & \dfrac{\partial \lambda_2}{\partial r} & \dfrac{\partial \lambda_3}{\partial z} & \dfrac{\partial \lambda_3}{\partial r}
\end{pmatrix}
\begin{pmatrix}
u_{1r} \\ u_{1z} \\ u_{2r} \\ u_{2z} \\ u_{3r} \\ u_{3z}
\end{pmatrix},
$$

which we write as

$$
\epsilon = \mathbf{B}_h \mathbf{U}. \tag{8.42}
$$

The element matrix can be expressed as

$$
\int_z \int_r \int_0^{2\pi} \mathbf{B}_h^T \mathbf{D} \mathbf{B}_h r \, d\theta dr dz = 2\pi \int_z \int_r \mathbf{B}_h^T \mathbf{D} \mathbf{B}_h r \, dr dz. \tag{8.43}
$$

Unlike the plane stress and plane strain cases, due to the presence of the term λ_i/r in matrix $\mathbf{B}_h$, the integrand in Eq. (8.43) is not constant. Hence, this integral needs to be evaluated. A simple approximation is to calculate

$\mathbf{B}_h$ at the centroid of a triangle K, $(\bar{r}, \bar{z})$, where $\bar{r} = (r_i + r_j + r_m)/3$ and $\bar{z} = (z_i + z_j + z_m)/3$. In this case, the element matrix becomes

$$2\pi\bar{r}\overline{\mathbf{B}}_h^T\mathbf{D}\overline{\mathbf{B}}_h|K|, \tag{8.44}$$

where $|K|$ is the area of the triangle K and $\overline{\mathbf{B}}_h$ is the matrix $\mathbf{B}_h$ evaluated at $(\bar{r}, \bar{z})$. Equation (8.44) holds for both isotropic and anisotropic materials.

8.4. Exercises

8.1. Show that if the Dirichlet boundary part Γ_D has a positive length, system (8.22) has a unique solution.

8.2. Show that if the boundary part Γ_D has a positive length, the discrete problem (8.23) has a unique solution.

8.3. We consider another two-dimensional version of the elastic model given by Eqs. (8.5), (8.8), (8.9), and (8.18). Let $\Omega \subset \mathbb{R}^2$ be a planar domain, and the elasticity body have the form $\Omega \times (-l, l)$, where l is a small real number. This problem corresponds to a thin elastic plate with a middle surface Ω subject to planar loads only (no transversal loads). Assuming a *plane stress state* (i.e., $\sigma_{i3} = 0$, $i = 1, 2, 3$), Eqs. (8.5), (8.8), (8.9), and (8.18) become

$$\epsilon_{ij} = \frac{1}{2}\left(\frac{\partial u_i}{\partial x_j} + \frac{\partial u_j}{\partial x_i}\right), \qquad i, j = 1, 2, \mathbf{x} \in \Omega,$$

$$\sum_{j=1}^{2} \frac{\partial \sigma_{ij}}{\partial x_j} + f_i = 0, \qquad i = 1, 2, \mathbf{x} \in \Omega,$$

$$\sigma_{ij} = 2\mu\epsilon_{ij}(\mathbf{u}) + \bar{\lambda}\big(\epsilon_{11}(\mathbf{u}) + \epsilon_{22}(\mathbf{u})\big)\delta_{ij}, \quad i, j = 1, 2, \mathbf{x} \in \Omega,$$

$$u_1 = u_2 = 0, \qquad\qquad\qquad \mathbf{x} \in \Gamma_D,$$

$$\sum_{j=1}^{2} \sigma_{ij}\nu_j = g_i, \qquad i = 1, 2, \mathbf{x} \in \Gamma_N,$$

where f_i and g_i are given forces and

$$\bar{\lambda} = \frac{E\nu}{1 - \nu^2}.$$

Write a Galerkin variational formulation for this problem in the displacement form, and formulate the corresponding finite element method using linear displacements on triangles (see Sec. 8.1.5).

Chapter 9

Application to Fluid Mechanics

The problems of solid and fluid mechanics are similar in that stresses occur and the material is displaced for both problems. There exists, however, a major difference between them. Fluids at rest cannot support any deviatory stresses. Hence, only a pressure or a mean compressive stress can be carried.

Deviatory stresses, in addition to pressure, can develop when fluids are in motion. These stresses are characterized by a quantity that is similar to the shear modulus of solid mechanics and is referred to as *dynamic viscosity* (molecular viscosity).

The other significant difference between solid and fluid problems is, even when the flow has a constant velocity (steady state), the effect of *convective acceleration* that makes the equations of fluid dynamics non-self-adjoint. Therefore, unless the velocity is so small that this effect is negligible, the treatment of fluid mechanics must be different from that of solid mechanics. As observed in Chapter 8, the standard finite element method produced an optimal solution for the solid mechanics equations. In the case of fluid mechanics, this method must be adjusted to give an optimal solution.

A simultaneous approximation of pressure and velocity needs to be carried out in a stable manner. The approximation spaces for these two variables must satisfy a *stability condition*. To overcome this difficulty, a wide variety of numerical methods, such as splitting, projection, and penalty methods, have been proposed to decouple the calculation of pressure from that of velocity (Glowinski, 2003; Zienkiewicz *et al.*, 2005). As an example, the splitting and projection methods are discussed here. In addition, the nonconforming and mixed finite element methods are also considered for the solution of the equations of fluid mechanics. These two methods are further described in Chapter 11. In Sec. 9.1, the basic governing equations of fluid dynamics are reviewed. Then, in Sec. 9.2, a characteristic-based splitting scheme, combined with the finite element method, is described. The projection approach, combined with the standard and nonconforming finite element methods, is introduced in Secs. 9.3 and 9.4, respectively. Finally, the mixed finite element method for the Stokes and Navier–Stokes equations is discussed in Secs. 9.5 and 9.6, respectively.

9.1. Equations of fluid dynamics

The motion of a continuous medium is governed by the fundamental principles of classical mechanics and thermodynamics for the *conservation of mass, momentum,* and *energy*. In particular, the application of the first two principles in a frame of reference leads to the following differential equations:

$$\frac{\partial \rho}{\partial t} + \nabla \cdot (\rho \mathbf{u}) = 0,$$

$$\frac{\partial}{\partial t}(\rho \mathbf{u}) + \nabla \cdot (\rho \mathbf{u}\mathbf{u} - \boldsymbol{\sigma}) = \mathbf{f},$$

(9.1)

where ρ, $\mathbf{u}$, and $\boldsymbol{\sigma}$ are, respectively, the density, velocity, and stress tensor of the continuous medium, and $\mathbf{f}$ is the force (per unit volume). These equations are in *divergence form*. Their *nondivergence form* is (see Exercise 9.1)

$$\frac{D\rho}{Dt} + \rho \nabla \cdot \mathbf{u} = 0,$$

$$\rho \frac{D\mathbf{u}}{Dt} - \nabla \cdot \boldsymbol{\sigma} = \mathbf{f},$$

(9.2)

where the *material derivative* is defined by

$$\frac{D}{Dt} = \frac{\partial}{\partial t} + \mathbf{u} \cdot \nabla.$$

These equations are based on the *Eulerian approach* for the description of continuum motion; i.e., the characteristic properties of the medium $(\rho, \mathbf{u}, \boldsymbol{\sigma})$ are treated as functions of time and space in the frame of reference. An alternative description is through the *Lagrangian approach* where the dependent variables are the characteristic properties of the material particles that are followed in motion. These properties are the functions of time and parameters used to identify the particles such as the particle coordinates at a fixed initial time. This approach, or more precisely the mixed

Lagrangian–Eulerian approach, is most effective for problems involving different media with interfaces. It is not as widely used in fluid mechanics as the Eulerian approach alone and thus it is not presented.

The basic unknowns in Eqs. (9.1) or (9.2) are $(\rho, \mathbf{u})$. A constitutive relationship is needed for the stress tensor $\boldsymbol{\sigma}$ as in the preceding chapter. A fluid is *Newtonian* if its stress tensor is a linear function of the velocity gradient. For this type of fluid, the *Newton law* (or *Navier–Stokes law*) applies:

$$\boldsymbol{\sigma} = -p\mathbf{I} + \boldsymbol{\tau}, \quad \boldsymbol{\tau} = \mu\left(\nabla\mathbf{u} + (\nabla\mathbf{u})^T\right) + \lambda\nabla \cdot \mathbf{u}\mathbf{I}, \tag{9.3}$$

where p and $\boldsymbol{\tau}$ are the pressure and viscous stress tensor, and μ and λ are the viscosity coefficients. These coefficients can be related by $\lambda = -2\mu/3$, e.g.

We substitute Eq. (9.3) into the second (momentum) equation in system (9.2) to obtain

$$\rho\frac{D\mathbf{u}}{Dt} + \nabla p = \mu\Delta\mathbf{u} + (\mu + \lambda)\nabla(\nabla \cdot \mathbf{u}) + \mathbf{f}$$
$$+ \nabla \cdot \mathbf{u}\nabla\lambda + \nabla\mu \cdot (\nabla\mathbf{u} + (\nabla\mathbf{u})^T). \tag{9.4}$$

In general, the viscosity coefficients depend on temperature; in the present case where the temperature is fixed, they are constant. Consequently, Eq. (9.4) becomes

$$\rho\frac{D\mathbf{u}}{Dt} + \nabla p = \mu\Delta\mathbf{u} + (\mu + \lambda)\nabla(\nabla \cdot \mathbf{u}) + \mathbf{f}. \tag{9.5}$$

An *incompressible flow* is characterized by the condition

$$\nabla \cdot \mathbf{u} = 0. \tag{9.6}$$

Using Eq. (9.6), the first (mass conservation) equation in system (9.2) becomes

$$\frac{D\rho}{Dt} = 0. \tag{9.7}$$

This equation implies that the density is constant along a fluid particle trajectory. In most cases, we can assume that ρ is constant so that Eq. (9.7) is satisfied everywhere.

Under condition (9.6), the momentum equation (9.5) becomes

$$\rho\left(\frac{\partial\mathbf{u}}{\partial t} + (\mathbf{u} \cdot \nabla)\mathbf{u}\right) + \nabla p - \mu\Delta\mathbf{u} = \mathbf{f}. \tag{9.8}$$

This equation is known as the Navier–Stokes equation. In the incompressible case, the unknown variables are the pressure and velocity field. They can be determined from Eqs. (9.6) and (9.8). The Navier–Stokes equation can be also presented in the *stream-function vorticity formulation*, which is not discussed in this chapter.

We observe that the Navier–Stokes equation is nonlinear. If we neglect the nonlinear term, we derive the *Stokes equation*:

$$\rho\frac{\partial \mathbf{u}}{\partial t} + \nabla p - \mu\Delta\mathbf{u} = \mathbf{f}. \tag{9.9}$$

Strictly speaking, the Stokes equation is valid only for a viscous Newtonian fluid over a limited range of flow rates where *turbulence, inertial,* and other high velocity effects are negligible. As the flow velocity is increased, deviations from the Stokes flow are observed. The generally accepted explanation is that, as the velocity is increased, deviations are due to inertial effects first, followed later by turbulent effects. Such a phenomenon can be characterized by the well-known *Reynolds number* that expresses the ratio between the inertial force and the viscous (frictional) force and can be defined by

$$Re = \frac{Lu^*}{\mu},$$

where L and u^* are some reference length and speed of a medium, respectively. This number can be used as a criterion to distinguish between *laminar flow* occurring at low velocities and *turbulent flow*. The critical number Re between these two types of flows in pipes is about 2100. In this chapter, we consider a low velocity flow of an incompressible Newtonian fluid. Turbulent flow is beyond the scope of this chapter.

9.2. A characteristic-based splitting method

A splitting method was proposed by Chorin (1968) for a finite difference solution of an incompressible flow problem. A similar method in the context of finite elements for this type of flow was developed by many researchers (Comini and Del Guidice, 1972; Zienkiewicz *et al.*, 2005). Due to the presence of convective acceleration in fluid mechanics noted above, the splitting in time can be carried out along the characteristics of the problem to produce an accurate solution. Characteristic-based finite element methods have been intensively studied (Douglas-Russell, 1982; Pironneau, 1982). These methods, which are further considered in Chapter 11, are somewhat complex in coding. In this section a simpler alternative is used, where programming difficulties are avoided at the expense of conditional stability (Zienkiewicz *et al.*, 1986).

9.2.1. An explicit characteristic-based method

To better understand the major idea of the characteristic-based finite element method considered, we focus on a one-dimensional *convection–diffusion problem* on the whole real line:

$$\frac{\partial p}{\partial t} + b(x,t)\frac{\partial p}{\partial x} - \frac{\partial}{\partial x}\left(a(x,t)\frac{\partial p}{\partial x}\right) = f(x,t), \quad -\infty < x < \infty,\ t > 0,$$

$$p(x,0) = p_0(x), \quad -\infty < x < \infty, \tag{9.10}$$

where p is the unknown variable and all other functions are given. Let the characteristic direction associated with the hyperbolic part of Eq. (9.10), $\partial p/\partial t + b\,\partial p/\partial x$, be denoted by $\tau(x)$, so

$$\frac{\partial}{\partial \tau(x)} = \frac{\partial}{\partial t} + b(x,t)\frac{\partial}{\partial x}.$$

Then, Eq. (9.10) can be rewritten as

$$\frac{\partial p}{\partial \tau} - \frac{\partial}{\partial x}\left(a(x,t)\frac{\partial p}{\partial x}\right) = f(x,t), \quad -\infty < x < \infty,\ t > 0,$$

$$p(x,0) = p_0(x), \quad -\infty < x < \infty. \tag{9.11}$$

The convective acceleration term disappears, and Eq. (9.11) can be now solved using the standard finite element method.

Let $0 = t^0 < t^1 < \cdots < t^n < \cdots$ be a partition in time, with time steps $\Delta t^n = t^n - t^{n-1}$. For a generic function v of time, set $v^n = v(t^n)$. For notational simplicity, assume that this partition is uniform: $\Delta t = \Delta t^n$, $n = 1, 2, \ldots$. The time discretization of Eq. (9.11) along the characteristics (Fig. 9.1) gives

$$\frac{1}{\Delta t}\left(p^{n+1} - p^n\Big|_{(x-\delta)}\right) \approx \theta\left(\frac{\partial}{\partial x}\left(a\frac{\partial p}{\partial x}\right) + f\right)^{n+1}$$

$$+ (1-\theta)\left(\frac{\partial}{\partial x}\left(a\frac{\partial p}{\partial x}\right) + f\right)^{n}\Bigg|_{(x-\delta)}, \tag{9.12}$$

where $\delta = \bar{b}^n \Delta t$ is the distance traveled by the particle in the x-direction, $\bar{b}^n$ is some average of b^n along the characteristics, $\theta = 0$ results in the explicit scheme, and the values of $0 < \theta \leq 1$ lead to semi implicit or fully implicit schemes in time. It is well known that the solution of Eq. (9.12) in moving

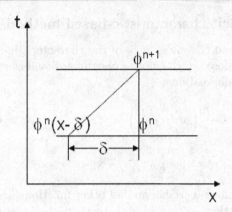

Figure 9.1   Approximate characteristics.

coordinates causes grid updating and presents difficulties in programming. These difficulties can be overcome through Taylor expansions:

$$p^n\Big|_{(x-\delta)} = p^n - \delta\frac{\partial p^n}{\partial x} + \frac{\delta^2}{2}\frac{\partial^2 p^n}{\partial x^2} + O\left(\delta^3\right),$$

$$\left(\frac{\partial}{\partial x}\left(a\frac{\partial p}{\partial x}\right)\right)^n\Big|_{(x-\delta)} = \left(\frac{\partial}{\partial x}\left(a\frac{\partial p}{\partial x}\right)\right)^n - \delta\frac{\partial}{\partial x}\left(\frac{\partial}{\partial x}\left(a\frac{\partial p}{\partial x}\right)\right)^n + O\left(\delta^2\right),$$

$$f^n\Big|_{(x-\delta)} = f^n - \delta\frac{\partial f^n}{\partial x} + O\left(\delta^2\right). \tag{9.13}$$

Different approximations of the average $\bar{b}$ of b along the characteristics generate different stablization terms. Here, a simple approximation is used:

$$\bar{b}^n = b^n - b^n\Delta t\frac{\partial b^n}{\partial x}. \tag{9.14}$$

Substitution of Eqs. (9.13) and (9.14) into Eq. (9.12), with $\theta = 1/2$, e.g., gives

$$p^{n+1} - p^n = -\Delta t\left(b^n\frac{\partial p^n}{\partial x} - \left(\frac{\partial}{\partial x}\left(a\frac{\partial p}{\partial x}\right)\right)^{n+1/2} - f^{n+1/2}\right)$$
$$+ \Delta t\left[\frac{\Delta t}{2}\frac{\partial}{\partial x}\left(b^2\frac{\partial p}{\partial x}\right) - \frac{\Delta t}{2}b\frac{\partial^2}{\partial x^2}\left(a\frac{\partial p}{\partial x}\right) - \frac{\Delta t}{2}b\frac{\partial f}{\partial x}\right]^n, \tag{9.15}$$

where

$$\left(\frac{\partial}{\partial x}\left(a\frac{\partial p}{\partial x}\right)\right)^{n+1/2} = \frac{1}{2}\left[\left(\frac{\partial}{\partial x}\left(a\frac{\partial p}{\partial x}\right)\right)^{n+1} + \left(\frac{\partial}{\partial x}\left(a\frac{\partial p}{\partial x}\right)\right)^{n}\right],$$

$$f^{n+1/2} = \frac{1}{2}(f^{n+1} + f^n).$$

When the terms with the superscript $n+1/2$ are replaced by those with n, the algorithm (9.15) becomes explicit in time. The additional terms in this algorithm are stabilization terms.

In the explicit case, algorithm (9.15) becomes

$$p^{n+1} - p^n = -\Delta t\left(b\frac{\partial p}{\partial x} - \frac{\partial}{\partial x}\left(a\frac{\partial p}{\partial x}\right) - f\right)^n$$

$$+ \frac{\Delta t^2}{2}\left[\frac{\partial}{\partial x}\left(b^2\frac{\partial p}{\partial x}\right) - b\frac{\partial^2}{\partial x^2}\left(a\frac{\partial p}{\partial x}\right) - b\frac{\partial f}{\partial x}\right]^n. \tag{9.16}$$

Generalization to multiple dimensions gives

$$p^{n+1} - p^n = -\Delta t\left[\mathbf{b}\cdot\nabla p - \nabla\cdot(\mathbf{a}\nabla p) - f\right]^n$$

$$+ \frac{\Delta t^2}{2}\left[\sum_{i,j=1}^{d}\frac{\partial}{\partial x_i}\left(b_i b_j\frac{\partial p}{\partial x_j}\right) - \mathbf{b}\cdot\nabla\left[\nabla\cdot(\mathbf{a}\nabla p)\right] - \mathbf{b}\cdot\nabla f\right]^n, \tag{9.17}$$

where $\mathbf{b} = (b_1, b_2, \ldots, b_d)$ is a vector function, $\mathbf{a}$ is a tensor, and $d = 2$ or 3 is the dimension number.

An alternative approximation for the average $\bar{b}$ is

$$b^n = \frac{1}{2}\left(b^{n+1} + b^n\bigg|_{(x-\delta)}\right),$$

where the following Taylor expansion can be used:

$$b^n\bigg|_{(x-\delta)} = b^n - \Delta t b^n\frac{\partial b^n}{\partial x} + O\left(\delta^2\right).$$

In this case, Eq. (9.15) becomes

$$p^{n+1} - p^n = -\Delta t\left(b\frac{\partial p}{\partial x} - \frac{\partial}{\partial x}\left(a\frac{\partial p}{\partial x}\right) - f\right)^n$$

$$+ \frac{\Delta t^2}{2}b^n\left[\frac{\partial}{\partial x}\left(b\frac{\partial p}{\partial x} - \frac{\partial}{\partial x}\left(a\frac{\partial p}{\partial x}\right) - f\right)\right]^n, \tag{9.18}$$

and the corresponding multidimensional counterpart is

$$p^{n+1} - p^n = -\Delta t \left[\nabla \cdot (\mathbf{b}p) - \nabla \cdot (\mathbf{a}\nabla p) - f \right]^n$$
$$+ \frac{\Delta t^2}{2} \mathbf{b}^n \cdot \nabla \left[\nabla \cdot (\mathbf{b}p) - \nabla \cdot (\mathbf{a}\nabla p) - f \right]^n . \tag{9.19}$$

Now, the standard finite element method can be used to solve Eqs. (9.17) or (9.19). The two different approximations for the average $\bar{b}$ produce different stabilization terms in these two equations. The difference is small, and if b is a constant, the equations are the same. In the subsequent application, the conservative form (9.19) is used.

As noted earlier, the explicit schemes (9.17) and (9.19) are only conditionally stable. For one-dimensional problems, when b is a constant and linear finite elements are used, the stability condition is

$$\Delta t \le \Delta t_c = \frac{h}{|b|}, \tag{9.20}$$

in the absence of the diffusion and source terms, where h is the spatial step size. If the diffusion $a > 0$ is present, this condition becomes

$$\Delta t \le \frac{\Delta t_c \Delta t_d}{\Delta t_c + \Delta t_d}, \tag{9.21}$$

where $\Delta t_d = h^2/(2a)$ is the diffusive limit for the critical one-dimensional time step. In multiple dimensions, the spatial step size h is taken as the minimum of the step sizes in all directions.

A fully implicit characteristic-based finite element method is computationally expensive. However, it is often possible to apply $1/2 \le \theta \le 1$ to the diffusion term only. The resulting scheme is nearly (or quasi) implicit, and as it is used, the stability condition (9.20) is retained.

9.2.2. Application to fluid mechanics

For notational convenience, we define the mass flow flux $\mathbf{U} = \rho \mathbf{u}$, and repeat the mass and momentum conservation equations in system (9.1), by using definition (9.3), as follows:

$$\frac{\partial \rho}{\partial t} = \frac{1}{c^2} \frac{\partial p}{\partial t} = -\nabla \cdot \mathbf{U},$$
$$\frac{\partial \mathbf{U}}{\partial t} = -\nabla \cdot (\mathbf{u}\mathbf{U} - \boldsymbol{\tau}) - \nabla p + \mathbf{f}, \tag{9.22}$$

where c is the speed of sound and depends on p, ρ, and E (the specific energy). Assuming a constant entropy, we see that

$$c^2 = \frac{\partial p}{\partial \rho} = \frac{\gamma p}{\rho},$$

where $\gamma = c_p/c_v$, with c_p and c_v the specific heat at constant pressure and constant volume, respectively. For a fluid with a small compressibility

$$c^2 = \frac{k}{\rho},$$

where k is the elastic bulk modulus. Depending on the application, we use an appropriate definition for c^2.

Two splitting approaches in time

The characteristic-based splitting method developed in the previous section is used to solve the momentum conservation equation; the pressure term is treated as a known (source type) quantity provided it can be calculated using an independent relation. With $\mathbf{U}$ and $\mathbf{u}$ replacing p and $\mathbf{b}$ in Eq. (9.19), respectively, the resulting equation becomes

$$\mathbf{U}^{n+1} - \mathbf{U}^n = \Delta t \left[-\nabla \cdot (\mathbf{u}\mathbf{U}) + \nabla \cdot \boldsymbol{\tau} + \mathbf{f} \right]^n - \Delta t \nabla p^{n+\theta_1}$$

$$+ \frac{\Delta t^2}{2} \mathbf{u}^n \cdot \nabla \left[\nabla \cdot (\mathbf{u}\mathbf{U}) - \nabla \cdot \boldsymbol{\tau} - \mathbf{f} \right]^n \qquad (9.23)$$

$$+ \frac{\Delta t^2}{2} \mathbf{u}^n \cdot \nabla \left(\nabla p^{n+\theta_1} \right),$$

where

$$\nabla p^{n+\theta_1} = (1 - \theta_1)\nabla p^n + \theta_1 \nabla p^{n+1} = \nabla p^n + \theta_1 \nabla(\bar{\Delta} p), \quad 0 \le \theta_1 \le 1,$$
$$(9.24)$$

with $\bar{\Delta} p = p^{n+1} - p^n$. During the time period $(t^n, t^{n+1}]$, the calculation of $\mathbf{U}^{n+1}$ is further split into two parts, *predictor* and *corrector*:

$$\mathbf{U}^{n+1} - \mathbf{U}^n = \bar{\Delta} \mathbf{U}^\star + \bar{\Delta} \mathbf{U}^{\star\star}.$$

Two splitting approaches are employed here. In the first approach, called *Split I*, all the pressure gradient terms are removed from Eq. (9.23), while in the second one, named *Split II*, these terms at the old time level p^n are retained. *Split I* is more suitable for steady-state problems, but *Split II* seems giving a slightly more accurate solution for transient problems (Zienkiewicz et al., 2005).

Split I. By neglecting the third-order stabilization term in Eq. (9.23), the predictor $\bar{\Delta}\mathbf{U}^{\star}$ is computed as

$$\bar{\Delta}\mathbf{U}^{\star} = \Delta t \left[-\nabla \cdot (\mathbf{u}\mathbf{U}) + \nabla \cdot \boldsymbol{\tau} + \mathbf{f}\right]^{n}$$
$$+ \frac{\Delta t^{2}}{2}\mathbf{u}^{n} \cdot \nabla \left[\nabla \cdot (\mathbf{u}\mathbf{U}) - \mathbf{f}\right]^{n}. \tag{9.25}$$

From Eq. (9.23), once the pressure increment is available, the corrector $\Delta\mathbf{U}^{\star\star}$ is calculated by

$$\bar{\Delta}\mathbf{U}^{\star\star} = -\Delta t\nabla p^{n+\theta_1} + \frac{\Delta t^{2}}{2}\mathbf{u}^{n} \cdot \nabla(\nabla p^{n}). \tag{9.26}$$

It follows from the mass conservation equation in system (9.22) that

$$\bar{\Delta}\rho = \left(\frac{1}{c^{2}}\right)^{n}\bar{\Delta}p = -\Delta t\nabla \cdot \mathbf{U}^{n+\theta_2} = -\Delta t(\nabla \cdot \mathbf{U}^{n} + \theta_2\nabla \cdot (\bar{\Delta}\mathbf{U})),$$

where $0 \leq \theta_2 \leq 1$. Replacing $\bar{\Delta}\mathbf{U}$ by $\bar{\Delta}\mathbf{U}^{\star} + \bar{\Delta}\mathbf{U}^{\star\star}$, applying Eqs. (9.24) and (9.26), and neglecting the third-order stabilization terms, we see that

$$\bar{\Delta}\rho = \left(\frac{1}{c^{2}}\right)^{n}\bar{\Delta}p = -\Delta t\{\nabla \cdot \mathbf{U}^{n} + \theta_2\nabla \cdot (\bar{\Delta}\mathbf{U}^{\star})$$
$$- \Delta t\theta_2 \left(\Delta p^{n} + \theta_1\Delta(\bar{\Delta}p)\right)\}, \tag{9.27}$$

where Δp^{n} and $\Delta(\bar{\Delta}p)$ are the Laplacians of p^{n} and $\bar{\Delta}p$, respectively. It is now obvious that the equations of fluid mechanics can be solved after spatial discretization in the following order: Equation (9.25) is solved for $\bar{\Delta}\mathbf{U}^{\star}$, Eq. (9.27) for $\bar{\Delta}p$, and Eq. (9.26) for $\Delta\mathbf{U}^{\star\star}$.

Split II. In this split, the known pressure gradients are retained in the evaluation of the predictor $\bar{\Delta}\mathbf{U}^{\star}$:

$$\bar{\Delta}\mathbf{U}^{\star} = \Delta t \left[-\nabla \cdot (\mathbf{u}\mathbf{U}) + \nabla \cdot \boldsymbol{\tau} - \nabla p + \mathbf{f}\right]^{n}$$
$$+ \frac{\Delta t^{2}}{2}\mathbf{u}^{n} \cdot \nabla \left[\nabla \cdot (\mathbf{u}\mathbf{U}) + \nabla p - \mathbf{f}\right]^{n}. \tag{9.28}$$

The corrector $\Delta\mathbf{U}^{\star\star}$ is given by

$$\bar{\Delta}\mathbf{U}^{\star\star} = -\Delta t\theta_1\nabla(\bar{\Delta}p). \tag{9.29}$$

Finally, the pressure increment is obtained:

$$\bar{\Delta}\rho = \left(\frac{1}{c^{2}}\right)^{n}\bar{\Delta}p = -\Delta t\{\nabla \cdot \mathbf{U}^{n} + \theta_2\nabla \cdot (\bar{\Delta}\mathbf{U}^{\star}) - \Delta t\theta_1\theta_2\Delta(\bar{\Delta}p)\}. \tag{9.30}$$

Boundary conditions

Different boundary conditions exist on the boundary Γ of a physical fluid domain Ω:

- *A solid boundary with a no-slip condition*: On such a boundary the fluid is assumed to stick or attach itself to the boundary; so, all velocity components are zero. Clearly, this condition is solely possible for viscous flow.

- *A solid boundary in viscid flow (slip condition)*: When the flow is viscid, the slipping boundary condition occurs, where only the normal velocity component is specified. This condition is zero if the boundary is stationary.

- *A prescribed traction boundary*: Tractions are given on this type of boundary, which includes zero traction in the case of free surfaces of fluid or any prescribed traction.

As an example, the no-slip boundary condition is used in the subsequent development:

$$\mathbf{u} = \mathbf{0} \quad \text{on } \Gamma. \tag{9.31}$$

The finite element formulation

It follows from Eq. (9.3) that the following relation is used in Eqs. (9.25) and (9.28):

$$\nabla \cdot \boldsymbol{\tau} = \mu \Delta \mathbf{u} + (\mu + \lambda)\nabla(\nabla \cdot \mathbf{u}).$$

To introduce the finite element formulation, we define the admissible function spaces:

$$\mathbf{V} = \Big\{ \text{Vector functions } \mathbf{v}: \mathbf{v} \text{ is continuous on } \Omega, \frac{\partial \mathbf{v}}{\partial x_1}, \frac{\partial \mathbf{v}}{\partial x_2},$$
$$\text{and } \frac{\partial \mathbf{v}}{\partial x_3} \text{ are piecewise continuous}$$
$$\text{and bounded on } \Omega, \text{ and } \mathbf{v} = \mathbf{0} \text{ on } \Gamma \Big\},$$

and

$$V = \Big\{ \text{Functions } v: v \text{ is continuous on } \Omega, \frac{\partial v}{\partial x_1}, \frac{\partial v}{\partial x_2},$$
$$\text{and } \frac{\partial v}{\partial x_3} \text{ are piecewise continuous}$$
$$\text{and bounded on } \Omega \Big\}.$$

The subsequent development is based on the first splitting approach *Split I*; a similar formulation can be established for the second approach *Split II*.

Applying Green's formula (2.9), Eq. (9.25) for the predictor $\bar{\Delta}\mathbf{U}^\star$ is written in the variational form:

$$\int_\Omega \bar{\Delta}\mathbf{U}^\star \cdot \mathbf{v} \, d\mathbf{x} = \Delta t \int_\Omega [-\nabla \cdot (\mathbf{u}U)\mathbf{v} + \boldsymbol{\tau} \cdot \nabla\mathbf{v} + \mathbf{f} \cdot \mathbf{v}]^n \, d\mathbf{x}$$
$$+ \frac{\Delta t^2}{2} \int_\Omega [\nabla \cdot (\mathbf{u}U) - \mathbf{f}]^n \, \nabla \cdot (\mathbf{u}^n\mathbf{v}) \, d\mathbf{x}, \quad \mathbf{v} \in \mathbf{V}, \tag{9.32}$$

where the homogeneous boundary condition in the space $\mathbf{V}$ is used. Note that, using Eqs. (9.26) and (9.31)

$$\{\mathbf{U}^n + \theta_2(\bar{\Delta}\mathbf{U}^\star) - \Delta t\theta_2(\nabla p^n + \theta_1\nabla(\bar{\Delta}p))\} \cdot \boldsymbol{\nu} = 0 \quad \text{on } \Gamma,$$

where $\boldsymbol{\nu}$ is the unit normal outward to the boundary Γ. As a result, the variational form for the pressure increment Eq. (9.27) is

$$\int_\Omega \bar{\Delta}\rho v \, d\mathbf{x} = \int_\Omega \left(\frac{1}{c^2}\right)^n \bar{\Delta}pv \, d\mathbf{x}$$
$$= \Delta t \int_\Omega \{\mathbf{U}^n + \theta_2(\bar{\Delta}\mathbf{U}^\star) - \Delta t\theta_2 \left(\nabla p^n + \theta_1\nabla(\bar{\Delta}p)\right)\} \cdot \nabla v \, d\mathbf{x},$$
$$v \in V. \tag{9.33}$$

Finally, the corrector $\bar{\Delta}\mathbf{U}^{\star\star}$ satisfies

$$\int_\Omega \bar{\Delta}\mathbf{U}^{\star\star} \cdot \mathbf{v} \, d\mathbf{x} = -\Delta t \int_\Omega \left(\nabla p^{n+\theta_1} \cdot \mathbf{v} + \frac{\Delta t}{2}\nabla \cdot (\mathbf{u}^n\mathbf{v}) \cdot \nabla p^n\right) d\mathbf{x}, \quad \mathbf{v} \in \mathbf{V}. \tag{9.34}$$

Let K_h be a triangulation of the problem domain Ω, and $\mathbf{V}_h \subset \mathbf{V}$ and $V_h \subset V$ be finite element spaces associated with K_h for the approximation of velocity and pressure, respectively. The triangulations for velocity and pressure do not need to be the same but are assumed to be for notational convenience here. The discrete versions of Eqs. (9.32)–(9.34) are

$$\int_\Omega \bar{\Delta}\mathbf{U}_h^\star \cdot v d\mathbf{x}$$
$$= \Delta t \int_\Omega [-\nabla \cdot (\mathbf{u}_h U_h)\mathbf{v} + \tau_h \cdot \nabla\mathbf{v} + \mathbf{f} \cdot \mathbf{v}]^n \, d\mathbf{x}$$
$$+ \frac{\Delta t^2}{2} \int_\Omega [\nabla \cdot (\mathbf{u}_h U_h) - \mathbf{f}]^n \, \nabla \cdot (\mathbf{u}_h^n\mathbf{v}) \, d\mathbf{x}, \quad \mathbf{v} \in \mathbf{V}_h, \tag{9.35}$$

$$\int_\Omega \left(\frac{1}{c^2}\right)^n \bar{\Delta} p_h v \, d\mathbf{x}$$

$$= \Delta t \int_\Omega \{\mathbf{U}_h^n + \theta_2(\bar{\Delta}\mathbf{U}_h^\star) - \Delta t \theta_2(\nabla p_h^n + \theta_1 \nabla(\bar{\Delta} p_h))\}$$

$$\cdot \nabla v \, d\mathbf{x}, \quad v \in V_h, \tag{9.36}$$

and

$$\int_\Omega \bar{\Delta}\mathbf{U}_h^{\star\star} \cdot \mathbf{v} d\mathbf{x} = -\Delta t \int_\Omega \left(\nabla p_h^{n+\theta_1} \cdot \mathbf{v} + \frac{\Delta t}{2} \nabla \cdot (\mathbf{u}_h^n \mathbf{v}) \cdot \nabla p_h^n\right) d\mathbf{x},$$

$$\mathbf{v} \in \mathbf{V}_h. \tag{9.37}$$

The sign ˜ below indicates a vector of nodal values. After the introduction of bases, $\{\boldsymbol{\varphi}_i\}$ and $\{\varphi_i\}$, in the respective finite dimensional spaces $\mathbf{V}_h$ and V_h, the discrete version (9.35) can be expressed in matrix form:

$$\mathbf{M}\bar{\Delta}\tilde{\mathbf{U}}^\star = \Delta t \left(-\mathbf{A}_u \tilde{\mathbf{U}} + \mathbf{A}_\tau \tilde{\mathbf{u}} + \mathbf{F} + \frac{\Delta t}{2}[\mathbf{A}_s \tilde{\mathbf{U}} + \mathbf{F}_s]\right)^n, \tag{9.38}$$

where the matrices and vectors are given by

$$\mathbf{M} = (M_{ij}), \quad M_{ij} = \int_\Omega \boldsymbol{\varphi}_i \cdot \boldsymbol{\varphi}_j d\mathbf{x},$$

$$\mathbf{A}_u = (a_{ij}^u), \quad a_{ij}^u = \int_\Omega \nabla \cdot (\mathbf{u}_h \boldsymbol{\varphi}_i) \boldsymbol{\varphi}_j d\mathbf{x},$$

$$\mathbf{A}_\tau = (a_{ij}^\tau), \quad a_{ij}^\tau = \int_\Omega (\mu \Delta \nabla \boldsymbol{\varphi}_i \cdot \nabla \boldsymbol{\varphi}_j + (\mu + \tau)\nabla \cdot \boldsymbol{\varphi}_i \nabla \cdot \boldsymbol{\varphi}_j) d\mathbf{x},$$

$$\mathbf{F} = (F_j), \quad F_j = \int_\Omega \mathbf{f} \cdot \boldsymbol{\varphi}_j \, d\mathbf{x},$$

$$\mathbf{A}_s = (a_{ij}^s), \quad a_{ij}^s = \int_\Omega \nabla \cdot (\mathbf{u}_h \boldsymbol{\varphi}_i)\nabla \cdot (\mathbf{u}_h \boldsymbol{\varphi}_j) d\mathbf{x},$$

$$\mathbf{F}_s = (F_j^s), \quad F_j^s = \int_\Omega \mathbf{f} \cdot \nabla \cdot (\mathbf{u}_h \boldsymbol{\varphi}_j) \, d\mathbf{x}.$$

Similarly, Eq. (9.36) is written as

$$(\mathbf{M}_p + \Delta t^2 \theta_1 \theta_2 \mathbf{A}_p)\bar{\Delta}\tilde{\mathbf{p}} = \Delta t(\mathbf{A}_{up}[\tilde{\mathbf{U}}^n + \theta_2 \bar{\Delta}\tilde{\mathbf{U}}^\star] - \Delta t \theta_2 \mathbf{A}_p \tilde{\mathbf{p}}^n), \tag{9.39}$$

where

$$\mathbf{M}_p = (M_{ij}^p), \quad M_{ij}^p = \int_\Omega \left(\frac{1}{c^2}\right)^n \varphi_i \varphi_j d\mathbf{x},$$

$$\mathbf{A}_p = (a_{ij}^p), \quad a_{ij}^p = \int_\Omega \nabla \varphi_i \cdot \nabla \varphi_j \, d\mathbf{x},$$

$$\mathbf{A}_{up} = (a_{ij}^{up}), \quad a_{ij}^{up} = \int_\Omega \varphi_i \cdot \nabla \varphi_j \, d\mathbf{x}.$$

Finally, Eq. (9.37) is of the matrix form

$$\mathbf{M}\bar{\Delta}\tilde{\mathbf{U}}^{\star\star} = -\Delta t \left(\mathbf{A}_{up}^T \left[\tilde{\mathbf{p}} + \theta_1 \bar{\Delta}\tilde{\mathbf{p}}\right] + \frac{\Delta t}{2}\mathbf{A}_{pu}\tilde{\mathbf{p}}\right)^n, \qquad (9.40)$$

where

$$\mathbf{A}_{pu} = (a_{ij}^{pu}), \quad a_{ij}^{pu} = \int_\Omega \nabla \varphi_i \cdot \nabla \cdot (\mathbf{u}_h \varphi_j) d\mathbf{x}.$$

All the global matrices in systems (9.38)–(9.40) can be assembled from local element matrices, as shown in Chapter 2. In addition, these matrix forms can be developed for the second splitting approach *Split II* (see Exercise 9.2).

9.2.3. Solution schemes in time

The first solution step in the two splitting approaches *Split I* and *Split II* is explicit for the predictor $\bar{\Delta}\mathbf{U}^\star$. The second step for the computation of pressure p, however, can be either explicit or implicit, depending on the choice of the parameter $0 \le \theta_1 \le 1$. In this section, we discuss a few options for the selection of θ_1 and their corresponding stability conditions. The parameter θ_2 used in the evaluation of the corrector $\bar{\Delta}\mathbf{U}^{\star\star}$ is always assumed to satisfy $1/2 \le \theta_2 \le 1$; so, stability does not depend on this parameter.

A fully explicit scheme

The choice of $\theta_1 = 0$ leads to a fully explicit scheme. When the viscosity effect is ignored and the linear finite elements are used, the time step size must be controlled for the explicit scheme to be stable:

$$\Delta t \le \frac{h}{c + |\mathbf{u}|}, \qquad (9.41)$$

where h is the element size. The fully explicit scheme is possible only for compressible flow unless an "artificial compressibility" is introduced (see Sec. 9.2.4).

A semi implicit scheme

The selection of $1/2 \leq \theta_1 \leq 1$ gives a semi implicit scheme. The permissible time step constraint mainly stems from the first solution step in the two splitting approaches:

$$\Delta t \leq \Delta t_u = \frac{h}{|\mathbf{u}|} \quad \text{and/or} \quad \Delta t \leq \Delta t_\nu = \frac{h^2}{2\nu}, \tag{9.42}$$

where $\nu = \mu/\rho$ is the *kinematic viscosity*. A more useful constraint incorporating these two limits is

$$\Delta t \leq \frac{\Delta t_u \Delta t_\nu}{\Delta t_u + \Delta t_\nu}. \tag{9.43}$$

Constraint (9.43) appears to give more reasonable time step limits with or without the dominance of viscosity.

A quasi-implicit scheme

To lessen the severe time step constraint caused by the diffusion terms, these terms can be handled in an implicit manner. This means that the viscous term for the predictor $\mathbf{U}^\star$ is solved implicitly. Hence, at each time step, coupled equations are solved in an iterative manner. This approach can be of great advantage in such cases as high viscous flow and low Mach number flow. In this case, the sole time step constraint is

$$\Delta t \leq \frac{h}{|\mathbf{u}|}. \tag{9.44}$$

The Mach number is the ratio of the relative speed of a source to a medium and the speed of sound in that medium.

9.2.4. Remarks on the splitting method

Artificial compressibility

If the real compressibility is small, the fluid can be treated as being incompressible, with the speed of sound approaching infinity. Even when this speed is finite, its value can be large and thus a very severe time step constraint must be imposed. In these cases, an artificial compressibility can be used to eliminate this constraint. This method is only possible if steady-state conditions exist and the transient term disappears in the limit. The pressure computation in Eq. (9.27) is now replaced by (Zienkiewicz *et al.*,

2005)

$$\left(\frac{1}{c^2}\right)^n \bar{\Delta}p \approx \left(\frac{1}{\beta^2}\right)^n \bar{\Delta}p = -\Delta t\{\nabla \cdot \mathbf{U}^n + \theta_2\nabla \cdot (\bar{\Delta}\mathbf{U}^\star) \tag{9.45}$$
$$- \Delta t\theta_2(\Delta p^n + \theta_1\Delta(\bar{\Delta}p))\},$$

where β is an artificial coefficient with the dimensions of speed. This coefficient can be either given as a constant throughout the entire domain or determined based on the convective and/or diffusive time step constraints. The latter option results in more manageable local and global time step sizes:

$$\beta = \max\{\epsilon, u_{\text{conv}}, u_{\text{diff}}\}, \tag{9.46}$$

where the small constant $\epsilon > 0$ ensures that β is bounded below by a positive constant, and u_{conv} and u_{diff} are the convective and diffusive velocities, respectively, given by

$$u_{\text{conv}} = |\mathbf{u}| = \left(\sum_{i=1}^{d} u_i^2\right)^{1/2}, \quad \mathbf{u} = (u_1, u_2, \ldots, u_d),$$
$$d = 2 \quad \text{or} \quad 3, \ u_{\text{diff}} = \nu/h. \tag{9.47}$$

From Eq. (9.41), the time step constraint for the artificial compressibility becomes

$$\Delta t \leq \frac{h}{\beta + |\mathbf{u}|}, \tag{9.48}$$

which includes the viscous effect.

Stability condition for finite element spaces

So far, we have not imposed any restriction on the nature of the finite element spaces $\mathbf{V}_h$ and V_h used for the approximation of velocity and pressure. If these spaces satisfy a stability (i.e., inf-sup) condition (see Sec. 9.5), steady-state incompressible flow problems can be handled without any difficulty by both *Split I* and *Split II* formulations. *Split I*, however, introduces an important formulation that does not require satisfaction of this stability condition by the velocity and pressure spaces.

For simplicity, we examine the Stokes equations of an incompressible fluid under the steady-state condition. In this case, systems (9.38)–(9.40) become

$$\mathbf{M}\bar{\Delta}\tilde{\mathbf{U}}^\star = \Delta t(\mathbf{A}_\tau\tilde{\mathbf{u}} + \mathbf{F})^n,$$
$$\Delta t^2\theta_1\theta_2\mathbf{A}_p\bar{\Delta}\tilde{\mathbf{p}} = \Delta t(\mathbf{A}_{up}[\tilde{\mathbf{U}}^n + \theta_2\bar{\Delta}\tilde{\mathbf{U}}^\star] - \Delta t\theta_2\mathbf{A}_p\tilde{\mathbf{p}}^n), \tag{9.49}$$
$$\mathbf{M}\bar{\Delta}\tilde{\mathbf{U}}^{\star\star} = -\Delta t\mathbf{A}_{up}^T[\tilde{\mathbf{p}} + \theta_1\bar{\Delta}\tilde{\mathbf{p}}]^n.$$

In the steady-state case, $\bar{\Delta}\tilde{U} = \bar{\Delta}\tilde{p} = 0$. After elimination of $\bar{\Delta}\tilde{U}^\star$ and $\bar{\Delta}\tilde{U}^{\star\star}$, the three systems in (9.49) reduce to (with the superscript n dropped)

$$A_\tau \tilde{u} + A_{up}^T \tilde{p} = F,$$
$$A_{up}\tilde{U} + \Delta t \theta_2 \left(A_{up}MA_{up}^T - A_p\right)\tilde{p} = 0. \tag{9.50}$$

The final system matrix for the unknown vectors $\tilde{U}$ and $\tilde{p}$ is

$$\begin{pmatrix} \dfrac{1}{\rho}A_\tau & A_{up}^T \\ A_{up} & \Delta t \theta_2 (A_{up}MA_{up}^T - A_p) \end{pmatrix},$$

which is positive definite (Codina *et al.*, 1995), and thus leads to a unique solution $\tilde{U}$ and $\tilde{p}$ for any pair of finite dimensional spaces V_h and V_h. This property is not preserved by *Split II* (see Exercise 9.3).

9.3. The finite element method

In the previous section, a characteristic-based splitting method was used to separate the computation of velocity from that of pressure. This method is very efficient for compressible, transient flow problems. For incompressible and steady-state flows, artificial compressibility and/or time stepping techniques must be used. In this section, a projection approach is introduced, which uses a different admissible function space for the velocity. To see the idea of the application of the finite element method to fluid mechanics using the projection approach, we concentrate on the stationary Stokes equations, together with the no-slip boundary condition, in a planar domain $\Omega \subset \mathbb{R}^2$:

$$\begin{aligned} -\mu\Delta\mathbf{u} + \nabla p &= \mathbf{f} \quad &\text{in } \Omega, \\ \nabla \cdot \mathbf{u} &= 0 \quad &\text{in } \Omega, \\ \mathbf{u} &= \mathbf{0} \quad &\text{on } \Gamma, \end{aligned} \tag{9.51}$$

where Γ is the boundary of Ω. We write Eq. (9.51) in a variational formulation. For this, we define the admissible function space

$$\mathbf{V} = \Big\{ \text{Vector functions } \mathbf{v}: \mathbf{v} \text{ is continuous on } \Omega, \frac{\partial \mathbf{v}}{\partial x_1} \text{ and } \frac{\partial \mathbf{v}}{\partial x_2}$$
$$\text{are piecewise continuous and bounded}$$
$$\text{on } \Omega, \nabla \cdot \mathbf{v} = 0 \text{ in } \Omega, \text{ and } \mathbf{v} = \mathbf{0} \text{ on } \Gamma \Big\}.$$

Then, using Green's formula (2.9), we are led to the variational formulation of Eq. (9.51): find $\mathbf{u} \in \mathbf{V}$ such that

$$\mu \int_\Omega \nabla \mathbf{u} \cdot \nabla \mathbf{v} dx = \int_\Omega \mathbf{f} \cdot \mathbf{v} \, dx, \quad \mathbf{v} \in \mathbf{V}. \tag{9.52}$$

It can be checked that Eq. (9.52) has a unique solution $\mathbf{u} \in \mathbf{V}$ (see Exercise 9.4). Note that the pressure p disappears in Eq. (9.52), which stems from the fact that we are working with the space $\mathbf{V}$ of velocities that satisfy the incompressibility condition.

To introduce a finite element method based on Eq. (9.52), we need to construct a finite element space $\mathbf{V}_h$ which is a subspace of the space $\mathbf{V}$. This is not an easy task because the elements $\mathbf{v}$ in $\mathbf{V}_h$ must satisfy the condition $\nabla \cdot \mathbf{v} = 0$, i.e., the *divergence free* condition. For simplicity, let Ω be a convex polygonal domain in the plane. It follows from a theorem in advanced calculus that if Ω does not contain any "holes," i.e., if Ω is simply connected, then $\nabla \cdot \mathbf{v} = 0$ if and only if there exists a smooth function w such that (Kaplan, 1991)

$$\mathbf{v} = \mathbf{rot} \; w \equiv \left(\frac{\partial w}{\partial x_2}, -\frac{\partial w}{\partial x_1} \right).$$

More precisely, it holds that

$$\mathbf{v} \in \mathbf{V} \text{ if and only if } \mathbf{v} = \mathbf{rot} \; w, \quad w \in \bar{V}, \tag{9.53}$$

where

$$\bar{V} = \left\{ \text{Functions } w: w \text{ and } \nabla w \text{ are continuous on } \Omega, \text{ the second} \right.$$
$$\text{partial derivatives of } w \text{ are piecewise continuous}$$
$$\left. \text{and bounded in } \Omega, \text{ and } w = \frac{\partial w}{\partial \nu} = 0 \text{ on } \Gamma \right\},$$

and ν is the outward unit normal to the boundary Γ. The function w is called the *stream function* associated with the velocity $\mathbf{v}$.

Let K_h be a regular triangulation of Ω into triangles as in Chapter 2. We define the finite element space

$$W_h = \left\{ \text{Functions } w: w \text{ and } \nabla w \text{ are continuous on } \Omega, w \text{ is a polynomial} \right.$$
$$\text{of degree 5 on each triangle in } K_h, \text{ and}$$
$$\left. w = \frac{\partial w}{\partial \nu} = 0 \text{ on } \Gamma \right\}.$$

As discussed in Example 3.5 in Chapter 3, because the first partial derivatives of functions in W_h are required to be continuous on Ω, there are at least six degrees of freedom on each interior edge in K_h. Thus, the polynomial degree of the finite element space W_h must be at least five. Each function in W_h is in $C^1(\bar{\Omega})$ (the set of functions that have continuous partial derivatives in Ω). Now, set

$$\mathbf{V}_h = \{\mathbf{v} \in \mathbf{V}: \mathbf{v} = \mathbf{rot}\ w,\ w \in W_h\}.$$

The Galerkin finite element method for Eq. (8.10) reads: find $\mathbf{u}_h \in \mathbf{V}_h$ such that

$$\mu \int_\Omega \nabla \mathbf{u}_h \cdot \nabla \mathbf{v}\, d\mathbf{x} = \int_\Omega \mathbf{f} \cdot \mathbf{v}\, d\mathbf{x}, \quad \mathbf{v} \in \mathbf{V}_h. \tag{9.54}$$

It possesses a unique solution (see Exercise 9.6). Furthermore, we have the error estimate between the exact solution $\mathbf{u}$ and its approximate solution $\mathbf{u}_h$ (Ciarlet, 1978; Chen, 2005):

$$\left(\int_\Omega |\nabla(\mathbf{u} - \mathbf{u}_h)|^2\, d\mathbf{x} \right)^{1/2} \leq Ch^4,$$

where the constant C depends on the fifth partial derivatives of $\mathbf{u}$. Note that the elements in $\mathbf{V}_h$ must satisfy the incompressibility condition exactly. To be able to satisfy this condition, we use the space W_h that consists of piecewise polynomials of degree five. To utilize a finite element space of polynomials of lower degree, we employ the nonconforming and mixed finite element methods introduced in the next two sections.

9.4. The nonconforming finite element method

As shown in the previous section, the simplest triangular element in the projected finite element method uses polynomials of degree five. The linear polynomials can be used at the expense of relaxing the element continuity across interelement boundaries; i.e., nonconforming elements are used. We develop the nonconforming finite element method for the solution of Eq. (9.51). It is further studied in Chapter 11.

The variational formulation is defined in Eq. (9.52). For a convex polygonal domain Ω, let K_h be a regular triangulation of Ω into triangles as in Chapter 2. We define the finite element space on triangles:

$$\mathbf{V}_h = \{\text{Functions } \mathbf{v}: \mathbf{v}|_K \text{ is linear, } K \in K_h; \mathbf{v} \text{ is continuous}$$

$$\text{at the midpoints of the interior edges and}$$

$$\text{is zero at the midpoints of the edges on } \Gamma;$$

$$\nabla \cdot \mathbf{v} = 0 \text{ on all } K \in K_h\}.$$

Note that functions in $\mathbf{V}_h$ are not required to be continuous over the whole domain Ω; so, the finite element space $\mathbf{V}_h$ is not a subspace of $\mathbf{V}$ (i.e., *nonconforming*). Namely, all the functions in $\mathbf{V}_h$ are the nonconforming P_1-elements that are divergence free on each triangle $K \in K_h$. The nonconforming method for system (9.51) is: find $\mathbf{u}_h \in \mathbf{V}_h$ such that

$$\mu \sum_{K \in K_h} \int_K \nabla \mathbf{u}_h \cdot \nabla \mathbf{v} \, d\mathbf{x} = \int_\Omega \mathbf{f} \cdot \mathbf{v} \, d\mathbf{x}, \quad \mathbf{v} \in \mathbf{V}_h. \qquad (9.55)$$

Existence and uniqueness of a solution to this problem can be shown exactly in the same way as for Eq. (1.10) in Sec. 1.2 (see Exercise 9.7). Furthermore, it can be proven (Crouzeix and Raviart, 1973) that the error between the exact solution $\mathbf{u}$ and its nonconforming finite element solution $\mathbf{u}_h$ is

$$\left(\int_\Omega |\mathbf{u} - \mathbf{u}_h|^2 \, d\mathbf{x} \right)^{1/2} \le Ch^2,$$

where the constant C depends on the second partial derivatives of $\mathbf{u}$ and the first partial derivatives of p.

A basis in $\mathbf{V}_h$ can be constructed as follows: using the divergence formula (2.7), we see that

$$0 = \int_K \nabla \cdot \mathbf{v} \, d\mathbf{x} = \int_{\partial K} \mathbf{v} \cdot \boldsymbol{\nu} \, d\ell = \sum_{i=1}^{3} \mathbf{v} \cdot \boldsymbol{\nu}(\mathbf{m}_K^i)|e_i|, \quad K \in K_h, \qquad (9.56)$$

where e_i is an edge of K, $\mathbf{m}_K^i$ is the midpoint of e_i, $|e_i|$ represents the length of e_i, and $\boldsymbol{\nu}$ is the outward unit normal to ∂K. The basis functions must satisfy property (9.56).

Let e be an edge in K_h. Denote by φ_e the piecewise linear function defined on K_h that is unity at the midpoint of e and zero at the midpoints of all other edges in K_h (see Sec. 2.2.3). We define $\boldsymbol{\varphi}_e = \varphi_e \mathbf{t}_e$, where e is an internal edge of K_h and $\mathbf{t}_e$ is a unit vector tangential to e. This basis function satisfies property (9.56).

Let $\mathbf{m}$ be an internal vertex and let $e_1, e_2, \ldots, e_l$ be the edges in K_h that have $\mathbf{m}$ as a common vertex. Define

$$\boldsymbol{\varphi}_\mathbf{m} = \sum_{i=1}^{l} \frac{\varphi_{e_i}}{|e_i|} \boldsymbol{\nu}_{e_i},$$

where $\boldsymbol{\nu}_{e_i}$ is a unit vector normal to e_i pointing in the counterclockwise direction (Fig. 9.2). Again, this function satisfies Eq. (9.56). It can be shown that a basis for $\mathbf{V}_h$ is given by the union of the two sets (see Exercise 9.8)

$$\{\boldsymbol{\varphi}_e\colon e \text{ is an internal edge in } K_h\}$$

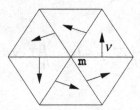

Figure 9.2   An illustration of the normal direction on triangles.

and

$$\{\varphi_{\mathbf{m}} : \mathbf{m} \text{ is an internal vertex in } K_h\}.$$

As an example, in this section, the nonconforming finite element method is described for triangles. It can be also defined using rectangular and three-dimensional elements (Chen, 2005).

9.5. The mixed finite element method

The equations of fluid mechanics can be solved more appropriately by the mixed finite element method, which is further considered in Chapter 11. As an example, we focus on the Stokes problem (9.51). Note that this problem determines the pressure p only up to an additive constant, which is usually fixed by enforcing the integral condition

$$\int_\Omega p \, d\mathbf{x} = 0.$$

We introduce the admissible function spaces

$$\mathbf{V} = \left\{ \text{Vector functions } \mathbf{v} \colon \mathbf{v} \text{ is continuous on } \Omega, \frac{\partial \mathbf{v}}{\partial x_1} \text{ and } \frac{\partial \mathbf{v}}{\partial x_2} \text{are} \right.$$
$$\left. \text{piecewise continuous and bounded on } \Omega, \text{ and } \mathbf{v} = \mathbf{0} \text{ on } \Gamma \right\},$$

and

$$W = \left\{ \text{Functions } w \colon w \text{ is defined and square integrable on } \Omega, \right.$$
$$\left. \text{i.e., } \int_\Omega w^2 \, d\mathbf{x} < \infty, \text{ and } \int_\Omega w \, d\mathbf{x} = 0 \right\}.$$

As in Chapter 2 (see Exercise 9.9), using Green's formula (2.9), problem (9.51) can be now written in a mixed formulation: find $\mathbf{u} \in \mathbf{V}$ and $p \in W$

such that

$$\mu \int_\Omega \nabla \mathbf{u} \cdot \nabla \mathbf{v} \, d\mathbf{x} - \int_\Omega \nabla \cdot \mathbf{v} p \, d\mathbf{x} = \int_\Omega \mathbf{f} \cdot \mathbf{v} \, d\mathbf{x}, \quad \mathbf{v} \in \mathbf{V},$$

$$\int_\Omega \nabla \cdot \mathbf{u} w \, d\mathbf{x} = 0, \qquad\qquad\qquad w \in W. \tag{9.57}$$

System (9.57) can be shown to have a unique solution $(\mathbf{u}, p) \in \mathbf{V} \times W$ (see Sec. 11.4). Moreover, the spaces $\mathbf{V}$ and W satisfy an inf-sup (stability) condition:

$$\inf_{w \in W} \sup_{\mathbf{v} \in \mathbf{V}} \frac{\int_\Omega \nabla \cdot \mathbf{v} w \, d\mathbf{x}}{\left(\int_\Omega |\nabla \mathbf{v}|^2 \, d\mathbf{x}\right)^{1/2} \left(\int_\Omega w^2 \, d\mathbf{x}\right)^{1/2}} \geq b_* > 0,$$

where b_* is a constant.

We now construct the mixed finite element method based on formulation (9.57). For the Stokes problem, the velocity has a derivative of higher order than the pressure. This suggests the rule of thumb that the degree of the piecewise polynomials used to approximate the velocity should be higher than that of the polynomials for the pressure. However, it is known that this rule does not suffice to guarantee stability, and the spaces $\mathbf{V}_h$ and W_h have to be constructed very carefully. In this section, we state three mixed finite element spaces that satisfy the discrete inf-sup condition:

$$\inf_{w \in W_h} \sup_{\mathbf{v} \in \mathbf{V}_h} \frac{\int_\Omega \nabla \cdot \mathbf{v} w \, d\mathbf{x}}{\left(\int_\Omega |\nabla \mathbf{v}|^2 \, d\mathbf{x}\right)^{1/2} \left(\int_\Omega w^2 \, d\mathbf{x}\right)^{1/2}} \geq b_{**} > 0, \tag{9.58}$$

where b_{**} is a constant, independent of h.

Example 9.1 Let K_h be a regular triangulation of Ω into triangles. We define

$$\mathbf{V}_h = \{\text{Functions } \mathbf{v} \in \mathbf{V} : \mathbf{v}|_K \in (P_2(K))^2, \ K \in K_h\},$$

$$W_h = \{\text{Functions } w \in W : w|_K \in P_0(K), \ K \in K_h\}.$$

Then, the mixed finite element method for problem (9.51) is: find $\mathbf{u}_h \in \mathbf{V}_h$ and $p_h \in W_h$ such that

$$\mu \int_\Omega \nabla \mathbf{u}_h \cdot \nabla \mathbf{v} \, d\mathbf{x} - \int_\Omega \nabla \cdot \mathbf{v} p_h \, d\mathbf{x} = \int_\Omega \mathbf{f} \cdot \mathbf{v} \, d\mathbf{x}, \qquad \mathbf{v} \in \mathbf{V}_h,$$

$$\int_\Omega \nabla \cdot \mathbf{u}_h w \, d\mathbf{x} = 0, \qquad\qquad\qquad w \in W_h. \tag{9.59}$$

System (9.59) has a unique solution $(\mathbf{u}_h, p_h)$ (see Sec. 11.4). Moreover, if the domain Ω is convex, this solution satisfies the error estimate (Girault-Raviart, 1981)

$$\left(\int_\Omega |\mathbf{u} - \mathbf{u}_h|^2 \, d\mathbf{x} \right)^{1/2} + h \left(\int_\Omega (p - p_h)^2 \, d\mathbf{x} \right)^{1/2} \leq Ch^2,$$

where the constant $C > 0$ depends on the second partial derivatives of $\mathbf{u}$ and the first derivatives of p.

Example 9.2 The second example is the so-called *MINI element* (Arnold *et al.*, 1984b). To introduce this element, let λ_1, λ_2, and λ_3 be the area (barycentric) coordinates of a triangle (they are x_1, x_2, and $1 - x_1 - x_2$ in the unit triangle with vertices $(0,0), (1,0)$, and $(0,1)$; see Sec. 5.2). We define

$$B_h = \{\text{Functions } v \colon v|_K \in \text{span}\{\lambda_1 \lambda_2 \lambda_3\}, \ K \in K_h\};$$

i.e., B_h is the space of *cubic bubble functions* in the sense that they are zero on the boundary of each element K. We define the space

$$V = \left\{ \text{Functions } v \colon v \text{ is a continuous function on } \Omega, \frac{\partial v}{\partial x_1} \text{ and } \frac{\partial v}{\partial x_2} \right.$$
$$\left. \text{are piecewise continuous and bounded on } \Omega \right\}.$$

Now, the mixed finite element spaces are

$$\mathbf{V}_h = \{\text{Functions } \mathbf{v} \in \mathbf{V} : \mathbf{v}|_K \in (P_1(K))^2, \ K \in K_h\} \oplus (B_h \cap V)^2,$$
$$W_h = \{\text{Functions } w \in W \cap V : w|_K \in P_1(K), \ K \in K_h\},$$

where the notation $\oplus$ denotes a direct sum. With these choices, the mixed method can be defined as in system (9.59). Moreover, if Ω is convex, the mixed finite element solution $(\mathbf{u}_h, p_h)$ satisfies (Chen, 2005)

$$\left(\int_\Omega |\mathbf{u} - \mathbf{u}_h|^2 \, d\mathbf{x} \right)^{1/2} + \left(\int_\Omega (p - p_h)^2 \, d\mathbf{x} \right)^{1/2} \leq Ch^2,$$

where the constant C depends on the second partial derivatives of both $\mathbf{u}$ and p. Without the cubic bubble space added to the space $\mathbf{V}_h$, the pair of $P_1 - P_1$ spaces does not satisfy the discrete inf-sup condition (9.58) unless some sort of stabilization term is inserted in the system (9.59). Recently, a class of stabilized mixed finite element methods have been introduced for the numerical solution of the Stokes and Navier–Stokes equations; these methods are based on two local Gauss integrals (Bochev *et al.*, 2006; Li *et al.*, 2007; Li and He, 2008).

Example 9.3 The third example is the *Taylor-Hood element*:

$$\mathbf{V}_h = \{\text{Functions } \mathbf{v} \in \mathbf{V} : \mathbf{v}|_K \in (P_2(K))^2, \ K \in K_h\},$$
$$W_h = \{\text{Functions } w \in W : w|_K \in P_1(K), \ K \in K_h\},$$

which satisfies the discrete inf-sup condition (9.58). In addition, the error estimate between the exact solution $(\mathbf{u}, p)$ and the approximate solution $(\mathbf{u}_h, p_h)$ is

$$\left(\int_\Omega |\mathbf{u} - \mathbf{u}_h|^2 \, d\mathbf{x} \right)^{1/2} + \left(\int_\Omega (p - p_h)^2 \, d\mathbf{x} \right)^{1/2} \leq Ch^2,$$

where the constant $C > 0$ depends on the second partial derivatives of both the solutions $\mathbf{u}$ and p.

9.6. The Navier–Stokes' equations

In this section, we make remarks on extensions of the standard, nonconforming, and mixed finite element methods developed in the previous three sections to the Navier–Stokes equations:

$$\begin{aligned}
-\mu \Delta \mathbf{u} + (\mathbf{u} \cdot \nabla)\mathbf{u} + \nabla p &= \mathbf{f} \quad && \text{in } \Omega, \\
\nabla \cdot \mathbf{u} &= 0 && \text{in } \Omega, \\
\mathbf{u} &= \mathbf{0} && \text{on } \Gamma.
\end{aligned} \tag{9.60}$$

As an example, we describe the mixed finite element method. The admissible function spaces $\mathbf{V}$ and W are defined as in the previous section. Then, in the same fashion as for system (9.57), problem (9.60) can be recast in the mixed formulation: find $\mathbf{u} \in \mathbf{V}$ and $p \in W$ such that

$$\int_\Omega (\mathbf{u} \cdot \nabla)\mathbf{u} \cdot \mathbf{v} \, d\mathbf{x} + \mu \int_\Omega \nabla \mathbf{u} \cdot \nabla \mathbf{v} \, d\mathbf{x} - \int_\Omega \nabla \cdot \mathbf{v} p \, d\mathbf{x} = \int_\Omega \mathbf{f} \cdot \mathbf{v} \, d\mathbf{x}, \quad \mathbf{v} \in \mathbf{V},$$
$$\int_\Omega \nabla \cdot \mathbf{u} w \, d\mathbf{x} = 0, \qquad\qquad w \in W. \tag{9.61}$$

System (9.61) can be shown to possess at least a solution $\mathbf{u} \in \mathbf{V}$ and $p \in W$. Proof of uniqueness of the solution requires some strong conditions on the function $\mathbf{f}$ and the viscosity μ (Girault and Raviart, 1981).

With an appropriate choice of the mixed finite element spaces $\mathbf{V}_h$ and W_h as in Sec. 9.5 that satisfy the discrete inf-sup condition (9.58), the mixed finite element method for the Navier–Stokes problem is: find $\mathbf{u}_h \in \mathbf{V}_h$ and

$p_h \in W_h$ such that

$$\int_\Omega (\mathbf{u}_h \cdot \nabla)\mathbf{u}_h \cdot \mathbf{v}\, dx + \mu \int_\Omega \nabla \mathbf{u}_h \cdot \nabla \mathbf{v}\, dx - \int_\Omega \nabla \cdot \mathbf{v} p_h\, dx$$

$$= \int_\Omega \mathbf{f} \cdot \mathbf{v}\, dx, \quad \mathbf{v} \in \mathbf{V}_h, \tag{9.62}$$

$$\int_\Omega \nabla \cdot \mathbf{u}_h w\, dx = 0, \quad w \in W_h.$$

Again, under suitable conditions, system (9.62) has a unique solution (Girault and Raviart, 1981). Note that this system is a nonlinear system and can be solved by using the Newton iteration method.

9.7. Exercises

9.1 Defining the material derivative

$$\frac{D}{Dt} = \frac{\partial}{\partial t} + \mathbf{u} \cdot \nabla,$$

derive system (9.2) from system (9.1) in detail.

9.2 Develop the finite element formulations and the corresponding matrix forms for Eqs. (9.28)–(9.30) used in the second splitting approach *Split II* as for Eqs. (9.25)–(9.27) in the first splitting approach *Split I*.

9.3 For the Stokes equations of an incompressible fluid under the steady-state condition, show that the final system matrix for the unknown vectors $\tilde{\mathbf{U}}$ and $\tilde{p}$ obtained in Exercise 9.2 is not positive definite.

9.4 Show that problem (9.52) has a unique solution $\mathbf{u} \in \mathbf{V}$, where the space $\mathbf{V}$ is defined as in Sec. 9.3.

9.5 Prove that the Stokes equation (9.51) in a simply connected domain $\Omega \subset \mathbb{R}^2$ can be written as the biharmonic problem (3.17) by introducing a suitable stream function as an unknown.

9.6 Show that the discrete problem (9.54) has a unique solution $\mathbf{u}_h \in \mathbf{V}_h$ (see Sec. 1.2), where the finite element space $\mathbf{V}_h$ is defined as in Sec. 9.3.

9.7 With the nonconforming finite element space $\mathbf{V}_h$ defined in Sec. 9.4, show that problem (9.55) possesses a unique solution $\mathbf{u}_h \in \mathbf{V}_h$.

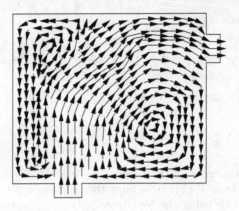

Figure 9.3   A numerical cavity problem.

9.8 Let $\mathbf{V}_h$ be the P_1-nonconforming finite element space defined in Sec. 9.4. Prove that a basis for $\mathbf{V}_h$ is given by the union of the two sets

$$\{\varphi_e \colon e \text{ is an internal edge in } K_h\}$$

and

$$\{\varphi_\mathbf{m} \colon \mathbf{m} \text{ is an internal vertex in } K_h\},$$

where the functions φ_e and $\varphi_\mathbf{m}$ are defined as in Sec. 9.4.

9.9 Derive system (9.57) from Eq. (9.51) in detail.

9.10 In this chapter, we have developed the finite element methods only for the stationary Stokes and Navier–Stokes equations. The corresponding transient equations can be treated using the techniques of this chapter with those in Sec. 7.1. Develop the nonconforming finite element method discussed in Sec. 9.4, using the backward Euler scheme (see Sec. 7.1) for the time derivative, for the transient Navier–Stokes equation

$$\frac{\partial \mathbf{u}}{\partial t} - \mu \Delta \mathbf{u} + (\mathbf{u} \cdot \nabla)\mathbf{u} + \nabla p = \mathbf{0} \quad \text{in } \Omega,$$

$$\nabla \cdot \mathbf{u} = 0 \quad \text{in } \Omega,$$

$$\mathbf{u}(\mathbf{x}, 0) = \mathbf{u}_0 \quad \text{in } \Omega,$$

where the domain $\Omega \subset \mathbb{R}^2$ is planar. A numerical cavity problem is presented in Fig. 9.3. The initial velocity $\mathbf{u}_0$ is zero, the inlet velocity is one, and the viscosity μ equals 10^{-3}. An example of the computed velocities after 10 time steps is given in this figure on a 20×10 grid (triangles constructed as in Fig. 2.6; see Johnson, 1994), where $\Delta t = h$ is used.

Chapter 10

Application to Porous Media Flow

This chapter is concerned with the application of the finite element method to fluid flows in porous media. In general, flows of these types involve multiple phases and multiple components that can transfer between the phases. The flows of a single phase and two immiscible phases through porous media are considered. For the treatment of a variety of compositional and thermal flows in porous media, the reader can refer to Chen $et\ al.$ (2006) and Chen (2007). Furthermore, the finite element method is concentrated on the numerical solution of the governing differential equations for the single-phase flow. For the finite element solution of two-phase flow, the reader can refer to Chen $et\ al.$ (2006). In Secs. 10.1 and 10.2, the governing differential equations for single- and two-phase flows are described. Then, in Sec. 10.3, the finite element method for the single-phase flow is presented.

10.1. Single-phase flow

10.1.1. Basic differential equations

We briefly decribe the governing differential equations for the flow and transport of a fluid in a porous medium. Denote by ϕ the $porosity$ of the porous medium, by ρ the density of the fluid per unit volume, by $\mathbf{u} = (u_1, u_2, u_3)$ the superficial $Darcy\ velocity$, and by q the external sources and sinks.

The $mass\ conservation\ equation$ for the fluid is

$$\frac{\partial(\phi\rho)}{\partial t} = -\nabla \cdot (\rho\mathbf{u}) + q, \qquad (10.1)$$

197

where we recall that $\nabla \cdot$ is the *divergence* operator:

$$\nabla \cdot (\rho \mathbf{u}) = \frac{\partial(\rho u_1)}{\partial x_1} + \frac{\partial(\rho u_2)}{\partial x_2} + \frac{\partial(\rho u_3)}{\partial x_3},$$

and q is negative for sinks and positive for sources.

The *formation volume factor B* of the fluid is

$$\rho = \frac{\rho_s}{B},$$

where ρ_s is the fluid density at standard conditions. Substituting ρ into Eq. (10.1), we have

$$\frac{\partial}{\partial t}\left(\frac{\phi}{B}\right) = -\nabla \cdot \left(\frac{1}{B}\mathbf{u}\right) + \frac{q}{\rho_s}. \tag{10.2}$$

Equations (10.1) and (10.2) are equivalent; they are the most general forms of the single phase mass conservation equation.

Darcy's law (Darcy, 1856) was originally a law for single-phase flow that relates the total volumetric flow rate of a fluid through a porous medium to the pressure gradient (∇p, or the potential gradient) and the properties of the fluid (viscosity, μ) and medium (permeability, $\mathbf{k}$):

$$\mathbf{u} = -\frac{1}{\mu}\mathbf{k}(\nabla p - \rho \wp \nabla z), \tag{10.3}$$

where $\mathbf{k}$ is the *absolute permeability tensor* of the porous medium, $\wp$ is the magnitude of the gravitational acceleration, z is the depth, and ∇ is the *gradient* operator:

$$\nabla p = \left(\frac{\partial p}{\partial x_1}, \frac{\partial p}{\partial x_2}, \frac{\partial p}{\partial x_3}\right).$$

The x_3-coordinate axis in Eq. (10.3) is in the vertical downward direction.

10.1.2. Units

In this chapter, the basic units are customary and metric. Here, we describe these units for the variables used in single phase flow (Table 10.1). One multiplies a customary unit by a conversion factor to obtain the corresponding metric unit. Standard barrel (STB) and standard cubic feet (SCF) are measured at 60°F and 14.696 psia, while std m^3 (at standard conditions) is measured at 15°C and 100 kPa. In Table 10.1, FVF stands for formation volume factor.

Table 10.1 Customary and metric units.

Quantity	Symbol	Customary	Metric	Conversion factor
Time	t	day	day	1.0
Length	x_i, z	ft	m	0.3048
Area	A	ft^2	m^2	0.09290304
Volume	V	ft^3	m^2	0.02831685
Porosity	ϕ	fraction	fraction	1.0
Permeability	$\mathbf{k}$	darcy	μm^2	0.9869233
Density	ρ	lbm/ft^3	kg/m^3	16.01846
Gravitational acceleration	$\wp$	32.174ft/s^2	9.8066352 m/s^2	0.3048
Fluid gravity	γ	psi/ft	kPa/m	22.62059
Pressure	p	psia	kPa	6.894757
Velocity	$\mathbf{u}$	ft/D	m/d	0.3048
Viscosity	μ	cp	Pa·s	0.001
Compressibility	c	psi^{-1}	kPa^{-1}	0.1450377
Compressibility factor	Z	dimensionless	dimensionless	1.0
Liquid flow rate	q	STB/D	stb m^3/d	0.1589873
Gas flow rate	q	SCF/D	stb m^3/d	0.0286364
Liquid FVF	B_w, B_o	RB/STB	m^3/stb m^3	1.0
Gas FVF	B_g	RB/SCF	m^3/stb m^3	5.5519314
Gravity conversion factor	γ_c	0.21584E-3	1.0E-3	—

10.1.3. Different forms of flow equations

Substituting Eq. (10.3) into Eq. (10.1) yields

$$\frac{\partial(\phi\rho)}{\partial t} = \nabla \cdot \left(\frac{\rho}{\mu}\mathbf{k}(\nabla p - \rho\wp\nabla z) \right) + q. \qquad (10.4)$$

An equation of state is expressed in terms of the *fluid compressibility* c_f:

$$c_f = -\frac{1}{V}\frac{\partial V}{\partial p}\bigg|_T = \frac{1}{\rho}\frac{\partial \rho}{\partial p}\bigg|_T, \qquad (10.5)$$

at a fixed temperature T, where V stands for the volume occupied by the fluid at reservoir conditions. Combining Eqs. (10.4) and (10.5) gives a closed system for the main unknown p or ρ. Simplified expressions such as a linear relationship between p and ρ for a *slightly compressible fluid* can be used.

It is sometimes convenient in mathematical analysis to write Eq. (10.4) in a form without the explicit appearance of gravity by the introduction of a *pseudo potential* (Hubbert, 1956):

$$\Phi' = \int_{p^o}^{p} \frac{1}{\rho(\xi)\wp} \, d\xi - z, \qquad (10.6)$$

where p^o is a reference pressure. Using relation (10.6), Eq. (10.4) reduces to

$$\frac{\partial(\phi\rho)}{\partial t} = \nabla \cdot \left(\frac{\rho^2 \wp}{\mu} \mathbf{k}\nabla\Phi'\right) + q. \tag{10.7}$$

In numerical simulation, we often use the usual *potential* (piezometric head)

$$\Phi = p - \rho\wp z,$$

which is related to Φ' (with, e.g., $p^o = 0$ and constant ρ) by

$$\Phi = \rho\wp\Phi'.$$

If we neglect the term $\wp z\nabla\rho$, in terms of Φ Eq. (10.4) becomes

$$\frac{\partial(\phi\rho)}{\partial t} = \nabla \cdot \left(\frac{\rho}{\mu} \mathbf{k}\nabla\Phi\right) + q. \tag{10.8}$$

Incompressible flow

When the rock and fluid are incompressible, the density ρ and porosity ϕ are assumed to be constant. In this case, Eq. (10.8) reduces to

$$\nabla \cdot \left(\frac{\rho}{\mu} \mathbf{k}\nabla\Phi\right) + q = 0, \tag{10.9}$$

which is an *elliptic equation* in Φ (see Sec. 10.2.2). For the flow of an incompressible fluid in a homogeneous and isotropic medium with a constant viscosity, Eq. (10.9) further becomes

$$\Delta\Phi = -\frac{\mu q}{\rho k}, \tag{10.10}$$

where the *Laplacian operator* Δ is recalled

$$\Delta\Phi = \frac{\partial^2\Phi}{\partial x_1^2} + \frac{\partial^2\Phi}{\partial x_2^2} + \frac{\partial^2\Phi}{\partial x_3^2}.$$

Equation (10.10) is the *Poisson equation* in Φ. If there is no external source/sink term (well), it is the *Laplace equation*.

Slightly compressible flow

It is sometimes possible to assume that the fluid compressibility c_f is constant over a certain range of pressures. Then, after integration, we write Eq. (10.5) as

$$\rho = \rho^o e^{c_f(p-p^o)}, \tag{10.11}$$

where ρ^o is the density at the reference pressure p^o. Using a Taylor series expansion, we see that

$$\rho = \rho^o \left\{ 1 + c_f(p - p^o) + \frac{1}{2!}c_f^2(p - p^o)^2 + \ldots \right\},$$

so, an approximation results

$$\rho \approx \rho^o \left(1 + c_f(p - p^o) \right). \qquad (10.12)$$

The *rock compressibility* is defined by

$$c_R = \frac{1}{\phi}\frac{d\phi}{dp}. \qquad (10.13)$$

After integration, it is given by

$$\phi = \phi^o e^{c_R(p-p^o)}, \qquad (10.14)$$

where ϕ^o is the porosity at p^o. Similarly, it is approximated by

$$\phi \approx \phi^o \left(1 + c_R(p - p^o) \right). \qquad (10.15)$$

Then, it follows that

$$\frac{d\phi}{dp} = \phi^o c_R. \qquad (10.16)$$

After carrying out the time differentiation in the left-hand side of Eq. (10.4), this equation becomes

$$\left(\phi \frac{\partial \rho}{\partial p} + \rho \frac{d\phi}{dp} \right) \frac{\partial p}{\partial t} = \nabla \cdot \left(\frac{\rho}{\mu}\mathbf{k} \left(\nabla p - \rho\wp\nabla z \right) \right) + q. \qquad (10.17)$$

Substituting Eqs. (10.5) and (10.16) into Eq. (10.17) gives

$$\rho(\phi c_f + \phi^o c_R)\frac{\partial p}{\partial t} = \nabla \cdot \left(\frac{\rho}{\mu}\mathbf{k} \left(\nabla p - \rho\wp\nabla z \right) \right) + q.$$

Defining the *total compressibility*

$$c_t = c_f + \frac{\phi^o}{\phi}c_R, \qquad (10.18)$$

we see that

$$\phi\rho c_t \frac{\partial p}{\partial t} = \nabla \cdot \left(\frac{\rho}{\mu}\mathbf{k} \left(\nabla p - \rho\wp\nabla z \right) \right) + q, \qquad (10.19)$$

which is a *parabolic equation* in p (see Sec. 10.2.2), with ρ given by Eq. (10.11).

Compressible flow

For gas flow, the compressibility of gas c_g is usually not assumed to be constant. In such a case, the general equation (10.17) applies; i.e.

$$c(p)\frac{\partial p}{\partial t} = \nabla \cdot \left(\frac{\rho}{\mu}\mathbf{k}(\nabla p - \rho\wp\nabla z) \right) + q, \tag{10.20}$$

where

$$c(p) = \phi\frac{\partial \rho}{\partial p} + \rho\frac{d\phi}{dp}. \tag{10.21}$$

A different form of Eq. (10.20) can be derived if we use the real *gas law* (the pressure–volume–temperature (PVT) relation)

$$\rho = \frac{pW}{ZRT}, \tag{10.22}$$

where W is the molecular weight, Z is the gas compressibility factor, and R is the *universal gas constant*. If pressure, temperature, and density are in atm, K, and g/cm^3 (physical unit system), respectively, the value of R is 82.057. For the English units (psia, R, and lbm/ft^3), $R = 10.73$; for the SI system (N/m^2, K, and kg/m^3), $R = 8,314$.

For a pure gas reservoir, the gravitational constant is usually small and neglected. We assume that the porous medium is isotropic; i.e., $\mathbf{k} = k\mathbf{I}$, where $\mathbf{I}$ is the identity tensor. Furthermore, we assume that ϕ and μ are constants. Then, substituting Eq. (10.22) into Eq. (10.4), we see that

$$\frac{\phi}{k}\frac{\partial}{\partial t}\left(\frac{p}{Z}\right) = \nabla \cdot \left(\frac{p}{\mu Z}\nabla p\right) + \frac{RT}{Wk}q. \tag{10.23}$$

Note that $2p\nabla p = \nabla p^2$, so Eq. (10.23) becomes

$$\frac{2\phi\mu Z}{k}\frac{\partial}{\partial t}\left(\frac{p}{Z}\right) = \Delta p^2 + 2pZ\frac{d}{dp}\left(\frac{1}{Z}\right)|\nabla p|^2 + \frac{2\mu ZRT}{Wk}q. \tag{10.24}$$

Because

$$c_g = \frac{1}{\rho}\frac{d\rho}{dp}\bigg|_T = \frac{1}{p} - \frac{1}{Z}\frac{dZ}{dp},$$

we have

$$\frac{\partial}{\partial t}\left(\frac{p}{Z}\right) = \frac{pc_g}{Z}\frac{\partial p}{\partial t}.$$

Inserting this equation into Eq. (10.24) and neglecting the term involving $|\nabla p|^2$ (often smaller than other terms in (10.24)), we obtain

$$\frac{\phi\mu c_g}{k}\frac{\partial p^2}{\partial t} = \Delta p^2 + \frac{2ZRT\mu}{Wk}q, \tag{10.25}$$

which is a parabolic equation in p^2.

There is another way to derive an equation similar to Eq. (10.25). We define a *pseudo pressure* by

$$\psi = 2 \int_{p^o}^{p} \frac{p}{Z\mu} \, dp.$$

Note that

$$\nabla\psi = \frac{2p}{Z\mu}\nabla p, \quad \frac{\partial\psi}{\partial t} = \frac{2p}{Z\mu}\frac{\partial p}{\partial t}.$$

Equation (10.23) becomes

$$\frac{\phi\mu c_g}{k}\frac{\partial\psi}{\partial t} = \Delta\psi + \frac{2RT}{Wk}q. \tag{10.26}$$

The derivation of Eq. (10.26) does not require us to neglect the second term in the right-hand side of Eq. (10.24).

10.1.4. Boundary and initial conditions

The mathematical model described so far for single-phase flow is not complete unless necessary *boundary* and *initial conditions* are specified. Below, we present three kinds of boundary conditions that are relevant to Eq. (10.4). We denote by Γ the external boundary or a boundary segment of the porous medium domain Ω under consideration.

Prescribed pressure. When the pressure is specified as a known function of position and time on Γ, the boundary condition is

$$p = g_1 \quad \text{on } \Gamma.$$

In the theory of partial differential equations, such a condition is termed a boundary condition of the *first kind*, or a *Dirichlet boundary condition* (see Sec. 1.3).

Prescribed mass flux. When the total mass flux is known on Γ, the boundary condition is

$$\rho\mathbf{u}\cdot\boldsymbol{\nu} = g_2 \quad \text{on } \Gamma,$$

where $\boldsymbol{\nu}$ indicates the outward unit normal to Γ. This condition is called a boundary condition of the *second kind*, or a *Neumann boundary condition*. For an *impervious boundary*, $g_2 = 0$ (i.e., a no-flow boundary condition).

Mixed boundary condition. A boundary condition of *mixed kind* (or *third kind*) takes the form

$$g_p p + g_u \rho\mathbf{u}\cdot\boldsymbol{\nu} = g_3 \quad \text{on } \Gamma,$$

where g_p, g_u, and g_3 are given functions. This condition is referred to as a *Robin* or *Dankwerts boundary condition*. Such a condition occurs when Γ is a semipervious boundary.

Finally, the initial condition can be defined in terms of p:

$$p(\mathbf{x}, 0) = p^0(\mathbf{x}), \quad \mathbf{x} \in \Omega.$$

In general, in reservoir simulation, an initial pressure is given only at a *datum level depth*. The pressure at other locations is determined by the *gravity equilibrium condition* (see Sec. 10.3.1).

10.2. Two-phase flow

In reservoir simulation, we are often interested in the *simultaneous flow* of two or more fluid phases within a porous medium. We now develop basic equations for multiphase flow in a porous medium. We consider two-phase flow where the fluids are *immiscible* and there is no mass transfer between the phases. One phase (e.g., water) wets the porous medium more than the other (e.g., oil), and is called the *wetting phase* and indicated by a subscript w. The other phase is termed the *nonwetting phase* and indicated by o. In general, water is the wetting fluid relative to oil and gas, while oil is the wetting fluid relative to gas.

10.2.1. Basic differential equations

Several new quantities peculiar to multiphase flow, such as *saturation, capillary pressure*, and *relative permeability*, must be introduced. The saturation of a fluid phase is defined as the fraction of the void volume of a porous medium filled by this phase. The fact that the two fluids jointly fill the voids implies the relation

$$S_{\mathrm{w}} + S_o = 1, \tag{10.27}$$

where S_{w} and S_o are the saturations of the wetting and nonwetting phases, respectively. Also, due to the *curvature* and *surface tension* of the interface between the two phases, the pressure in the wetting fluid is less than that in the nonwetting fluid. The pressure difference is given by the capillary pressure

$$p_c = p_o - p_{\mathrm{w}}, \tag{10.28}$$

where p_{w} and p_o are the respective pressures of the wetting and nonwetting phases. Empirically, the capillary pressure is a function of saturation S_{w}.

Mass is conserved within each phase:

$$\frac{\partial(\phi \rho_\alpha S_\alpha)}{\partial t} = -\nabla \cdot (\rho_\alpha \mathbf{u}_\alpha) + q_\alpha, \quad \alpha = w, o, \tag{10.29}$$

where each phase also has its own Darcy's velocity $\mathbf{u}_\alpha$ and mass flow rate q_α.

Darcy's law for single-phase flow can be directly extended to multiphase flow. In the present case, it relates the total volumetric flow rate of each fluid phase through a porous medium to its pressure gradient and the properties of the fluid (viscosity, μ_w or μ_o) and medium (*effective permeability*, k_w or k_o):

$$\mathbf{u}_\alpha = -\frac{1}{\mu_\alpha}\mathbf{k}_\alpha(\nabla p_\alpha - \rho_\alpha \wp \nabla z), \quad \alpha = w, o, \tag{10.30}$$

where $\wp$ is the magnitude of the gravitational acceleration and z is the depth.

Since the simultaneous flow of two fluids causes each to interfere with the other, the effective permeabilities are not greater than the absolute permeability $\mathbf{k}$ of the porous medium. The *relative permeabilities* $k_{r\alpha}$ are widely used in reservoir simulation:

$$\mathbf{k}_\alpha = k_{r\alpha}\mathbf{k}, \quad \alpha = w, o. \tag{10.31}$$

The function $k_{r\alpha}$ indicates the tendency of phase α to wet the porous medium. For typical functions of $p_c(S_w)$ and $k_{r\alpha}(S_w)$, the reader can refer to Chen *et al.* (2006).

10.2.2. Alternative differential equations

In this section, we derive several alternative formulations for the two-phase flow differential equations.

Formulation in phase pressures

Assume that the capillary pressure p_c has a unique inverse function:

$$S_w = p_c^{-1}(p_o - p_w).$$

We use p_w and p_o as the main unknowns. Then, it follows from Eqs. (10.27)–(10.30) that

$$
\begin{aligned}
\nabla \cdot \left(\frac{\rho_w}{\mu_w}\mathbf{k}_w(\nabla p_w - \rho_w \wp \nabla z)\right) &= \frac{\partial(\phi \rho_w p_c^{-1})}{\partial t} - q_w, \\
\nabla \cdot \left(\frac{\rho_o}{\mu_o}\mathbf{k}_o(\nabla p_o - \rho_o \wp \nabla z)\right) &= \frac{\partial\left(\phi \rho_o (1 - p_c^{-1})\right)}{\partial t} - q_o.
\end{aligned}
\tag{10.32}
$$

This system was employed in the *simultaneous solution* (SS) scheme in petroleum reservoirs (Douglas *et al.*, 1959). The equations in this system are strongly nonlinear and coupled.

Formulation in phase pressure and saturation

We use p_o and S_w as the main variables. Applying Eqs. (10.27), (10.28), and (10.30), Eq. (10.29) can be rewritten:

$$\nabla \cdot \left(\frac{\rho_w}{\mu_w} \mathbf{k}_w \left(\nabla p_o - \frac{dp_c}{dS_w} \nabla S_w - \rho_w \wp \nabla z \right) \right) = \frac{\partial(\phi \rho_w S_w)}{\partial t} - q_w,$$

$$\nabla \cdot \left(\frac{\rho_o}{\mu_o} \mathbf{k}_o (\nabla p_o - \rho_o \wp \nabla z) \right) = \frac{\partial(\phi \rho_o (1 - S_w))}{\partial t} - q_o. \tag{10.33}$$

Carrying out the time differentiation in system (10.33), dividing the first and second equations by ρ_w and ρ_o, respectively, and adding the resulting equations, we obtain

$$\frac{1}{\rho_w} \nabla \cdot \left(\frac{\rho_w}{\mu_w} \mathbf{k}_w \left(\nabla p_o - \frac{dp_c}{dS_w} \nabla S_w - \rho_w \wp \nabla z \right) \right)$$

$$+ \frac{1}{\rho_o} \nabla \cdot \left(\frac{\rho_o}{\mu_o} \mathbf{k}_o (\nabla p_o - \rho_o \wp \nabla z) \right) \tag{10.34}$$

$$= \frac{S_w}{\rho_w} \frac{\partial(\phi \rho_w)}{\partial t} + \frac{1 - S_w}{\rho_o} \frac{\partial(\phi \rho_o)}{\partial t} - \frac{q_w}{\rho_w} - \frac{q_o}{\rho_o}.$$

Note that if the saturation S_w in Eq. (10.34) is explicitly evaluated, we can use this equation to solve for p_o. After computing this pressure, the second equation in system (10.33) can be used to calculate S_w. This is the *implicit pressure-explicit saturation* (IMPES) scheme and has been widely exploited for two-phase flow in petroleum reservoirs.

Simplifications for incompressible fluids

We now develop three alternative formulations under the assumption that the two fluids are incompressible, which is physically reasonable for water and oil. The following three formulations also have similar counterparts for compressible fluids (Chen and Ewing, 1997a).

Phase formulation. Introduce the *phase mobilities*

$$\lambda_\alpha = \frac{k_{r\alpha}}{\mu_\alpha}, \quad \alpha = w, \, o,$$

and the *total mobility*

$$\lambda = \lambda_w + \lambda_o.$$

Also, define the *fractional flow* functions

$$f_\alpha = \frac{\lambda_\alpha}{\lambda}, \quad \alpha = w, \, o.$$

The oil pressure and water saturation are used as the *primary variables*

$$p = p_o, \quad S = S_w. \tag{10.35}$$

Define the total velocity

$$\mathbf{u} = \mathbf{u}_w + \mathbf{u}_o. \tag{10.36}$$

Under the assumption that the fluids are incompressible, we apply Eqs. (10.27) and (10.36) to Eq. (10.29) to see that

$$\nabla \cdot \mathbf{u} = \tilde{q}(p, S) \equiv \tilde{q}_w(p, S) + \tilde{q}_o(p, S), \tag{10.37}$$

and Eqs. (10.28) and (10.36) to Eq. (10.30) to obtain

$$\mathbf{u} = -\mathbf{k}\big[\lambda(S)\nabla p - \lambda_w(S)\nabla p_c - \big(\lambda_w \rho_w + \lambda_o \rho_o\big)\wp \nabla z\big], \tag{10.38}$$

where $\tilde{q}_w = q_w/\rho_w$ and $\tilde{q}_o = q_o/\rho_o$. Substituting Eq. (10.38) into Eq. (10.37) yields the *pressure equation*

$$-\nabla \cdot (\mathbf{k}\lambda\nabla p) = \tilde{q} - \nabla \cdot \big(\mathbf{k}(\lambda_w \nabla p_c + (\lambda_w \rho_w + \lambda_o \rho_o)\wp \nabla z)\big). \tag{10.39}$$

The phase velocities $\mathbf{u}_w$ and $\mathbf{u}_o$ are related to the total velocity $\mathbf{u}$ by

$$\mathbf{u}_w = f_w \mathbf{u} + \mathbf{k}\lambda_o f_w \nabla p_c + \mathbf{k}\lambda_o f_w (\rho_w - \rho_o)\wp \nabla z,$$

$$\mathbf{u}_o = f_o \mathbf{u} - \mathbf{k}\lambda_w f_o \nabla p_c + \mathbf{k}\lambda_w f_o (\rho_o - \rho_w)\wp \nabla z.$$

Similarly, we apply Eqs. (10.28), (10.36), and (10.38) to Eqs. (10.29) and (10.30) with $\alpha = $ w to obtain the *saturation equation*

$$\phi \frac{\partial S}{\partial t} + \nabla \cdot \left\{ \mathbf{k} f_w(S)\lambda_o(S) \left(\frac{dp_c}{dS}\nabla S - (\rho_o - \rho_w)\wp \nabla z \right) \right.$$
$$\left. + f_w(S)\mathbf{u} \right\} = \tilde{q}_w(p, S), \tag{10.40}$$

where, for notational convenience, we assume that $\phi = \phi(\mathbf{x})$.

Weighted formulation. We introduce a pressure that is smoother than the phase pressure:

$$p = S_w p_w + S_o p_o. \tag{10.41}$$

Even if a phase disappears (i.e., either S_w or S_o is zero), there is still a nonzero smooth variable p. Applying the same algebraic manipulations as in deriving the phase formulation, we obtain

$$\mathbf{u} = -\mathbf{k}\big\{ \lambda(S)\nabla p + \big(S\lambda(S) - \lambda_w(S)\big)\nabla p_c$$
$$+ \lambda(S)p_c \nabla S - \big(\lambda_w \rho_w + \lambda_o \rho_o\big)\wp \nabla z\big\}. \tag{10.42}$$

Equations (10.37) and (10.40) remain the same.

Global formulation. Note that the capillary pressure p_c appears in both Eqs. (10.38) and (10.42). To remove it, we define a global pressure (Antontsev, 1972; Chavent-Jaffré, 1978):

$$p = p_o - \int^S \left(f_w \frac{dp_c}{dS} \right)(\xi) \, d\xi. \tag{10.43}$$

Using this pressure, the total velocity becomes

$$\mathbf{u} = -\mathbf{k}\big(\lambda(S)\nabla p - (\lambda_w \rho_w + \lambda_o \rho_o)\wp\nabla z\big). \tag{10.44}$$

It follows from Eqs. (10.28) and (10.43) that

$$\lambda\nabla p = \lambda_w \nabla p_w + \lambda_o \nabla p_o,$$

which implies that the global pressure is the pressure that would produce a flow of a fluid (with mobility λ) equal to the sum of the flows of fluids w and o. Equations (10.37) and (10.40) remain the same.

The coupling between the pressure and saturation equations in the global formulation is less than that in the phase and weighted formulations, and the nonlinearity is weakened as well. This formulation is most suitable for a mathematical analysis for two-phase flow (Antontsev, 1972; Chavent-Jaffré, 1978; Chen, 2001; Chen, 2002). When the capillary effect is neglected, the three formulations are the same. In this case, the saturation equation becomes the well-known *Buckley-Leverett equation*.

Classification of differential equations

There are basically three types of second-order partial differential equations: elliptic, parabolic, and hyperbolic. We must be able to distinguish among these types when numerical methods for their solution are devised.

If two independent variables (either (x_1, x_2) or (x_1, t)) are considered, then second-order partial differential equations have the form, with $x = x_1$

$$a\frac{\partial^2 p}{\partial x^2} + b\frac{\partial^2 p}{\partial t^2} = f\left(\frac{\partial p}{\partial x}, \frac{\partial p}{\partial t}, p\right).$$

This equation is (1) elliptic if $ab > 0$; (2) parabolic if $ab = 0$; and (3) hyperbolic if $ab < 0$.

The simplest elliptic equation is the *Poisson equation*:

$$\frac{\partial^2 p}{\partial x_1^2} + \frac{\partial^2 p}{\partial x_2^2} = f(x_1, x_2).$$

A typical parabolic equation is the *heat conduction equation*:

$$\phi \frac{\partial p}{\partial t} = \frac{\partial^2 p}{\partial x_1^2} + \frac{\partial^2 p}{\partial x_2^2}.$$

Finally, the prototype hyperbolic equation is the *wave equation*:

$$\frac{1}{v^2} \frac{\partial^2 p}{\partial t^2} = \frac{\partial^2 p}{\partial x_1^2} + \frac{\partial^2 p}{\partial x_2^2},$$

where v is a wave speed. In the one-dimensional case, this equation can be "factorized" into two first-order parts:

$$\left(\frac{1}{v} \frac{\partial}{\partial t} - \frac{\partial}{\partial x} \right) \left(\frac{1}{v} \frac{\partial}{\partial t} + \frac{\partial}{\partial x} \right) p = 0.$$

The second part gives the first-order hyperbolic equation:

$$\frac{\partial p}{\partial t} + v \frac{\partial p}{\partial x} = 0.$$

We now return to the two-phase flow equations. While the phase mobilities λ_w and λ_o can be zero (Chen *et al.*, 2006), the total mobility λ is always positive, so the pressure equation (10.39) is elliptic in p. If one of the densities varies, this equation becomes parabolic. In general, $-\mathbf{k}\lambda_o f_w dp_c/dS$ is semi positive definite, so the saturation equation (10.40) is a parabolic equation in S, which is *degenerate* in the sense that the diffusion can be zero. This equation becomes hyperbolic if the capillary pressure is ignored. The total velocity is used in the global pressure formulation. This velocity is smoother than the phase velocities. It can be also used in the formulations (10.32) and (10.33) (Chen-Ewing, 1997b). Finally, with $p_c = 0$, Eq. (10.40) becomes the known *Buckley-Leverett equation* whose flux function f_w is generally nonconvex over the range of saturation values where this function is nonzero, as illustrated in Fig. 10.1; see the next section for the formulation in hyperbolic form.

Formulation in hyperbolic form

Assume that $p_c = 0$ and rock compressibility is neglected. Then, Eq. (10.40) becomes

$$\phi \frac{\partial S}{\partial t} + \nabla \cdot \left(f_w \mathbf{u} - \lambda_o f_w (\rho_o - \rho_w) \wp \mathbf{k} \nabla z \right) = \frac{q_w}{\rho_w}. \tag{10.45}$$

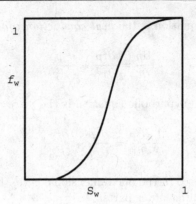

Figure 10.1 A flux function f_w.

Using Eq. (10.37) and the fact that $f_w + f_o = 1$, this equation can be manipulated into

$$\phi\frac{\partial S}{\partial t} + \left(\frac{df_w}{dS}\mathbf{u} - \frac{d(\lambda_o f_w)}{dS}(\rho_o - \rho_w)\wp\mathbf{k}\nabla z\right) \cdot \nabla S = \frac{f_o q_w}{\rho_w} - \frac{f_w q_o}{\rho_o},$$

$$(10.46)$$

which is a hyperbolic equation in S. Finally, if we neglect the gravitational term, we obtain

$$\phi\frac{\partial S}{\partial t} + \frac{df_w}{dS}\mathbf{u} \cdot \nabla S = \frac{f_o q_w}{\rho_w} - \frac{f_w q_o}{\rho_o}, \qquad (10.47)$$

which is the familiar form of a waterflooding equation, i.e., the *Buckley-Leverett equation*. The source term in Eq. (10.47) is zero for production since

$$\frac{q_w}{\rho_w} = f_w\left(\frac{q_w}{\rho_w} + \frac{q_o}{\rho_o}\right),$$

by Darcy's law. For injection, this term may not be zero since it equals $(1 - f_w)q_w/\rho_w \neq 0$ in this case.

10.2.3. Boundary conditions

As for single-phase flow in the previous section, the mathematical model described so far for two-phase flow is not complete unless necessary boundary and initial conditions are specified. Below, we present boundary conditions of three kinds that are relevant to systems (10.32), (10.33), (10.39) ((10.42) or (10.44)), and (10.40). We denote by Γ the external boundary or a boundary segment of the porous medium domain Ω considered.

Boundary conditions for system (10.32)

The symbol α, as a subscript, with $\alpha = w, o$, is used to indicate a considered phase. When a phase pressure is specified as a known function of position and time on Γ, the boundary condition reads

$$p_\alpha = g_{\alpha,1} \quad \text{on } \Gamma. \tag{10.48}$$

When the mass flux of phase α is known on Γ, the boundary condition is

$$\rho_\alpha \mathbf{u}_\alpha \cdot \boldsymbol{\nu} = g_{\alpha,2} \quad \text{on } \Gamma, \tag{10.49}$$

where $\boldsymbol{\nu}$ indicates the outward unit normal to Γ and $g_{\alpha,2}$ is given. For an impervious boundary for the α phase, $g_{\alpha,2} = 0$ (no flow for this phase).

When Γ is a semipervious boundary for the α phase, a boundary condition of mixed kind occurs:

$$g_{\alpha,p} p_\alpha + g_{\alpha,u} \rho_\alpha \mathbf{u}_\alpha \cdot \boldsymbol{\nu} = g_{\alpha,3} \quad \text{on } \Gamma, \tag{10.50}$$

where $g_{\alpha,p}$, $g_{\alpha,u}$, and $g_{\alpha,3}$ are given functions.

Initial conditions specify the values of the main unknowns p_w and p_o over the entire domain at some initial time, usually taken at $t = 0$:

$$p_\alpha(\mathbf{x}, 0) = p_\alpha^0(\mathbf{x}), \quad \alpha = w, o,$$

where $p_\alpha^0(\mathbf{x})$ are known functions.

Boundary conditions for system (10.33)

Boundary conditions for system (10.33) can be imposed as for system (10.32); i.e., Eqs. (10.48)–(10.50) are applicable to system (10.33). The only difference between the boundary conditions for these two systems is that a prescribed saturation is sometimes given on the boundary Γ for system (10.33):

$$S_w = g_4 \quad \text{on } \Gamma.$$

In practice, this prescribed saturation boundary condition seldom occurs. However, a condition $g_4 = 1$ does take place when a medium is in contact with a body of this wetting phase. The condition $S_w = 1$ can be exploited on the bottom of a water pond on the ground surface, e.g. An initial saturation is also specified:

$$S_w(\mathbf{x}, 0) = S_w^0(\mathbf{x}),$$

where $S_w^0(\mathbf{x})$ is given.

Boundary conditions for (10.39) ((10.42) or (10.44)) and (10.40)

Boundary conditions are usually specified in terms of phase quantities like those in Eqs. (10.48)–(10.50). These conditions can be transformed into the boundary conditions in terms of the global quantities introduced in Eqs. (10.43) and (10.44). For the prescribed pressure boundary condition in Eq. (10.48), the corresponding boundary condition is given by

$$p = g_1 \quad \text{on } \Gamma,$$

where the pressure p is defined by Eq. (10.43) and g_1 is determined by

$$g_1 = g_{o,1} - \int^{g_{o,1}-g_{w,1}} f_{\mathrm{w}}\left(p_{\mathrm{c}}^{-1}(\xi)\right)d\xi.$$

Also, when the total mass flux is known on the boundary Γ, it follows from Eq. (10.49) that

$$\mathbf{u} \cdot \boldsymbol{\nu} = g_2 \quad \text{on } \Gamma,$$

where

$$g_2 = \frac{g_{o,2}}{\rho_o} + \frac{g_{w,2}}{\rho_w}.$$

For an impervious boundary for the total flow, $g_2 = 0$.

10.3. Finite element solution of single-phase flow

The development of the finite element method will be carried out for the general single-phase flow equation (10.20), which applies to all the cases: incompressible, slightly compressible, and compressible. The discretization of these cases will be different, particularly in the treatment of transmissibility terms, and the difference will be pointed out as appropriate. Most noticeably, the flow equation in the incompressible case is linear and does not contain the accumulation term, while the equations in other two cases are nonlinear and involve the accumulation term; nonlinearity in the slightly compressible case is weaker than that in the compressible case.

We recall the general equation (10.20), with appropriate boundary and initial conditions, for the primary unknown p:

$$
\begin{aligned}
c(p)\frac{\partial p}{\partial t} &= \nabla \cdot \left(\frac{\rho}{\mu}\mathbf{k}(\nabla p - \rho \wp \nabla z)\right) + q && \text{in } \Omega \times J, \\
\frac{\rho}{\mu}\mathbf{k}\left(\nabla p - \rho \wp \nabla z\right) \cdot \boldsymbol{\nu} &= 0 && \text{on } \Gamma_N \times J, \\
p &= g && \text{on } \Gamma_D \times J, \\
p(\cdot,0) &= p^0 && \text{in } \Omega,
\end{aligned}
$$

$$(10.51)$$

where the no-flow boundary part Γ_N and the pressure-specified boundary part Γ_D are disjoint, $\Gamma = \Gamma_N \cup \Gamma_D$, ν is the outward unit normal to the boundary Γ_N, and $J = (0, T]$ $(T > 0)$ is the time interval of interest.

10.3.1. Treatment of initial conditions

In reservoir simulation, in general, an initial pressure p_d^0 is given at a *datum level depth* z_d. The pressure at other location z is determined by the *gravity equilibrium condition*:

$$p^0|_z = p_d^0 + \gamma(z - z_d), \qquad (10.52)$$

where $\gamma = \rho \wp$ is the fluid density in terms of pressure per distance (usually called *fluid gravity*). For slightly compressible and compressible flow, the fluid density γ depends on p. Hence, an iteration procedure may be required in the application of Eq. (10.52).

10.3.2. The finite element method

The admissible function space V is

$$V = \left\{ \text{Functions } v \colon v \text{ is a continuous function on } \Omega, \frac{\partial v}{\partial x_1}, \frac{\partial v}{\partial x_2}, \text{ and } \frac{\partial v}{\partial x_3} \right.$$
$$\left. \text{are piecewise continuous and bounded on } \Omega, \text{ and } v = 0 \text{ on } \Gamma_D \right\}.$$

Using Green's formula (2.9), problem (10.51) can be written in the variational form, with $p = g$ on Γ_D,

$$\int_\Omega c(p) \frac{\partial p}{\partial t} v \, d\mathbf{x} + \int_\Omega \frac{\rho}{\mu} \mathbf{k} (\nabla p - \gamma \nabla z) \cdot \nabla v \, d\mathbf{x}$$
$$= \int_\Omega q v \, d\mathbf{x} \quad \forall v \in V, \ t \in J, \qquad (10.53)$$
$$p(\mathbf{x}, 0) = p^0(\mathbf{x}) \quad \forall \mathbf{x} \in \Omega.$$

Let V_h be a finite element subspace of V (see Chapters 4–6). The finite element version of system (10.53) is: find $p_h : J \to V_h + g$ such that

$$\int_\Omega c(p_h) \frac{\partial p_h}{\partial t} v \, d\mathbf{x} + \int_\Omega \frac{\rho(p_h)}{\mu} \mathbf{k} (\nabla p_h - \gamma(p_h) \nabla z) \cdot \nabla v \, d\mathbf{x}$$
$$= \int_\Omega q v \, d\mathbf{x} \quad \forall v \in V_h, \qquad (10.54)$$
$$\int_\Omega p_h(x, 0) v(x) \, d\mathbf{x} = \int_\Omega p^0 v \, d\mathbf{x} \quad \forall v \in V_h,$$

where $V_h + g$ means that $p_h = g$ on Γ_D and $p_h - g \in V_h$. As for system (7.28), after the introduction of basis functions in V_h, system (10.54) can be stated in matrix form (see Exercise 10.4):

$$\mathbf{C}(\mathbf{p})\frac{d\mathbf{p}}{dt} + \mathbf{A}(\mathbf{p})\mathbf{p} = \mathbf{f}(\mathbf{p}), \quad t \in J,$$

$$\mathbf{B}\mathbf{p}(0) = \mathbf{p}^0.$$
(10.55)

Under the assumption that the coefficient $c(p)$ is bounded below by a positive constant, this nonlinear system of ODEs possesses a unique solution (at least locally). In fact, if assumption (7.25) is satisfied for c, ρ, and γ (and possibly for q), the solution $\mathbf{p}(t)$ exists for all t. Several approaches for solving system (10.55) are discussed next, in a similar fashion as for the solution of the system (7.28).

Linearization approach

Let $0 = t^0 < t^1 < t^2 < \cdots < t^N$ be a partition of J, and set $\Delta t^n = t^n - t^{n-1}$, $n = 1, 2, \ldots, N$. The nonlinear system (10.55) can be linearized by allowing the nonlinearities to lag one time step behind. Thus, the modified backward Euler method for system (10.54) takes the form: find $p_h^n \in V_h + g^n$, $n = 1, 2, \ldots, N$, such that

$$\int_\Omega c(p_h^{n-1})\frac{p_h^n - p_h^{n-1}}{\Delta t^n}v\, d\mathbf{x} + \int_\Omega \frac{\rho(p_h^{n-1})}{\mu}\mathbf{k}(\nabla p_h^n - \gamma(p_h^{n-1})\nabla z) \cdot \nabla v\, d\mathbf{x}$$

$$= \int_\Omega q^{n-1}v\, d\mathbf{x} \quad \forall v \in V_h,$$
(10.56)

$$\int_\Omega p_h^0 v\, d\mathbf{x} = \int_\Omega p^0 v\, d\mathbf{x} \quad \forall v \in V_h.$$

In matrix form it is given by

$$\mathbf{C}\left(\mathbf{p}^{n-1}\right)\frac{\mathbf{p}^n - \mathbf{p}^{n-1}}{\Delta t^n} + \mathbf{A}(\mathbf{p}^{n-1})\mathbf{p}^n = \mathbf{f}(\mathbf{p}^{n-1}),$$

$$\mathbf{B}\mathbf{p}(0) = \mathbf{p}^0.$$
(10.57)

For slightly compressible flow, the pressure-dependent fluid properties, ρ (or B) and μ, are weakly nonlinear and can be evaluated at the previous time level.

Note that system (10.57) is a system of linear equations in $\mathbf{p}^n$, which can be solved using iterative algorithms. When V_h is the finite element space of piecewise linear functions, the error $p^n - p_h^n$ ($0 \leq n \leq N$) in the norm $(\int_\Omega |p^n - p_h^n|^2 d\mathbf{x})^{1/2}$ is asymptotically of order $\mathcal{O}(\Delta t + h^2)$ under appropriate smoothness assumptions on the solution p and for Δt small enough (Thomée, 1984; Chen and Douglas, 1991), where the quantity $\Delta t = \max_{1 \leq n \leq N} \Delta t^n$. We may use the Crank-Nicholson discretization method in Eq. (10.56). However, the linearization decreases the order of the time discretization error to $\mathcal{O}(\Delta t)$, giving $\mathcal{O}(\Delta t + h^2)$ overall, which is true for any higher-order time discretization method with the present linearization technique. This drawback can be overcome by using an *extrapolation technique* in the linearization of the coefficients (see Sec. 7.2.2).

Implicit time approximation

We now consider a fully *implicit time approximation* scheme for problem (10.54): find $p_h^n \in V_h + g^n$, $n = 1, 2, \ldots, N$, such that

$$\int_\Omega c(p_h^n) \frac{p_h^n - p_h^{n-1}}{\Delta t^n} v \, d\mathbf{x} + \int_\Omega \frac{\rho(p_h^n)}{\mu} \mathbf{k} \left(\nabla p_h^n - \gamma(p_h^n) \nabla z \right) \cdot \nabla v \, d\mathbf{x}$$

$$= \int_\Omega q^n v \, d\mathbf{x} \quad \forall v \in V_h, \tag{10.58}$$

$$\int_\Omega p_h^0 v \, d\mathbf{x} = \int_\Omega p^0 v \, d\mathbf{x} \quad \forall v \in V_h.$$

Its matrix form is

$$\mathbf{C}(\mathbf{p}^n) \frac{\mathbf{p}^n - \mathbf{p}^{n-1}}{\Delta t^n} + \mathbf{A}(\mathbf{p}^n)\mathbf{p}^n = \mathbf{f}(\mathbf{p}^n),$$

$$\mathbf{B}\mathbf{p}(0) = \mathbf{p}^0. \tag{10.59}$$

Now, system (10.59) is a system of nonlinear equations in $\mathbf{p}^n$, which must be solved at each time step via an iteration method. Let us consider *Newton's method* (or the Newton–Raphson method). Note that the first equation in system (10.59) can be rewritten:

$$\left(\mathbf{A}(\mathbf{p}^n) + \frac{1}{\Delta t^n} \mathbf{C}(\mathbf{p}^n) \right) \mathbf{p}^n - \frac{1}{\Delta t^n} \mathbf{C}(\mathbf{p}^n) \mathbf{p}^{n-1} - \mathbf{f}(\mathbf{p}^n) = \mathbf{0}.$$

We express this equation as

$$\mathbf{F}(\mathbf{p}^n) = \mathbf{0}. \tag{10.60}$$

Newton's method for system (10.60) is

$$\text{Set } \mathbf{v}^0 = \mathbf{p}^{n-1};$$

$$\text{Iterate } \mathbf{v}^k = \mathbf{v}^{k-1} + \mathbf{d}^k, \quad k = 1, 2, \ldots,$$

where $\mathbf{d}^k$ satisfies the equation

$$\mathbf{G}(\mathbf{v}^{k-1})\mathbf{d}^k = -\mathbf{F}(\mathbf{v}^{k-1}),$$

with $\mathbf{G}$ the Jacobian matrix of the vector function $\mathbf{F}$:

$$\mathbf{G} = \left(\frac{\partial F_i}{\partial p_j}\right)_{i,j=1,2,\ldots,M},$$

where M is the dimension number of $\mathbf{p}$. Recall that if the matrix $\mathbf{G}(\mathbf{p}^n)$ is non singular and the second partial derivatives of $\mathbf{F}$ are bounded, Newton's method converges quadratically in a neighborhood of $\mathbf{p}^n$; i.e., there are constants $\epsilon > 0$ and C such that if $|\mathbf{v}^{k-1} - \mathbf{p}^n| \leq \epsilon$, then

$$|\mathbf{v}^k - \mathbf{p}^n| \leq C|\mathbf{v}^{k-1} - \mathbf{p}^n|^2.$$

Explicit time approximation

We conclude with a remark about the application of a *forward, explicit time approximation* method to problem (10.54): find $p_h^n \in V_h + g^n$, $n = 1, 2, \ldots, N$, such that

$$\int_\Omega c(p_h^{n-1})\frac{p_h^n - p_h^{n-1}}{\Delta t^n}v\,d\mathbf{x} + \int_\Omega \frac{\rho(p_h^{n-1})}{\mu}\mathbf{k}(\nabla p_h^{n-1} - \gamma(p_h^{n-1})\nabla z)\cdot\nabla v\,d\mathbf{x}$$

$$= \int_\Omega q^{n-1}v\,d\mathbf{x} \qquad \forall v \in V_h, \tag{10.61}$$

$$\int_\Omega p_h^0 v\,d\mathbf{x} = \int_\Omega p^0 v\,d\mathbf{x} \quad \forall v \in V_h.$$

In matrix form, it is written

$$\mathbf{C}(\mathbf{p}^n)\frac{\mathbf{p}^n - \mathbf{p}^{n-1}}{\Delta t^n} + \mathbf{A}(\mathbf{p}^{n-1})\mathbf{p}^{n-1} = \mathbf{f}(\mathbf{p}^{n-1}),$$
$$\mathbf{B}\mathbf{p}(0) = \mathbf{p}_0. \tag{10.62}$$

Note that the only nonlinearity is in matrix $\mathbf{C}$. With an appropriate *mass lumping* (a diagonalization technique; off-diagonal quantities are placed in the right-hand side of Eq. (10.62)) in this matrix, the first equation in system (10.62) represents M scalar nonlinear equations of the form

$$\mathcal{F}(p_i^n) = 0, \quad i = 1, 2, \ldots, M. \tag{10.63}$$

Each single equation in system (10.63) can be easily solved via any standard method (Ostrowski, 1973; Rheinboldt, 1998).

For the explicit method (10.61) to be stable in the sense defined in Sec. 7.1.3, a *stability condition* of the following type must be satisfied:

$$\Delta t^n \leq Ch^2, \quad n = 1, 2, \ldots, N, \tag{10.64}$$

where C now depends on c and $\mathbf{a}$ (see constraint (7.22)). Unfortunately, this condition on the time steps is very restrictive for long-time integration, as noted earlier.

As in Sec. 7.2, we have developed linearization, implicit, and explicit time approximation approaches for numerically solving the single-phase problem (10.54). In terms of computational effort, the explicit approach is the simplest at each time step; however, it requires an impractical stability restriction. The linearization approach is more practical, but it reduces the order of accuracy in time for high-order time discretization methods (unless extrapolations are exploited). An efficient and accurate method is the fully implicit approach; the extra cost involved at each time step for this implicit method is usually more than compensated for by the fact that larger time steps may be taken, particularly when Newton's method with a good initial guess is employed. Modified implicit methods such as *semi implicit methods* (Aziz and Settari, 1979) can be applied; for a given physical problem, the linearization approach should be applied for weak nonlinearity (e.g., the dependence of viscosity μ on pressure p), while the implicit one should be used for strong nonlinearity (e.g., the dependence of density ρ on p). For more details, the reader can refer to Chen *et al.* (2000c).

10.3.3. Practical issues

In this section, we address a couple of issues related to the computation of the system matrix $\mathbf{A}$, which appeared in system (10.55) (also see Eqs. (10.57), (10.59), and (10.62)).

The system matrix

The simplest three-dimensional continuum element is a tetrahedron. Let K_h be a partition of a polygonal domain $\Omega \subset \mathbb{R}^3$ into non overlapping *tetrahedra* such that no vertex of any tetrahedron lies in the interior of an edge or face of another tetrahedron. The finite element space is

$$V_h = \{\text{Functions } v : v \text{ is a continuous function on } \Omega, v \text{ is linear}$$

$$\text{on each tetrahedron } K \in K_h, \text{ and } v = 0 \text{ on } \Gamma_D\}.$$

We recall the finite element system (10.54): find $p_h : J \to V_h + g$ such that

$$\int_\Omega c(p_h)\frac{\partial p_h}{\partial t} v \; d\mathbf{x} + \int_\Omega \frac{\rho(p_h)}{\mu}\mathbf{k}(\nabla p_h - \gamma(p_h)\nabla z) \cdot \nabla v \; d\mathbf{x}$$

$$= \int_\Omega qv \; d\mathbf{x} \quad \forall v \in V_h, \tag{10.65}$$

$$\int_\Omega p_h(x,0)v(x) \; d\mathbf{x} = \int_\Omega p^0 v \; d\mathbf{x} \quad \forall v \in V_h.$$

As an example, we show how to calculate the term

$$\int_\Omega \frac{\rho(p_h)}{\mu}\mathbf{k}\nabla p_h \cdot \nabla v \; d\mathbf{x}; \tag{10.66}$$

other terms in system (10.65) are easier to evaluate.

On each tetrahedron $K \in K_h$, the solution p_h is represented using the volume coordinates λ_i (see Sec. 6.2 and Fig. 10.2):

$$p_h = \sum_{l=1}^{4} p_l \lambda_l = p_1\lambda_1 + p_2\lambda_2 + p_3\lambda_3 + p_4\lambda_4,$$

where p_i indicates the value of p_h at the ith node of tetrahedron K. As a result, with $v = \lambda_i$, the term (10.66) on the tetrahedron K is written as

$$\sum_{l=1}^{4} \int_K \frac{\rho}{\mu}\mathbf{k}\nabla \lambda_l \cdot \nabla \lambda_i \; d\mathbf{x} \; p_l, \quad i = 1, 2, 3, 4.$$

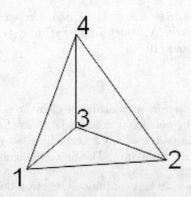

Figure 10.2   A tetrahedron.

Since $\nabla \lambda_l$ is constant, this equation becomes

$$\sum_{l=1}^{4} |K| \overline{\left(\frac{\rho}{\mu}\mathbf{k}\right)} \nabla \lambda_l \cdot \nabla \lambda_i \, p_l, \qquad i = 1, 2, 3, 4, \qquad (10.67)$$

where $|K|$ is the volume of tetrahedron K and $\overline{(\frac{\rho}{\mu}\mathbf{k})}$ is some average of $\frac{\rho}{\mu}\mathbf{k}$ over K.

Note that (see Sec. 6.2)

$$\lambda_1 + \lambda_2 + \lambda_3 + \lambda_4 = 1.$$

Consequently, Eq. (10.67) equals

$$\sum_{l=1}^{4} |K| \overline{\left(\frac{\rho}{\mu}\mathbf{k}\right)} \nabla \lambda_l \cdot \nabla \lambda_i \, p_l = -T_{ij}(p_j - p_i) - T_{ik}(p_k - p_i) - T_{im}(p_m - p_i),$$

$$(10.68)$$

where the *transmissibility coefficients* T_{ij}, T_{ik}, and T_{im} are

$$T_{ij} = -|K| \overline{\left(\frac{\rho}{\mu}\mathbf{k}\right)} \nabla \lambda_j \cdot \nabla \lambda_i, \quad T_{ik} = -|K| \overline{\left(\frac{\rho}{\mu}\mathbf{k}\right)} \nabla \lambda_k \cdot \nabla \lambda_i,$$

$$T_{im} = -|K| \overline{\left(\frac{\rho}{\mu}\mathbf{k}\right)} \nabla \lambda_m \cdot \nabla \lambda_i,$$

and $\{i, j, k, m\}$ are cyclic permutations of the indices $\{1, 2, 3, 4\}$.

We now consider the assembly of the corresponding *global transmissibility matrix*. Each connection between any two adjacent nodes $\mathbf{m}_i$ and $\mathbf{m}_j$ includes the contributions from tetrahedra $\{K_l\}$ that share the common edge with endpoints $\mathbf{m}_i$ and $\mathbf{m}_j$. The transmissibility between $\mathbf{m}_i$ and $\mathbf{m}_j$ is now

$$T_{ij} = -\sum_{l} \left(|K| \overline{\left(\frac{\rho}{\mu}\mathbf{k}\right)} \nabla \lambda_j \cdot \nabla \lambda_i \right) \Big|_{K_l}. \qquad (10.69)$$

Then, the part of the linear system associated with the term (10.66) at node i is

$$-\sum_{j \in \Omega_i} T_{ij} \left(p_j - p_i \right), \qquad (10.70)$$

where Ω_i is the set of all neighboring nodes of node $\mathbf{m}_i$.

Treatment of block transmissibility

On each element K, some sort of average is used in the transmissibility:

$$\overline{\left(\frac{\rho}{\mu} \mathbf{k} \right)}.$$

It contains two distinct quantities: $\mathbf{k}$ represents the rock property and ρ/μ represents the fluid properties. These rock and fluid properties are often given only at the block centers, but the transmissibilities are computed at the gridblock boundaries or boundary nodes due to the use of certain numerical quadrature formulas (see Sec. 6.7). Hence, some sort of averaging techniques must be utilized to estimate these properties between two adjacent gridblocks.

Because the fluid properties in the transmissibility term for the single-phase flow do not change much from block to block (they are slowly varying functions of pressure only), the usual arithmetic average can be used for them. As shown in Chen (2007), the harmonic average must be used for the rock property $\mathbf{k}$ in the transmissibility to produce an accurate, physically meaning solution (see Sec. 11.1.4).

We end this chapter with a remark that the fluid properties are constant for incompressible flow. It is in the slightly compressible and compressible cases that they must be carefully evaluated. For multiphase flow, the relative permeabilities can change a great deal from block to block, which will require use of a different averaging, such as *upstream weighting* (Chen, 2007).

10.4. Exercises

10.1. Write Eq. (10.19) in the cylindrical coordinates (r, θ, x_3), where $x_1 = r \cos \theta$, $x_2 = r \sin \theta$, and $x_3 = x_3$.

10.2. Derive Eq. (10.40) in detail.

10.3. Derive Eq. (10.44) in detail.

10.4. Let V_h be a finite element subspace of the admissible function space V defined in Sec. 10.3. After the introduction of a set of basis functions in V_h and appropriate matrices and vectors, write system (10.54) in the matrix form (10.55) in detail.

Chapter 11

Other Finite Element Methods

The development and study of the standard finite element method is the major focus of this book. There exist a wide variety of variations of this method that are more often used for some of the practical problems considered in the previous three chapters. For example, the nonconforming and mixed finite element methods have been more widely applied in solid and fluid mechanics, and the control volume finite element (CVFE) method has been more recently used in porous media flow. This chapter is devoted to a brief introduction of these methods, along with some other useful methods such as the multipoint flux approximation (MPFA), discontinuous, characteristic, adaptive, and multiscale finite element methods. To see more details of these methods, the reader can refer to the books by Chen (2005) and Chen *et al.* (2006). The CVFE, MPFA, nonconforming, mixed, discontinuous, characteristic, adaptive, and multiscale finite element methods are, respectively, described in Secs. 11.1–11.8. The characteristic, nonconforming, and mixed finite element methods were briefly discussed in Chapter 9 for fluid mechanics.

11.1. The CVFE method

The classical finite difference method is locally conservative, but it is not flexible in the treatment of complex geometry. On the other hand, the standard finite element method described has greater flexibility, but it is not conservative on local elements (e.g., on triangles). The standard finite element method conserves mass or energy on the entire physical domain, which makes it globally conservative. In this section, we introduce a variation of

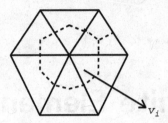

Figure 11.1 A control volume.

the finite element method so that it becomes locally conservative on each *control volume*. Control volumes can be formed around grid nodes by joining the midpoints of the edges of a triangle with a point inside the triangle, for example (Fig. 11.1). Different locations of the point give rise to different forms of the flux term between grid nodes. When it is the barycenter of the triangle, the resulting grid is CVFE type, and the resulting finite element method is the CVFE method. This method was first introduced by Lemonnier (1979) for reservoir simulation. The CVFE grids are different from the perpendicular bisection (PEBI) grids (also called *Voronoi* grids (Heinrich, 1987)) in that the latter are locally orthogonal. The CVFE grids are more flexible in grid requirements. In this section, we formulate the CVFE method based on triangles; it can be also formulated using quadrilateral grids in two and three dimensions (Aavatsmark, 2002) and tetrahedral grids (Verma and Aziz, 1996).

11.1.1. The basic CVFE method

To sketch the CVFE idea, we focus on linear triangular elements in two space dimensions. As an example, we consider the stationary problem

$$-\nabla \cdot (\mathbf{a}\nabla p) = f(x_1, x_2) \quad \text{in } \Omega, \tag{11.1}$$

where Ω is a bounded domain in the plane and p is the sought solution.

Let V_i be a control volume. Replacing p by $p_h \in V_h$ (the finite element space of continuous piecewise linear functions on Ω; see Chapters 4–6) in Eq. (11.1) and integrating over V_i, we see that

$$-\int_{V_i} \nabla \cdot (\mathbf{a}\nabla p_h) \, d\mathbf{x} = \int_{V_i} f \, d\mathbf{x}.$$

The divergence theorem implies

$$-\int_{\partial V_i} \mathbf{a}\nabla p_h \cdot \boldsymbol{\nu} \, d\ell = \int_{V_i} f \, d\mathbf{x}, \tag{11.2}$$

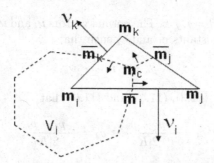

Figure 11.2   A base triangle for CVFE.

where $\boldsymbol{\nu}$ is the unit normal outward to the surface ∂V_i of the control volume V_i. Note that $\nabla p_h \cdot \boldsymbol{\nu}$ is continuous across each segment of ∂V_i (that lies inside a triangle). Thus, if the coefficient **a** is continuous across that segment, so is the flux $\mathbf{a}\nabla p_h \cdot \boldsymbol{\nu}$. Therefore, the flux is continuous across the edges of the control volume V_i. Furthermore, Eq. (11.2) indicates that the CVFE method is locally (i.e., on each control volume) conservative. If **a** is discontinuous across an interface of the control volume, some special averaging technique will be used to guarantee the flux continuity (see Sec. 11.1.4).

Given a triangle K with vertices $\mathbf{m}_i$, $\mathbf{m}_j$, and $\mathbf{m}_k$, edge midpoints $\overline{\mathbf{m}}_i$, $\overline{\mathbf{m}}_j$, and $\overline{\mathbf{m}}_k$, and center $\mathbf{m}_c$ (Fig. 11.2), we define the "backward" vertex integers

$$m(i) = k, \quad m(j) = i, \quad m(k) = j,$$

and the "forward" vertex integers

$$q(i) = j, \quad q(j) = k, \quad q(k) = i.$$

Let $\boldsymbol{\nu}_l$ be the vector normal to the edge $\mathbf{m}_l\mathbf{m}_{q(l)}$ (outward to the triangle K; Fig. 11.2), with the same length as this edge, $l = i, j, k$. With $|K|$ the area of the triangle K, we see that

$$-2|K| = \mathbf{m}_i\mathbf{m}_j \cdot \boldsymbol{\nu}_k = \mathbf{m}_i\mathbf{m}_k \cdot \boldsymbol{\nu}_i. \tag{11.3}$$

We consider the computation of the left-hand side of Eq. (11.2) on the interface segment $\overline{\mathbf{m}}_i\mathbf{m}_c\overline{\mathbf{m}}_k$ (Fig. 11.2):

$$f_i \equiv -\int_{\overline{\mathbf{m}}_i\mathbf{m}_c + \mathbf{m}_c\overline{\mathbf{m}}_k} \mathbf{a}\nabla p_h \cdot \boldsymbol{\nu}\, d\ell. \tag{11.4}$$

Since p_h is linear on the triangle K

$$p_j = p_i + \mathbf{m}_i\mathbf{m}_j \cdot \nabla p_h, \quad p_k = p_i + \mathbf{m}_i\mathbf{m}_k \cdot \nabla p_h, \tag{11.5}$$

where $p_l = p_h(\mathbf{m}_l)$, $l = i, j, k$. The normal vectors $\boldsymbol{\nu}_i$ and $\boldsymbol{\nu}_k$ are independent so that there are constants c_1 and c_2 such that

$$\nabla p_h = c_1 \boldsymbol{\nu}_i + c_2 \boldsymbol{\nu}_k. \tag{11.6}$$

It follows from Eqs. (11.3), (11.5), and (11.6) that

$$c_1 = -\frac{p_k - p_i}{2|K|}, \quad c_2 = -\frac{p_j - p_i}{2|K|},$$

which are substituted into expression (11.6) to have

$$\nabla p_h = -\frac{p_k - p_i}{2|K|} \boldsymbol{\nu}_i - \frac{p_j - p_i}{2|K|} \boldsymbol{\nu}_k. \tag{11.7}$$

Let the coefficient $\mathbf{a}$ be a constant tensor $\mathbf{a}_i$ on the control volume V_i. Then, on the interface segments $\overline{\mathbf{m}}_i \mathbf{m}_c$ and $\mathbf{m}_c \overline{\mathbf{m}}_k$ (Fig. 11.2), it follows from Eqs. (11.4) and (11.7) that

$$\begin{aligned}
f_i &= -\int_{\overline{\mathbf{m}}_i \mathbf{m}_c + \mathbf{m}_c \overline{\mathbf{m}}_k} \mathbf{a} \nabla p_h \cdot \boldsymbol{\nu} \, d\ell \\
&= \left(\frac{p_k - p_i}{2|K|} \mathbf{a}_i \boldsymbol{\nu}_i + \frac{p_j - p_i}{2|K|} \mathbf{a}_i \boldsymbol{\nu}_k \right) \cdot \boldsymbol{\nu}_{\overline{\mathbf{m}}_i \mathbf{m}_c} \\
&\quad + \left(\frac{p_k - p_i}{2|K|} \mathbf{a}_i \boldsymbol{\nu}_i + \frac{p_j - p_i}{2|K|} \mathbf{a}_i \boldsymbol{\nu}_k \right) \cdot \boldsymbol{\nu}_{\mathbf{m}_c \overline{\mathbf{m}}_k}.
\end{aligned} \tag{11.8}$$

where $\boldsymbol{\nu}_{\overline{\mathbf{m}}_i \mathbf{m}_c}$ and $\boldsymbol{\nu}_{\mathbf{m}_c \overline{\mathbf{m}}_k}$ are the normal vectors to the interface segments $\overline{\mathbf{m}}_i \mathbf{m}_c$ and $\mathbf{m}_c \overline{\mathbf{m}}_k$, respectively, with their respective length.

We write the flux f_i as

$$f_i = -T_{ij}(p_j - p_i) - T_{ik}(p_k - p_i), \tag{11.9}$$

where the *transmissibility coefficients* T_{ij} and T_{ik} are

$$T_{ij} = -\frac{1}{2|K|} \mathbf{a}_i \boldsymbol{\nu}_k \cdot (\boldsymbol{\nu}_{\overline{\mathbf{m}}_i \mathbf{m}_c} + \boldsymbol{\nu}_{\mathbf{m}_c \overline{\mathbf{m}}_k}),$$

$$T_{ik} = -\frac{1}{2|K|} \mathbf{a}_i \boldsymbol{\nu}_i \cdot (\boldsymbol{\nu}_{\overline{\mathbf{m}}_i \mathbf{m}_c} + \boldsymbol{\nu}_{\mathbf{m}_c \overline{\mathbf{m}}_k}).$$

We now consider the assembly of the *global transmissibility matrix*. Each connection between any two adjacent nodes $\mathbf{m}_i$ and $\mathbf{m}_j$ includes the contributions from two triangles K_1 and K_2 that share the common edge with endpoints $\mathbf{m}_i$ and $\mathbf{m}_j$ (Fig. 11.3). The transmissibility between $\mathbf{m}_i$ and

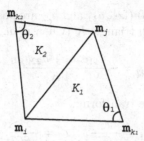

Figure 11.3 Two adjacent triangles.

$\mathbf{m}_j$, where at least one of them is not on the external boundary, consists of the contributions from these two triangles:

$$T_{ij} = -\sum_{l=1}^{2} \frac{1}{2|K_l|} \mathbf{a}_i \nu_k \cdot (\nu_{\overline{\mathbf{m}}_i \mathbf{m}_c} + \nu_{\mathbf{m}_c \overline{\mathbf{m}}_k}) \bigg|_{K_l}. \qquad (11.10)$$

Applying Eqs. (11.2) and (11.9), we obtain the linear system on the control volume V_i in terms of pressure values at the vertices of triangles (the centers of control volumes):

$$-\sum_{j \in \Omega_i} T_{ij} (p_j - p_i) = F_i, \qquad (11.11)$$

where Ω_i is the set of all neighboring nodes of $\mathbf{m}_i$ and $F_i = \int_{V_i} f \, d\mathbf{x}$.

If ∂V_i contains part of the Neumann boundary, then the flux on that part is given; if it contains part of the Dirichlet boundary, the pressure on the corresponding part is given. The third boundary condition can be also incorporated as in Chapter 2. Since linear elements are used, an *error estimate* as in (2.51) holds for the CVFE method considered. Finally, the CVFE method can be extended to transient problems as in Chapter 7.

11.1.2. Positive transmissibilities

The *transmissibility coefficient* T_{ij} defined in Eq. (11.10) must be positive. Positive transmissibilities or *positive flux linkages* always yield a direction of the discrete flux in the physical direction. Negative transmissibilities are not physically meaningful, and generate unsatisfactory solutions.

For the triangle K with vertices $\mathbf{m}_i$, $\mathbf{m}_j$, and $\mathbf{m}_k$ (Fig. 11.2), we define

$$a_i = m_{j,2} - m_{k,2}, \quad b_i = -(m_{j,1} - m_{k,1}),$$

where $\mathbf{m}_i = (m_{i,1}, m_{i,2})$ are the coordinates of the ith node and $\{i, j, k\}$ is cyclically permuted. For simplicity, consider a *homogeneous anisotropic* medium: $\mathbf{a} = \text{diag}(a_{11}, a_{22})$ (i.e., a_{11} and a_{22} are positive constants). In this case, T_{ij} restricted to each triangle K (Chen *et al.*, 2006) is

$$T_{ij} = -\frac{a_{11}a_j a_i + a_{22}b_j b_i}{4|K|}.$$

We introduce a coordinate transform:

$$x_1' = \frac{x_1}{\sqrt{a_{11}}}, \quad x_2' = \frac{x_2}{\sqrt{a_{22}}}.$$

Under this transform, the area of the transformed triangle K' is

$$|K'| = \frac{|K|}{\sqrt{a_{11}a_{22}}}.$$

Consequently, T_{ij} becomes

$$T_{ij} = \sqrt{a_{11}a_{22}}\frac{|\mathbf{m}_{k'}\mathbf{m}_{j'}||\mathbf{m}_{k'}\mathbf{m}_{i'}|\cos\theta_{k'}}{4|K'|} = \sqrt{a_{11}a_{22}}\frac{\cot\theta_{k'}}{2},$$

where $\theta_{k'}$ is the angle of the triangle at node $\mathbf{m}_{k'}$ in the transformed plane. Because each global transmissibility consists of the contributions from two adjacent triangles, the global T_{ij} between nodes $\mathbf{m}_i$ and $\mathbf{m}_j$ (Fig. 11.3) is

$$T_{ij} = \sqrt{a_{11}a_{22}}\left(\frac{\cot\theta_{k_1'} + \cot\theta_{k_2'}}{2}\right), \tag{11.12}$$

where $\theta_{k_1'}$ and $\theta_{k_2'}$ are the opposite angles of the two triangles. Thus, the requirement $T_{ij} > 0$ is equivalent to

$$\theta_{k_1'} + \theta_{k_2'} < \pi. \tag{11.13}$$

For an edge on the external boundary, the requirement for the angle opposite this edge is

$$\theta_{k'} < \frac{\pi}{2}. \tag{11.14}$$

Note that all these angles are measured in the (x_1', x_2') coordinate plane.

11.1.3. The CVFE grid construction

It is interesting to note that condition (11.13) is related to a *Delaunay triangulation*. A Delaunay triangulation satisfies the *empty circle criterion*: the circumcircle of each triangle must not contain any other nodes in its interior. Given a shape-regular triangulation K_h (inequality (2.52)) of a convex

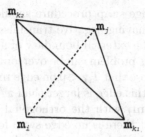

Figure 11.4   An edge swap.

domain, K_h can be converted to a Delaunay triangulation in a sequence of local edge swaps as follows (Joe, 1986; D'Azevedo and Simpson, 1989): Each internal edge in K_h is examined. If it is a part of a convex quadrilateral (Fig. 11.3), then the circumcircles of the two triangles are checked. If one of the circumcircles contains the fourth vertex of the quadrilateral, then the diagonal of this quadrilateral is swapped (Fig. 11.4). The resulting local triangulation then satisfies the empty circle criterion, i.e., the *local optimality condition* (Joe, 1986; D'Azevedo and Simpson, 1989). A sequence of local edge swaps eventually converges so that every internal edge is locally optimal. All internal edges of a triangulation are locally optimal if and only if it is a Delaunay triangulation (Joe, 1986).

On the other hand, the local optimal condition is equivalent to condition (11.13) (D'Azevedo and Simpson, 1989). Hence, the edge-swapping procedure can be given geometrically: given a regular triangulation K_h, if the sum of the two angles opposite edge $\mathbf{m}_i\mathbf{m}_j$ (Fig. 11.3) is larger than π, then this edge is replaced by edge $\mathbf{m}_{k_1}\mathbf{m}_{k_2}$. This edge swap can only be carried out if the quadrilateral is convex. If it is not, then condition (11.13) must necessarily be true. For a convex domain, no addition or movement of nodes is required to convert K_h to a Delaunay triangulation.

For problem (11.1), the edge-swapping procedure can be generalized (Forsyth, 1991): Each edge $\mathbf{m}_i\mathbf{m}_j$ is examined, and the transmissibility T_{ij} is computed using Eq. (11.10). If T_{ij} is negative, then this edge is replaced by $\mathbf{m}_{k_1}\mathbf{m}_{k_2}$. If the solution domain is convex and $\mathbf{a}$ is constant, this procedure is equivalent to establishing a Delaunay triangulation in the (x'_1, x'_2) plane where $\mathbf{a}'$ is the identity tensor. The equivalence of positive transmissibilities with a Delaunay triangulation is true only for internal edges in the transformed plane when $\mathbf{a}$ is constant. In general, a Delaunay triangulation of the physical plane cannot ensure positive transmissibilities, even for internal edges. Most domains that arise in practical applications can be treated as a union of convex regions with a constant permeability tensor **a**, and the local edge swap procedure should tend to minimize the number of internal edges having negative transmissibilities.

In general, edges on the external boundary of a domain can have negative transmissibilities. This problem can be overcome by adding a boundary node as in Fig. 11.5. Suppose that $T_{ij} < 0$ on edge $\mathbf{m}_i\mathbf{m}_j$; i.e., in the (x'_1, x'_2) plane, the angle opposite this edge is larger than $\pi/2$. A new node is added at the intersection of $\mathbf{m}_i\mathbf{m}_j$ with the orthogonal line segment to $\mathbf{m}_i\mathbf{m}_j$ drawn from $\mathbf{m}_k$. Note that there is no edge swap for a boundary edge.

In three dimensions, the Delaunay empty sphere criterion is not equivalent to positive transmissibilities (Letniowski, 1992).

11.1.4. Flux continuity

Harmonic average

In general, the coefficient **a** is given at the centers of control volumes. However, in the computation of the integral equation (11.2), this coefficient needs to be evaluated at the boundary of each control volume. For the flux to be continuous across each internal interface, an appropriate average must be used for **a** at the interface.

We consider the calculation of a transmissibility at the interface between two cells 1 and 2 in one dimension (Fig. 11.6). Let the coefficient a (e.g., permeability) in cell i be a_i, and let h_i be the length of this cell,

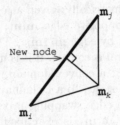

Figure 11.5 An addition of a new boundary node.

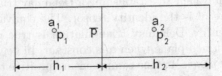

Figure 11.6 Two adjacent cells.

$i = 1, 2$. Equating the flux at each side of the interface gives

$$-a_1\frac{\bar{p} - p_1}{h_1/2} = -a_2\frac{p_2 - \bar{p}}{h_2/2}, \tag{11.15}$$

where $\bar{p}$ represents the value of the approximate solution p_h at the interface. Setting $T_i = a_i/h_i$, $i = 1, 2$, we can solve for $\bar{p}$ in Eq. (11.15):

$$\bar{p} = \frac{T_1 p_1 + T_2 p_2}{T_1 + T_2}.$$

Substituting this expression back into the left-hand side of Eq. (11.15) yields the flux equation

$$f_1 = -\frac{2}{1/T_1 + 1/T_2}(p_2 - p_1). \tag{11.16}$$

This equation shows that the transmissibility at the interface between two cells is approximated by the *harmonic average* of the cell transmissibilities:

$$T_{\text{har}} = \frac{2}{1/T_1 + 1/T_2} = \frac{2T_1 T_2}{T_1 + T_2}. \tag{11.17}$$

In multiple dimensions, if the permeability $\mathbf{a}$ is a tensor, this harmonic average is used for each component of $\mathbf{a}$ (or the transmissibility) and the result is denoted by $\mathbf{a}_{\text{har}}$.

It can be seen that for an inactive node (i.e., the node where $\mathbf{a} = \mathbf{0}$), the harmonic average gives the correct value (i.e., $\mathbf{a} = \mathbf{0}$), in contrast with the arithmetic average.

The upstream weighted CVFE

The basic idea of *upstream weighting* is to choose the value of a property coefficient according to the upstream direction of a flux.

Suppose that Eq. (11.1) is of the form

$$-\nabla \cdot (\lambda \mathbf{a} \nabla p) = f(x_1, x_2) \quad \text{in } \Omega, \tag{11.18}$$

where $\mathbf{a}$ and λ can be a permeability tensor and a mobility coefficient, respectively, e.g. For this problem, a CVFE method analog to (11.11) can be derived.

For the mobility coefficient λ, in practice, upstream weighting must be used to maintain stability for the CVFE methods. As a result, the transmissibility between nodes $\mathbf{m}_i$ and $\mathbf{m}_j$ restricted to each triangle K becomes (Fig. 11.2)

$$T_{ij} = -|K|\,\lambda_{ij}^{\text{up}}\mathbf{a}_{\text{har}}\nabla\lambda_j \cdot \nabla\lambda_i, \tag{11.19}$$

where the *potential-based upstream weighting* scheme is defined by

$$\lambda_{ij}^{\mathrm{up}} = \begin{cases} \lambda(\mathbf{m}_i) & \text{if } p_i > p_j, \\ \lambda(\mathbf{m}_j) & \text{if } p_i < p_j. \end{cases} \qquad (11.20)$$

In fact, it is a pressure-based approach in the current context. The name *potential based* is due to the fact that potentials are usually used in place of p in reservoir simulation. This potential-based upstream weighting scheme is easy to implement. However, it violates the important flux continuity property across the interfaces between control volumes (Chen *et al.*, 2006).

An alternative approach is the flux-based approach where the upstream direction is determined by the sign of a flux. The flux on edge $\mathbf{m}_a \mathbf{m}_c$ at node $\mathbf{m}_i$ is indicated by $f_{i,\mathbf{m}_a \mathbf{m}_c}$ (Fig. 11.2). Then, the upstream weighting between nodes $\mathbf{m}_i$ and $\mathbf{m}_j$ is defined by

$$\lambda_{ij}^{\mathrm{up}} = \begin{cases} \lambda(\mathbf{m}_i) & \text{if } f_{i,\mathbf{m}_a \mathbf{m}_c} > 0, \\ \lambda(\mathbf{m}_j) & \text{if } f_{i,\mathbf{m}_a \mathbf{m}_c} < 0. \end{cases} \qquad (11.21)$$

The flux-based upstream weighted CVFE method has a continuous flux across the internal interfaces of the control volumes. ·

The CVFE methods can be generalized in a variety of ways. The simplest generalization is to finite elements of higher order as in Chapters 4–6, piecewise polynomials of higher degree. They can also be generalized to nonpolynomial functions, such as *spline functions* (Chen *et al.*, 2006).

11.2. Multipoint flux approximations

In this section, control volume discretizations using MPFA are briefly touched up on. They are designed to give a more accurate flux approximation across the interfaces of control volumes. As in the previous section, the discretizations are based on triangular grids. The major difference between CVFE and MPFA is that a linear approximation is used to approximate the pressure p on each triangle in the triangulation K_h in the CVFE method, while this linear approximation is used on each control volume in the MPFA method.

11.2.1. Definition of MPFA

For each control volume V_i, Eq. (11.2) is repeated as follows:

$$-\int_{\partial V_i} \mathbf{a} \nabla p_h \cdot \boldsymbol{\nu}_i \, d\ell = \int_{V_i} f \, d\mathbf{x}, \qquad (11.22)$$

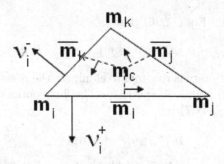

Figure 11.7 A base triangle for MPFA.

where $\boldsymbol{\nu}_i$ is the unit normal outward to the boundary ∂V_i of V_i. Now, the linear approximation p_h to the pressure on the volume V_i is

$$p_h(\mathbf{x}) = p_i + (\mathbf{x} - \mathbf{m}_i) \cdot \nabla p_h, \quad \mathbf{x} \in V_i, \qquad (11.23)$$

where $p_i = p(\mathbf{m}_i)$ and ∇p_h is constant on V_i.

To calculate the constant ∇p_h, let $\boldsymbol{\nu}_i^+$ be the vector orthogonal to the edge $\mathbf{m}_i \overline{\mathbf{m}}_i$, with the length $|\mathbf{m}_i \overline{\mathbf{m}}_i|$, and $\boldsymbol{\nu}_i^-$ be the vector orthogonal to the edge $\mathbf{m}_i \overline{\mathbf{m}}_{m(i)}$, with the length $|\mathbf{m}_i \overline{\mathbf{m}}_{m(i)}|$ (Fig. 11.7), where we recall that $m(i)$ is the backward integer introduced in Sec. 11.1.1. In addition, let $|K_i|$ be the area of the triangle with vertices $\mathbf{m}_i$, $\overline{\mathbf{m}}_i$, and $\overline{\mathbf{m}}_{m(i)}$. Then, restricted to this smaller triangle, in the same way as for Eq. (11.7) it follows that

$$\nabla p_h = -\frac{\overline{p}_{m(i)} - p_i}{2|K_i|}\boldsymbol{\nu}_i^+ - \frac{\overline{p}_i - p_i}{2|K_i|}\boldsymbol{\nu}_i^-, \quad \mathbf{x} \in V_i, \qquad (11.24)$$

where $\overline{p}_{m(i)} = p_h(\overline{\mathbf{m}}_{m(i)})$ and $\overline{p}_i = p_h(\overline{\mathbf{m}}_i)$.

Fix an interface l, $l = i, j, k$ (Fig. 11.7). From Eq. (11.22), the flux through this interface is approximated in the volume V_i by

$$f_{li} = -\boldsymbol{\nu}_l^T \mathbf{a}_i \nabla p_i, \qquad (11.25)$$

where $\mathbf{a}_i$ is the restriction of the tensor $\mathbf{a}$ to the volume V_i, $\nabla p_i = \nabla p_h|_{V_i}$, and the length of the normal vector $\boldsymbol{\nu}_l$ is the length of the corresponding interface. Substituting expression (11.24) into Eq. (11.25) gives

$$f_{li} = \frac{\overline{p}_{m(i)} - p_i}{2|K_i|}\boldsymbol{\nu}_l^T \mathbf{a}_i \boldsymbol{\nu}_i^+ + \frac{\overline{p}_i - p_i}{2|K_i|}\boldsymbol{\nu}_l^T \mathbf{a}_i \boldsymbol{\nu}_i^-, \quad i = l, q(l). \qquad (11.26)$$

We define

$$\omega_{li}^{\pm} = \frac{1}{2|K_i|}\boldsymbol{\nu}_l^T \mathbf{a}_i \boldsymbol{\nu}_i^{\pm}.$$

Then, it follows from Eq. (11.26) that

$$f_{li} = (\overline{p}_{m(i)} - p_i)\omega_{li}^+ + (\overline{p}_i - p_i)\omega_{li}^-, \quad i = l, \ q(l). \tag{11.27}$$

Additional continuity is required to eliminate the pressure values at the middle (or dividing) points, $\overline{p}_i$ and $\overline{p}_{m(i)}$. The flux continuity at each of the three interfaces (Fig. 11.7) is

$$f_{ll} = f_{lq(l)}, \quad l = i, j, k,$$

which, together with Eq. (11.27), yields the linear system

$$\omega_{ii}^+(\overline{p}_k - p_i) + \omega_{ii}^-(\overline{p}_i - p_i) = \omega_{ij}^+(\overline{p}_i - p_j) + \omega_{ij}^-(\overline{p}_j - p_j),$$

$$\omega_{jj}^+(\overline{p}_i - p_j) + \omega_{jj}^-(\overline{p}_j - p_j) = \omega_{jk}^+(\overline{p}_j - p_k) + \omega_{jk}^-(\overline{p}_k - p_k), \tag{11.28}$$

$$\omega_{kk}^+(\overline{p}_j - p_k) + \omega_{kk}^-(\overline{p}_k - p_k) = \omega_{ki}^+(\overline{p}_k - p_i) + \omega_{ki}^-(\overline{p}_i - p_i).$$

We define

$$\overline{\mathbf{p}} = (\overline{p}_i, \overline{p}_j, \overline{p}_k)^T, \quad \mathbf{p} = (p_i, p_j, p_k)^T.$$

Then, system (11.28) is expressed in matrix form

$$\mathbf{A}\overline{\mathbf{p}} - \mathbf{B}\mathbf{p} = \mathbf{C}\overline{\mathbf{p}} - \mathbf{D}\mathbf{p}, \tag{11.29}$$

where the matrices $\mathbf{A}$, $\mathbf{B}$, $\mathbf{C}$, and $\mathbf{D}$ are

$$\mathbf{A} = \begin{pmatrix} \omega_{ii}^- & 0 & \omega_{ii}^+ \\ \omega_{jj}^- & \omega_{jj}^+ & 0 \\ 0 & \omega_{kk}^- & \omega_{kk}^+ \end{pmatrix}, \quad \mathbf{B} = \begin{pmatrix} \omega_{ii}^- + \omega_{ii}^+ & 0 & 0 \\ 0 & \omega_{jj}^- + \omega_{jj}^+ & 0 \\ 0 & 0 & \omega_{kk}^- + \omega_{kk}^+ \end{pmatrix},$$

$$\mathbf{C} = \begin{pmatrix} \omega_{ij}^+ & \omega_{ij}^- & 0 \\ 0 & \omega_{jk}^+ & \omega_{jk}^- \\ \omega_{ki}^- & 0 & \omega_{ki}^+ \end{pmatrix}, \quad \mathbf{D} = \begin{pmatrix} 0 & \omega_{ij}^+ + \omega_{ij}^- & 0 \\ 0 & 0 & \omega_{jk}^+ + \omega_{jk}^- \\ \omega_{ki}^+ + \omega_{ki}^- & 0 & 0 \end{pmatrix}.$$

Consequently, it follows from Eq. (11.29) that

$$\overline{\mathbf{p}} = (\mathbf{A} - \mathbf{C})^{-1}(\mathbf{B} - \mathbf{D})\mathbf{p}. \tag{11.30}$$

Finally, using Eqs. (11.27) and (11.30), we obtain the fluxes on the three interfaces within the triangle with vertices $\mathbf{m}_i$, $\mathbf{m}_j$, and $\mathbf{m}_k$ (Fig. 11.7):

$$\mathbf{f} = \begin{pmatrix} f_{ii} \\ f_{jj} \\ f_{kk} \end{pmatrix} = \mathbf{A}\overline{\mathbf{p}} - \mathbf{B}\mathbf{p} = \left\{ \mathbf{A}(\mathbf{A} - \mathbf{C})^{-1}(\mathbf{B} - \mathbf{D}) - \mathbf{B} \right\} \mathbf{p}. \tag{11.31}$$

The transmissibility matrix $\mathbf{T}$ is defined by

$$\mathbf{T} = \mathbf{A}(\mathbf{A} - \mathbf{C})^{-1}(\mathbf{B} - \mathbf{D}) - \mathbf{B}$$

and the rows in $\mathbf{T}$ contain the coefficients associated with the flux interfaces i, j, and k, respectively. To calculate all transmissibilities for a grid, one can loop over every triangle, since all transmissibilities associated with a triangle can be calculated simultaneously. Finally, assembling the fluxes of the subinterfaces gives the transmissibility coefficients for all interfaces.

11.2.2. A-orthogonal grids

The flux expression developed for MPFA generally involves more than two cell values. A two-point flux can be obtained for a class of grids termed a-orthogonal grids. Let the direction and magnitude of the principal axes of the permeability tensor $\mathbf{a}$ be a_{11} and a_{22}, and a triangle K have the vertices $\mathbf{m}_i$, $\mathbf{m}_j$, and $\mathbf{m}_k$ and the edge midpoints $\overline{\mathbf{m}}_i$, $\overline{\mathbf{m}}_j$, and $\overline{\mathbf{m}}_k$ (Fig. 11.8). Moreover, let $\mathbf{m}_c$ be a point inside the triangle K.

Suppose that for each l the line $\mathbf{m}_c\overline{\mathbf{m}}_l$ is $\mathbf{a}^{-1}$-orthogonal to the triangle edge $\mathbf{m}_l\mathbf{m}_{q(l)}$:

$$(\mathbf{m}_{q(l)} - \mathbf{m}_l)^T \mathbf{a}^{-1}(\overline{\mathbf{m}}_l - \mathbf{m}_c) = 0, \quad l = i, j, k. \tag{11.32}$$

Since $\overline{\mathbf{m}}_l$ is the midpoint of the edge $\mathbf{m}_l\mathbf{m}_{q(l)}$, it follows from (11.32) that

$$([\mathbf{m}_{q(l)} - \mathbf{m}_c] - [\mathbf{m}_l - \mathbf{m}_c])^T \mathbf{a}^{-1}([\mathbf{m}_{q(l)} - \mathbf{m}_c] + [\mathbf{m}_l - \mathbf{m}_c])$$
$$= (\mathbf{m}_{q(l)} - \mathbf{m}_c)^T \mathbf{a}^{-1}(\mathbf{m}_{q(l)} - \mathbf{m}_c) - (\mathbf{m}_l - \mathbf{m}_c)^T \mathbf{a}^{-1}(\mathbf{m}_l - \mathbf{m}_c) = 0,$$

which indicates that the lines that are $\mathbf{a}^{-1}$-orthogonal to the three edges and run through the edge midpoints intersect at the center of the circumscribed a-ellipse of the triangle (Fig. 11.8):

$$(\mathbf{m}_l - \mathbf{m}_c)^T \mathbf{a}^{-1}(\mathbf{m}_l - \mathbf{m}_c) = 0, \quad l = i, j, k. \tag{11.33}$$

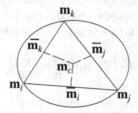

Figure 11.8 A triangle with a circumscribed a-ellipse.

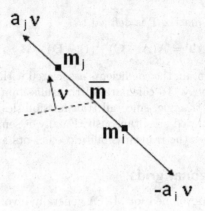

Figure 11.9   A two-point flux in an **a**-orthogonal polygonal grid.

The ratio between the half axes of the ellipse equals the square root of the anisotropy ratio, $\sqrt{a_{22}/a_{11}}$.

Clearly, the center of the **a**-ellipse does not always lie in the triangle. We introduce a transformed coordinate system, where the ellipse reduces to a circle. The center of the ellipse will lie inside the triangle only if the corner angles in the transformed triangle are less than $\pi/2$ (see Sec. 11.1.3).

A grid where the volume edge normals $\boldsymbol{\nu}_l$ are **a**-orthogonal to the normals $\boldsymbol{\nu}_l^+$ and $\boldsymbol{\nu}_{q(l)}^-$ of the corresponding triangle edges, i.e.,

$$\boldsymbol{\nu}_l^T \mathbf{a}_l \boldsymbol{\nu}_l^+ = \boldsymbol{\nu}_l^T \mathbf{a}_{q(l)} \boldsymbol{\nu}_{q(l)}^- = 0, \quad l = i, j, k, \tag{11.34}$$

is called an **a**-orthogonal grid. For **a**-orthogonal grids, a simple expression can be obtained for the transmissibilities. Consider two adjacent volumes V_i and V_j, with a common flux interface (Fig. 11.9). The interface has a normal vector $\boldsymbol{\nu}$, directed from V_i to V_j and defined with a length equal to the length of the flux interface.

In an **a**-orthogonal grid, the vector $\mathbf{a}_l\boldsymbol{\nu}$ aligns with the line $\overline{\mathbf{m}}\mathbf{m}_l$ ($l = i, j$), where $\overline{\mathbf{m}}$ is the midpoint between $\mathbf{m}_i$ and $\mathbf{m}_j$. Thus, the flux through the flux interface is proportional to the directional derivative along the line $\overline{\mathbf{m}}\mathbf{m}_l$ and can be obtained for each side of this interface as follows:

$$\boldsymbol{\nu}^T \mathbf{a}_l \nabla p_l = (-1)^{\delta_{il}} |\mathbf{a}_l \boldsymbol{\nu}| \frac{p_i - \overline{p}}{|\mathbf{m}_l - \overline{\mathbf{m}}|}, \quad l = i, j, \tag{11.35}$$

where $\overline{p} = p_h(\overline{\mathbf{m}})$, δ_{il} is the Kronecker symbol, and $| \cdot |$ stands for the magnitude of a vector. The flux continuity condition can be used to eliminate $\overline{p}$ in the usual manner, and the flux can then be expressed by a two-point scheme

$$f_{ij} = \overline{a}(p_i - p_j), \qquad (11.36)$$

where the transmissibility $\overline{a}$ for the interface is the harmonic average:

$$\frac{1}{\overline{a}} = \frac{|\mathbf{m}_i - \overline{\mathbf{m}}|}{|\mathbf{a}_i \boldsymbol{\nu}|} + \frac{|\mathbf{m}_j - \overline{\mathbf{m}}|}{|\mathbf{a}_j \boldsymbol{\nu}|}.$$

In the **a**-orthogonal case, the fluxes over the interfaces of the control volumes have the same form as those derived for the CVFE method in the previous section.

11.3. The nonconforming finite element method

In the development of the finite element method for a second-order differential equation problem in Chapters 1–9 and the previous two sections of this chapter, piecewise polynomials in a finite element space V_h were required to be *continuous* throughout the entire domain Ω. Due to this continuity requirement, the resulting method is called the H^1-*conforming finite element method* (in short, the conforming method). For the discretization of a fourth-order problem, functions in the discrete space V_h and their first derivatives were required to be continuous on Ω. In this case, the finite element method is termed the H^2-*conforming method* or conforming method. In this section, we introduce the *nonconforming finite element method* in which functions in a finite element space V_h for the discretization of a second-order problem are not required continuous on Ω; for a fourth-order problem, their derivatives (and even the functions themselves in some cases) are not required continuous on Ω.

Compared with the conforming finite element spaces introduced so far, the finite element spaces used in the nonconforming method (i.e., *nonconforming spaces*) employ fewer degrees of freedom, particularly for a fourth-order differential equation problem. The nonconforming method was initially introduced in the early 1960s (Adini and Clough, 1961). Since then, it has been widely used in computational mechanics and structural engineering. In this section, as an example, we discuss its application to second-order partial differential equation problems. For its application to fourth-order problems, the reader can refer to Chen (2005).

11.3.1. Second-order partial differential problems

For the purpose of introduction, we consider a stationary problem for the unknown variable p:

$$-\nabla \cdot (\mathbf{a}\nabla p) = f \quad \text{in } \Omega,$$
$$p = g \qquad \text{on } \Gamma, \tag{11.37}$$

where $\Omega \subset \mathbb{R}^2$ (the plane) is a bounded two-dimensional domain with boundary Γ, the diffusion tensor $\mathbf{a}$ is assumed to be bounded, symmetric, and uniformly positive definite in $\mathbf{x}$, and f and g are given real-valued piecewise continuous bounded functions in Ω and Γ, respectively. A typical problem in the form (11.37) is heat conduction where one seeks the temperature distribution p in an inhomogeneous plate Ω with conductivity tensor $\mathbf{a}$.

We recall the admissible function space

$$V = \left\{ \text{Functions } v: v \text{ is continuous on } \Omega, \frac{\partial v}{\partial x_1} \text{ and } \frac{\partial v}{\partial x_2} \text{ are} \right.$$

$$\left. \text{piecewise continuous and bounded on } \Omega, \text{ and } v = 0 \text{ on } \Gamma \right\}.$$

Multiplying the first equation in system (11.37) by a test function $v \in V$ and integrating the resulting equation over Ω, we see that

$$-\int_\Omega \nabla \cdot (\mathbf{a}\nabla p)v \, d\mathbf{x} = \int_\Omega fv \, d\mathbf{x}.$$

Applying Green's formula (2.9) to this equation and using the boundary condition in the definition of the space V, we have

$$\int_\Omega (\mathbf{a}\nabla p) \cdot \nabla v \, d\mathbf{x} = \int_\Omega fv \, d\mathbf{x} \quad \forall v \in V, \tag{11.38}$$

where $p|_\Gamma = g$. As observed in Chapter 2, the variational form (11.38) is equivalent to a minimization problem (see Sec. 2.2.2). In the next section, as an example, we construct the finite element method for solving problem (11.37) using the simplest *nonconforming elements* on triangles.

11.3.2. Nonconforming finite elements on triangles

Let Ω be a polygonal domain in the plane, and let K_h be a triangulation of Ω into nonoverlapping (open) triangles K:

$$\bar{\Omega} = \bigcup_{K \in K_h} \bar{K},$$

such that no vertex of one triangle lies in the interior of an edge of another triangle, where $\bar{\Omega}$ and $\bar{K}$ represent the closure of Ω and K (i.e., $\bar{\Omega} = \Omega \cup \Gamma$ and $\bar{K} = K \cup \partial K$, where ∂K is the boundary of K), respectively. The mesh parameters h_K and h are defined as in Chapter 2:

$$h_K = \text{diam}(K) \quad \text{and} \quad h = \max_{K \in K_h} h_K,$$

where $\text{diam}(K)$ is the length of the longest edge of $\bar{K}$.

Now, we introduce the finite element spaces on triangles

$$\tilde{V}_h = \{\text{Functions } v: v|_K \text{ is linear}, \ K \in K_h; v \text{ is continuous}$$
$$\text{at the midpoints of interior edges}\},$$

and

$$V_h = \{\text{Functions } v: v|_K \text{ is linear}, \ K \in K_h; v \text{ is continuous}$$
$$\text{at the midpoints of interior edges and}$$
$$\text{is zero at the midpoints of edges on } \Gamma\}.$$

It can be shown that the degrees of freedom (i.e., the function values at the midpoints of edges) for V_h (Fig. 11.10) are legitimate. Namely, a linear function v on each $K \in K_h$ is uniquely determined by them. In fact, by connecting the midpoints of the edges on each triangle $K \in K_h$, we obtain a smaller triangle (Fig. 11.10) on which $v \in P_1$ (the linear function space) vanishes at the vertices (see Exercise 11.2).

In Chapters 1–9, functions in the finite element spaces are required to be continuous across interelement boundaries. In contrast, the functions here are continuous only at the midpoints of interior edges, so $V_h \not\subset V$ (not a subspace). In this case, the finite element space V_h is referred to as a *nonconforming finite element space*. Because of the nonconformity, we introduce the mesh-dependent bilinear form $a_h(\cdot, \cdot): V_h \times V_h \to \mathbb{R}$ (the set

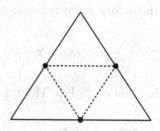

Figure 11.10   The degrees of freedom for the Crouzeix–Raviart element.

of real numbers)

$$a_h(v, w) = \sum_{K \in K_h} \int_K \mathbf{a} \nabla v \cdot \nabla w \, d\mathbf{x}, \quad v, \ w \in V_h.$$

Then, the *nonconforming finite element method* for problem (11.37) is formulated as follows:

$$\text{Find } p_h \in \tilde{V}_h \text{ such that } a_h(p_h, v) = \int_\Omega fv \, d\mathbf{x} \quad \forall v \in V_h, \qquad (11.39)$$

where p_h equals g at the midpoints of the edges on the boundary Γ.

Existence and uniqueness of a solution to problem (11.39) can be easily checked. In fact, set $f = g = 0$. Then, Eq. (11.39) becomes

$$a_h(p_h, v) = 0 \quad \forall v \in V_h.$$

With $v = p_h$ in this equation, we see that p_h is a constant on each $K \in K_h$. Due to the continuity of p_h at interior midpoints, p_h is a constant on Ω. Consequently, the zero boundary condition implies $p_h = 0$. Uniqueness also yields existence since system (11.39) is a finite-dimensional system.

Denote the midpoints (nodes) of edges in the triangulation K_h by $\mathbf{x}_1, \mathbf{x}_2, \ldots, \mathbf{x}_{\tilde{M}}$, where $\tilde{M}$ is the number of the midpoints. The basis (shape) functions φ_i in $\tilde{V}_h$, $i = 1, 2, \ldots, \tilde{M}$, are defined as follows:

$$\varphi_i(\mathbf{x}_j) = \begin{cases} 1 & \text{if } i = j, \\ 0 & \text{if } i \neq j. \end{cases}$$

The *support* of φ_i, i.e., the set where $\varphi_i(\mathbf{x}) \neq 0$, consists of the triangles with the common node $\mathbf{x}_i$ (Fig. 11.11). In the present case, the support of each φ_i consists of at most two triangles.

Let M ($M < \tilde{M}$) be the number of interior nodes in K_h. For notational convenience, the interior nodes are chosen to be the first M nodes. Then, any function $v \in V_h$ has the unique representation

$$v(\mathbf{x}) = \sum_{i=1}^M v_i \varphi_i(\mathbf{x}), \quad \mathbf{x} \in \Omega,$$

where $v_i = v(\mathbf{x}_i)$. Also, the solution to Eq. (11.39) is given by

$$p_h = \sum_{i=1}^M p_i \varphi_i + \sum_{k=M+1}^{\tilde{M}} g_k \varphi_k, \qquad (11.40)$$

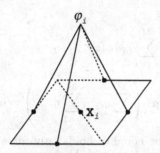

Figure 11.11 A nonconforming basis function in two dimensions.

where $p_i = p_h(\mathbf{x}_i)$ and $g_k = g(\mathbf{x}_k)$.

For each j, we take $v = \varphi_j$ in Eq. (11.39) to see that

$$a_h(p_h, \varphi_j) = \int_\Omega f\varphi_j \, d\mathbf{x}, \quad j = 1, 2, \ldots, M.$$

Substituting expression (11.40) into this equation, we have

$$\sum_{i=1}^M a_h(\varphi_i, \varphi_j)p_i = \int_\Omega f\varphi_j \, d\mathbf{x} - \sum_{k=M+1}^{\tilde{M}} a_h(\varphi_k, \varphi_j)g_k, \quad j = 1, 2, \ldots, M.$$

This is a linear system of M algebraic equations in the M unknowns $p_1, p_2, \ldots, p_M$. It can be written in matrix form

$$\mathbf{Ap} = \mathbf{f}, \tag{11.41}$$

where the matrix $\mathbf{A}$ and vectors $\mathbf{p}$ and $\mathbf{f}$ are given by

$$\mathbf{A} = (a_{ij}), \quad \mathbf{p} = (p_j), \quad \mathbf{f} = (f_j),$$

with, $i, j = 1, 2, \ldots, M$,

$$a_{ij} = a_h(\varphi_i, \varphi_j), \quad f_j = \int_\Omega f\varphi_j \, d\mathbf{x} - \sum_{k=M+1}^{\tilde{M}} a_h(\varphi_k, \varphi_j)g_k.$$

Symmetry of $\mathbf{A}$ can be seen from the definition of a_{ij}: $a_{ij} = a_{ji}$. Positive definiteness can be checked as in Chapter 1: with

$$\eta = \sum_{i=1}^M \eta_i\varphi_i \in V_h, \quad \boldsymbol{\eta}^T = (\eta_1, \eta_2, \ldots, \eta_M),$$

we see that

$$
\sum_{i,j=1}^{M} \eta_i a_{ij} \eta_j = \sum_{i,j=1}^{M} \eta_i a_h \left(\varphi_i, \varphi_j \right) \eta_j
$$

$$
= a_h \left(\sum_{i=1}^{M} \eta_i \varphi_i, \sum_{j=1}^{M} \eta_j \varphi_j \right) = a_h \left(\eta, \eta \right) \geq 0,
$$

so, as for Eq. (11.39), the equality holds only for $\eta \equiv 0$ since a constant function η must be zero because of the boundary condition in the space V_h. Particularly, positive definiteness implies that $\mathbf{A}$ is nonsingular. As a result, system (11.41) has a unique solution. This is another way to show that Eq. (11.39) has a unique solution. System (11.41) can be solved using linear system solution techniques, e.g.

We have considered a Dirichlet boundary value problem in this section. A Neumann or more general boundary value problem can be treated in a similar manner as in Chapter 2. An error analysis for the nonconforming finite element method (11.39) is delicate. A general theory for this method exists (Ciarlet, 1978; Chen, 2005). Here, we just state the error estimate: If the solution p to Eq. (11.37) is sufficiently smooth (e.g., its second partial derivatives are bounded), then

$$
\left(\int_\Omega |p - p_h|^2 \, d\mathbf{x} \right)^{1/2} + h \left(\sum_{K \in K_h} \int_K |\nabla(p - p_h)|^2 \, d\mathbf{x} \right)^{1/2} \leq Ch^2,
$$

$$
(11.42)
$$

where p_h is the solution of Eq. (11.39), C is a constant independent of h and dependent on the second partial derivatives of p, and the triangulation K_h is assumed to be shape regular (see the definition of regularity on a triangulation in inequality (2.52)).

The nonconforming finite element under consideration is the linear Crouzeix and Raviart (1973) element, which is the simplest nonconforming element on triangles (also called the P_1-nonconforming element). For a quadratic nonconforming element on triangles, refer to Fortin and Soulie (1983). For general high-order nonconforming elements on triangles, see Arbogast and Chen (1995). For rectangular and three-dimensional nonconforming elements, refer to Chen (2005).

11.4. The mixed finite element method

In this section, we study the *mixed finite element method*, which generalizes the finite element method discussed in Chapters 1–9. This method was initially introduced by engineers in the 1960s (Fraeijs de Veubeke, 1965; Hellan, 1967; Hermann, 1967) for solving problems in solid continua. Since then, it has been applied to many areas, particularly to solid and fluid mechanics. Here, we discuss its applications to second-order partial differential equation problems. The main reason for using the mixed method is that in some applications a vector variable (e.g., a fluid velocity, see Chapter 10) is the primary variable of interest. This method is developed to approximate both this variable and a scalar variable (e.g., pressure) simultaneously and to give a high-order approximation of both variables. Instead of a single finite element space used in the standard finite element method, the *mixed* finite element method employs two different spaces. These two spaces must satisfy an inf-sup condition for the mixed method to be stable. Raviart and Thomas (1977) introduced the first family of mixed finite element spaces for second-order elliptic problems in two dimensions. Somewhat later, Nédélec (1980) extended these spaces to three-dimensional problems. Motivated by these two papers, there are now many mixed finite element spaces available in the literature; see Brezzi *et al.* (1985, 1987a,b) and Chen and Douglas (1989).

11.4.1. A one-dimensional model problem

As in Chapter 1, for the purpose of demonstration, we consider a stationary problem for the unknown p in one dimension

$$-\frac{d^2p}{dx^2} = f(x), \quad 0 < x < 1,$$

$$p(0) = p(1) = 0,$$

(11.43)

where the function f is given. The first admissible function space is

$$W = \left\{ \text{Functions } w \colon w \text{ is defined and square integrable} \right.$$

$$\left. \text{on } (0,1); \text{ i.e., } \int_0^1 w^2(x) \, dx < \infty \right\}.$$

Observe that the functions in W are not required to be continuous on the interval $(0, 1)$. We also define the second admissible function space

$$V = \left\{ \text{Functions } v \colon v \text{ is continuous on } [0, 1], \text{ and its first} \right.$$

$$\left. \frac{dv}{dx} \text{ derivative is piecewise continuous and bounded on } (0, 1) \right\}.$$

After introducing the variable

$$u = -\frac{dp}{dx},$$ (11.44)

the first equation in system (11.43) can be recast in the form

$$\frac{du}{dx} = f.$$ (11.45)

Multiplying Eq. (11.44) by any function $v \in V$ and integrating over $(0, 1)$, we see that

$$\int_0^1 uv \, dx = -\int_0^1 \frac{dp}{dx} v \, dx.$$

Application of integration by parts to the right-hand side of this equation leads to

$$\int_0^1 uv \, dx = \int_0^1 p\frac{dv}{dx} \, dx,$$

where we use the boundary conditions $p(0) = p(1) = 0$ from system (11.43). Also, we multiply Eq. (11.45) by any function $w \in W$ and integrate over $(0, 1)$ to give

$$\int_0^1 \frac{du}{dx} w \, dx = \int_0^1 fw \, dx.$$

Therefore, we see that the pair of functions u and p satisfies the system

$$\begin{aligned} \int_0^1 uv \, dx - \int_0^1 p\frac{dv}{dx} \, dx &= 0, \quad v \in V, \\ \int_0^1 \frac{du}{dx} w \, dx &= \int_0^1 fw \, dx, \qquad w \in W. \end{aligned}$$ (11.46)

This system is referred to as a *mixed variational* (or *weak*) *formulation* of problem (11.43). If the pair of functions (u, p) is a solution to Eqs. (11.44) and (11.45), then this pair also satisfies system (11.46). The converse also holds if the solution p is sufficiently smooth (e.g., if the second derivative of p is bounded in (0,1); see Exercise 11.6).

We introduce the functional $F: V \times W \to \mathbb{R}$ by

$$F(v, w) = \frac{1}{2} \int_0^1 v^2 \, dx - \int_0^1 \frac{dv}{dx} w \, dx + \int_0^1 fw \, dx, \quad v \in V, \ w \in W,$$

where $\mathbb{R}$ is the set of real numbers. It can be shown (Chen, 2005) that problem (11.46) is equivalent to the *saddle point problem*: Find $u \in V$ and $p \in W$ such that

$$F(u, w) \le F(u, p) \le F(v, p) \quad \forall v \in V, \ w \in W.$$ (11.47)

For this reason, problem (11.46) is also referred to as a *saddle point problem*.

To construct the mixed finite element method for solving problem (11.43), for a positive integer M let $0 = x_1 < x_2 < \cdots < x_M = 1$ be a partition of $(0,1)$ into a set of subintervals $I_{i-1} = (x_{i-1}, x_i)$, with length $h_i = x_i - x_{i-1}$, $i = 2, 3, \ldots, M$. Set $h = \max\{h_i, 2 \le i \le M\}$. Define the *mixed finite element spaces*

$$V_h = \{\text{Functions } v \colon v \text{ is a continuous function on } [0,1]$$

$$\text{and linear on each subinterval } I_i\},$$

$$W_h = \{\text{Functions } w \colon w \text{ is constant on each subinterval } I_i\}.$$

Note that $V_h \subset V$ and $W_h \subset W$ (subspaces). Now, the *mixed finite element method* for problem (11.43) is defined:

Find $u_h \in V_h$ and $p_h \in W_h$ such that

$$\int_0^1 u_h v \, dx - \int_0^1 p_h \frac{dv}{dx} \, dx = 0, \quad v \in V_h, \tag{11.48}$$

$$\int_0^1 \frac{du_h}{dx} w \, dx = \int_0^1 fw \, dx, \qquad w \in W_h.$$

To show that system (11.48) has a unique solution, let $f = 0$; take $v = u_h$ and $w = p_h$ in system (11.48), and add the resulting equations to give

$$\int_0^1 u_h^2 \, dx = 0,$$

so that $u_h = 0$. Consequently, it follows from the first equation of system (11.48) that

$$\int_0^1 \frac{dv}{dx} p_h \, dx = 0, \quad v \in V_h.$$

Choose $v \in V_h$ such that $dv/dx = p_h$ (thanks to the definition of the spaces V_h and W_h) in this equation to see that $p_h = 0$. Hence, the solution of Eq. (11.48) is unique. Uniqueness also yields existence since system (11.48) is equivalent to a finite-dimensional linear system.

In the same fashion as for the equivalence between system (11.46) and inequality (11.47), problem (11.48) is equivalent to the saddle point problem: find $u_h \in V_h$ and $p_h \in W_h$ such that

$$F(u_h, w) \le F(u_h, p_h) \le F(v, p_h) \quad \forall v \in V_h, \ w \in W_h. \tag{11.49}$$

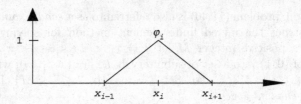

Figure 11.12 A basis function φ_i in one dimension.

We introduce the *basis functions* (shape functions) $\varphi_i \in V_h$, $i = 1$, $2, \ldots, M$ (Fig. 11.12)

$$\varphi_i(x_j) = \begin{cases} 1 & \text{if } i = j, \\ 0 & \text{if } i \neq j, \end{cases}$$

and the basis functions $\psi_i \in W_h$, $i = 1, 2, \ldots, M - 1$,

$$\psi_i(x) = \begin{cases} 1 & \text{if } x \in I_i, \\ 0 & \text{otherwise.} \end{cases}$$

The functions ψ_i are *characteristic* functions. Now, functions $v \in V_h$ and $w \in W_h$ have the unique representations

$$v(x) = \sum_{i=1}^{M} v_i \varphi_i(x), \quad w(x) = \sum_{i=1}^{M-1} w_i \psi_i(x), \quad 0 \le x \le 1,$$

where $v_i = v(x_i)$ and $w_i = w|_{I_i}$. Take v and w in system (11.48) to be these basis functions to see that

$$\int_0^1 u_h \varphi_j \, dx - \int_0^1 \frac{d\varphi_j}{dx} p_h \, dx = 0, \quad j = 1, 2, \ldots, M,$$

$$\int_0^1 \frac{du_h}{dx} \psi_j \, dx = \int_0^1 f \psi_j \, dx, \qquad j = 1, 2, \ldots, M - 1. \tag{11.50}$$

Set

$$u_h(x) = \sum_{i=1}^{M} u_i \varphi_i(x), \quad u_i = u_h(x_i),$$

and

$$p_h(x) = \sum_{k=1}^{M-1} p_k \psi_k(x), \quad p_k = p_h|_{I_k}.$$

Substitute these expressions into system (11.50) to give

$$\sum_{i=1}^{M} \int_0^1 \varphi_i \varphi_j \, dx \, u_i - \sum_{k=1}^{M-1} \int_0^1 \frac{d\varphi_j}{dx} \psi_k \, dx \, p_k = 0, \quad j = 1, \dots, M,$$

$$\sum_{i=1}^{M} \int_0^1 \frac{d\varphi_i}{dx} \psi_j \, dx \, u_i = \int_0^1 f \psi_j \, dx, \qquad\qquad j = 1, \dots, M-1.$$

$$(11.51)$$

We introduce the matrices and vectors

$$\mathbf{A} = (a_{ij})_{i,j=1,2,\dots,M}, \quad \mathbf{B} = (b_{jk})_{j=1,2,\dots,M,k=1,2,\dots,M-1},$$
$$\mathbf{U} = (u_i)_{i=1,2,\dots,M}, \quad \mathbf{p} = (p_k)_{k=1,2,\dots,M-1}, \quad \mathbf{f} = (f_j)_{j=1,2,\dots,M-1},$$

where

$$a_{ij} = \int_0^1 \varphi_i \varphi_j \, dx, \quad b_{jk} = -\int_0^1 \frac{d\varphi_j}{dx} \psi_k \, dx, \quad f_j = \int_0^1 f \psi_j \, dx.$$

With the notation, system (11.51) can be written in matrix form

$$\begin{pmatrix} \mathbf{A} & \mathbf{B} \\ \mathbf{B}^T & 0 \end{pmatrix} \begin{pmatrix} \mathbf{U} \\ \mathbf{p} \end{pmatrix} = \begin{pmatrix} 0 \\ -\mathbf{f} \end{pmatrix}, \tag{11.52}$$

where $\mathbf{B}^T$ is the transpose of matrix $\mathbf{B}$. Note that system (11.52) is symmetric, but *indefinite*. It can be shown that the matrix $\mathbf{M}$ defined by

$$\mathbf{M} = \begin{pmatrix} \mathbf{A} & \mathbf{B} \\ \mathbf{B}^T & 0 \end{pmatrix} \tag{11.53}$$

has both positive and negative eigenvalues (see Exercise 11.11).

The matrix $\mathbf{A}$ is symmetric and positive definite (see Sec. 1.2.2). It is also sparse. In the one-dimensional case, it is tridiagonal. It follows from the definition of the basis functions φ_i that

$$a_{ij} = \int_0^1 \varphi_i \varphi_j \, dx = 0 \quad \text{if } |i - j| \geq 2,$$

so that

$$a_{11} = \frac{h_2}{3}, \quad a_{MM} = \frac{h_M}{3},$$

and, for $i = 2, 3, \dots, M - 1$,

$$a_{i-1,i} = \frac{h_i}{6}, \quad a_{ii} = \frac{h_i}{3} + \frac{h_{i+1}}{3}, \quad a_{i,i+1} = \frac{h_{i+1}}{6}.$$

It can be also seen that

$$b_{jj} = 1, \quad b_{j+1,j} = -1, \quad j = 1, 2, \ldots, M - 1;$$

all other entries of $\mathbf{B}$ are zero. The $M \times (M - 1)$ matrix $\mathbf{B}$ is *bidiagonal*

$$\mathbf{B} = \begin{pmatrix} 1 & 0 & 0 & \cdots & 0 & 0 \\ -1 & 1 & 0 & \cdots & 0 & 0 \\ 0 & -1 & 1 & \cdots & 0 & 0 \\ \vdots & \vdots & \vdots & \ddots & \vdots & \vdots \\ 0 & 0 & 0 & \cdots & 1 & 0 \\ 0 & 0 & 0 & \cdots & -1 & 1 \\ 0 & 0 & 0 & \cdots & 0 & -1 \end{pmatrix}.$$

In the case where the partition is uniform, i.e., $h = h_i$,

$$\mathbf{A} = \frac{h}{6} \begin{pmatrix} 2 & 1 & 0 & \cdots & 0 & 0 \\ 1 & 4 & 1 & \cdots & 0 & 0 \\ 0 & 1 & 4 & \cdots & 0 & 0 \\ \vdots & \vdots & \vdots & \ddots & \vdots & \vdots \\ 0 & 0 & 0 & \cdots & 4 & 1 \\ 0 & 0 & 0 & \cdots & 1 & 2 \end{pmatrix}.$$

Even for the one-dimensional problem considered, an error analysis for the mixed finite element method (11.48) is delicate. An error estimate of the following type can be obtained for system (11.48):

$$\left(\int_0^1 |p - p_h|^2 \, dx \right)^{1/2} + \left(\int_0^1 |u - u_h|^2 \, dx \right)^{1/2} \leq Ch, \qquad (11.54)$$

where (u, p) and (u_h, p_h) are the respective solutions of systems (11.46) and (11.48) and the constant C depends on the second derivative of p. When u is sufficiently smooth (e.g., its second derivative is bounded in $(0, 1)$), we can show the error estimate (Brezzi and Fortin, 1991; Chen, 2005)

$$\left(\int_0^1 |u - u_h|^2 \, dx \right)^{1/2} \leq Ch^2. \qquad (11.55)$$

Error bounds (11.54) and (11.55) are optimal for the solutions p and u.

11.4.2. A two-dimensional model problem

We extend the mixed finite element method in the previous section to a stationary problem in two dimensions:

$$-\Delta p = f \quad \text{in } \Omega,$$
$$p = 0 \quad \text{on } \Gamma, \tag{11.56}$$

where Ω is a bounded domain in the plane with boundary Γ and f is a given function. Again, we define the first admissible function space

$$W = \Big\{ \text{Functions } w \colon w \text{ is defined and square integrable}$$

$$\text{on } \Omega; \text{ i.e., } \int_\Omega w^2(\mathbf{x}) \, d\mathbf{x} < \infty \Big\}.$$

The admissible space $\mathbf{V}$ is defined slightly differently from the previous section:

$$\mathbf{V} = \Big\{ \text{Vector functions } \mathbf{v} \colon \mathbf{v} \text{ and } \nabla \cdot \mathbf{v} \text{ are defined and square}$$

$$\text{integrable on } \Omega; \text{ i.e., } \int_\Omega \left(|\mathbf{v}|^2 + |\nabla \cdot \mathbf{v}|^2 \right)(\mathbf{x}) \, d\mathbf{x} < \infty \Big\},$$

where we recall that, for $\mathbf{v} = (v_1, v_2)$,

$$\nabla \cdot \mathbf{v} = \frac{\partial v_1}{\partial x_1} + \frac{\partial v_2}{\partial x_2}.$$

It can be checked (see Exercise 11.12) that for any decomposition of Ω into subdomains such that the interiors of these subdomains are pairwise disjoint, the space $\mathbf{V}$ consists of those functions whose normal components are continuous across the interior edges in this decomposition.

Introduce the (e.g., velocity) vector

$$\mathbf{u} = -\nabla p. \tag{11.57}$$

The first equation in system (11.56) becomes

$$\nabla \cdot \mathbf{u} = f. \tag{11.58}$$

Multiply Eq. (11.57) by $\mathbf{v} \in \mathbf{V}$ and integrate over Ω to see that

$$\int_\Omega \mathbf{u} \cdot \mathbf{v} \, d\mathbf{x} = - \int_\Omega \mathbf{v} \cdot \nabla p \, d\mathbf{x}.$$

Applying Green's formula (2.9) to the right-hand side of this equation, we have

$$\int_\Omega \mathbf{u} \cdot \mathbf{v} \, d\mathbf{x} = \int_\Omega \nabla \cdot \mathbf{v} p \, d\mathbf{x},$$

where we use the boundary condition in system (11.56). Also, multiplying Eq. (11.58) by any $w \in W$, we get

$$\int_\Omega \nabla \cdot \mathbf{u} w \, d\mathbf{x} = \int_\Omega f w \, d\mathbf{x}.$$

Thus, we have the system for $\mathbf{u}$ and p:

$$\int_\Omega \mathbf{u} \cdot \mathbf{v} \, d\mathbf{x} - \int_\Omega \nabla \cdot \mathbf{v} p \, d\mathbf{x} = 0, \quad \mathbf{v} \in \mathbf{V},$$

$$\int_\Omega \nabla \cdot \mathbf{u} w \, d\mathbf{x} = \int_\Omega f w \, d\mathbf{x}, \qquad w \in W. \tag{11.59}$$

This is the mixed variational formulation of problem (11.56). If $\mathbf{u}$ and p satisfy Eqs. (11.57) and (11.58), they also satisfy system (11.59). The converse also holds if p is sufficiently smooth (e.g., if the second partial derivatives of p are bounded in Ω; see Exercise 11.13). In a similar fashion as for system (11.46) and inequality (11.47), system (11.59) can be written as a saddle point problem.

For a polygonal domain Ω, let K_h be a partition of Ω into nonoverlapping (open) triangles such that no vertex of one triangle lies in the interior of an edge of another triangle. Define the mixed finite element spaces

$$W_h = \{\text{Functions } w \colon w \text{ is constant on each triangle in } K_h\},$$

and

$$\mathbf{V}_h = \{\text{Vector functions } \mathbf{v} \colon \mathbf{v}\big|_K = (b_K x_1 + a_K, b_K x_2 + c_K), \ K \in K_h, a_K,$$
$$b_K, c_K \in \mathbb{R}, \text{ and the normal components of } \mathbf{v} \text{ are continuous}$$
$$\text{across the interior edges in } K_h\},$$

where we recall that $\mathbb{R}$ is the set of real numbers. Note that $\mathbf{V}_h \subset \mathbf{V}$ and $W_h \subset W$. The mixed finite element method for problem (11.56) is defined: Find $\mathbf{u}_h \in \mathbf{V}$ and $p_h \in W_h$ such that

$$\int_\Omega \mathbf{u}_h \cdot \mathbf{v} \, d\mathbf{x} - \int_\Omega \nabla \cdot \mathbf{v} p_h \, d\mathbf{x} = 0, \quad \mathbf{v} \in \mathbf{V}_h,$$

$$\int_\Omega \nabla \cdot \mathbf{u}_h w \, d\mathbf{x} = \int_\Omega f w \, d\mathbf{x}, \qquad w \in W_h. \tag{11.60}$$

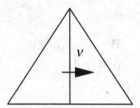

Figure 11.13 An illustration of the unit normal $\boldsymbol{\nu}$.

As for system (11.48) it can be proven that system (11.60) has a unique solution.

Let $\{\mathbf{x}_i\}$ be the set of the midpoints of edges in the triangulation K_h, $i = 1, 2, \ldots, M$. With each point $\mathbf{x}_i$, we associate a unit normal vector $\boldsymbol{\nu}_i$. For $\mathbf{x}_i \in \Gamma$, $\boldsymbol{\nu}_i$ is just the outward unit normal to the boundary Γ; for $\mathbf{x}_i \in e = \bar{K}_1 \cap \bar{K}_2$, $K_1, K_2 \in K_h$, let $\boldsymbol{\nu}_i$ be a (prefixed) unit vector orthogonal to the edge e (Fig. 11.13). We now define the basis functions of the space $\mathbf{V}_h$, $i = 1, 2, \ldots, M$, by

$$(\boldsymbol{\varphi}_i \cdot \boldsymbol{\nu}_i)(\mathbf{x}_j) = \begin{cases} 1 & \text{if } i = j, \\ 0 & \text{if } i \neq j. \end{cases}$$

Any function $\mathbf{v} \in \mathbf{V}_h$ has the unique representation

$$\mathbf{v}(\mathbf{x}) = \sum_{i=1}^{M} v_i \boldsymbol{\varphi}_i(\mathbf{x}), \quad \mathbf{x} \in \Omega,$$

where $v_i = (\mathbf{v} \cdot \boldsymbol{\nu}_i)(\mathbf{x}_i)$. Also, the basis functions $\psi_i \in W_h$, $i = 1, 2, \ldots, N$, can be defined as in the previous section; i.e.,

$$\psi_i(\mathbf{x}) = \begin{cases} 1 & \text{if } \mathbf{x} \in K_i, \\ 0 & \text{otherwise}, \end{cases}$$

where $\bar{\Omega} = \bigcup_{i=1}^{N} \bar{K}_i$ and N is the number of triangles in the partition K_h. Any function $w \in W_h$ has the representation

$$w(\mathbf{x}) = \sum_{i=1}^{N} w_i \psi_i(\mathbf{x}), \quad \mathbf{x} \in \Omega, \ w_i = w\big|_{K_i}.$$

In the same manner as in the previous section, system (11.60) can be recast in matrix form (see Exercise 11.14):

$$\begin{pmatrix} \mathbf{A} & \mathbf{B} \\ \mathbf{B}^T & 0 \end{pmatrix} \begin{pmatrix} \mathbf{U} \\ \mathbf{p} \end{pmatrix} = \begin{pmatrix} \mathbf{0} \\ -\mathbf{f} \end{pmatrix}, \tag{11.61}$$

where

$$\mathbf{A} = (a_{ij})_{i,j=1,2,\dots,M}, \quad \mathbf{B} = (b_{jk})_{j=1,2,\dots,M,\ k=1,2,\dots,N},$$
$$\mathbf{U} = (u_i)_{i=1,2,\dots,M}, \quad \mathbf{p} = (p_k)_{k=1,2,\dots,N}, \quad \mathbf{f} = (f_j)_{j=1,2,\dots,N},$$

with

$$a_{ij} = \int_\Omega \boldsymbol{\varphi}_i \cdot \boldsymbol{\varphi}_j \ d\mathbf{x}, \quad b_{jk} = -\int_\Omega \nabla \cdot \boldsymbol{\varphi}_j \psi_k \ d\mathbf{x}, \quad f_j = \int_\Omega f\psi_j \ d\mathbf{x}.$$

Again, the matrix $\mathbf{M}$ defined by

$$\mathbf{M} = \begin{pmatrix} \mathbf{A} & \mathbf{B} \\ \mathbf{B}^T & 0 \end{pmatrix}$$

has both positive and negative eigenvalues. The matrix $\mathbf{A}$ is symmetric, positive definite, and sparse. In fact, it has at most five nonzero entries per row in the present case (see Exercise 11.14). The matrix $\mathbf{B}$ is also sparse, with two nonzero entries in each row.

Let $(\mathbf{u}, p)$ and $(\mathbf{u}_h, p_h)$ be the respective solutions of systems (11.59) and (11.60). Then, the following error estimate holds (Brezzi and Fortin, 1991; Chen, 2005):

$$\left(\int_\Omega |p - p_h|^2 \ d\mathbf{x}\right)^{1/2} + \left(\int_\Omega |\mathbf{u} - \mathbf{u}_h|^2 \ d\mathbf{x}\right)^{1/2} \leq Ch, \qquad (11.62)$$

where C depends on the size of the second partial derivatives of the solution p. This estimate is optimal for this pair of mixed finite element spaces.

11.4.3. Extension to boundary conditions of other kinds

A Neumann boundary condition

In the previous section, we considered the Dirichlet boundary condition in system (11.56). We now extend the mixed finite element method to the stationary problem with the *homogeneous Neumann boundary condition*:

$$\begin{aligned} -\Delta p &= f \quad \text{in } \Omega, \\ \frac{\partial p}{\partial \boldsymbol{\nu}} &= 0 \quad \text{on } \Gamma, \end{aligned} \qquad (11.63)$$

where $\partial p/\partial \boldsymbol{\nu}$ is the derivative of p normal to boundary Γ.

Application of Green's formula (2.9) to problem (11.63) yields

$$\int_\Omega f \, d\mathbf{x} = 0.$$

This is a *compatibility condition*. In this case, the solution p is unique up to an additive constant.

The admissible function spaces are accordingly modified:

$$W = \Big\{ \text{Functions } w \colon w \text{ is defined and square integrable on } \Omega$$

$$\Big(\text{i.e., } \int_\Omega w^2 \, d\mathbf{x} < \infty \Big), \quad \text{and} \quad \int_\Omega w \, d\mathbf{x} = 0 \Big\},$$

and

$$\mathbf{V} = \Big\{ \text{Vector functions } \mathbf{v} \colon \mathbf{v} \text{ and } \nabla \cdot \mathbf{v} \text{ are defined and square integrable}$$

$$\text{on } \Omega \Big(\text{i.e., } \int_\Omega (|\mathbf{v}|^2 + |\nabla \cdot \mathbf{v}|^2) d\mathbf{x} < \infty \Big) \text{ and } \mathbf{v} \cdot \boldsymbol{\nu} = 0 \text{ on } \Gamma \Big\}.$$

Recall that the use of the condition $\int_\Omega w \, d\mathbf{x} = 0$ is to rectify the up-to-a-constant uniqueness problem for the solution p.

With the choice of these two spaces, the mixed variational formulation of problem (11.63) is

Find $\mathbf{u} \in \mathbf{V}$ and $p \in W$ such that

$$\int_\Omega \mathbf{u} \cdot \mathbf{v} \, d\mathbf{x} - \int_\Omega \nabla \cdot \mathbf{v} p \, d\mathbf{x} = 0, \quad \mathbf{v} \in \mathbf{V}, \tag{11.64}$$

$$\int_\Omega \nabla \cdot \mathbf{u} w \, d\mathbf{x} = \int_\Omega f w \, d\mathbf{x}, \qquad w \in W.$$

Note that the Neumann boundary condition becomes the *essential* condition that must be incorporated into the definition of the space $\mathbf{V}$. In contrast, the Dirichlet boundary condition is the essential condition in the standard finite element method (see Sec. 2.3.2).

Let K_h be a partition of Ω into non overlapping triangles, as defined in the previous section. We define the mixed finite element spaces

$$W_h = \Big\{ \text{Functions } w \colon w \text{ is constant on each triangle in } K_h$$

$$\text{and} \quad \int_\Omega w \, d\mathbf{x} = 0 \Big\},$$

and

$$\mathbf{V}_h = \{\text{Vector functions } \mathbf{v}: \mathbf{v}\big|_K = (b_K x_1 + a_K, b_K x_2 + c_K),\ K \in K_h,$$
$$a_K, b_K, c_K \in \mathbb{R},\ \text{the normal components of } \mathbf{v} \text{ are continuous}$$
$$\text{across the interior edges in } K_h, \text{ and } \mathbf{v} \cdot \boldsymbol{\nu} = 0 \text{ on } \Gamma\}.$$

Again, $\mathbf{V}_h \subset \mathbf{V}$ and $W_h \subset W$. The mixed finite element method for problem (11.63) reads as follows:

Find $\mathbf{u}_h \in \mathbf{V}_h$ and $p_h \in W_h$ such that
$$\int_\Omega \mathbf{u}_h \cdot \mathbf{v}\, d\mathbf{x} - \int_\Omega \nabla \cdot \mathbf{v} p_h\, d\mathbf{x} = 0, \quad \mathbf{v} \in \mathbf{V}_h, \tag{11.65}$$
$$\int_\Omega \nabla \cdot \mathbf{u}_h w\, d\mathbf{x} = \int_\Omega f w\, d\mathbf{x}, \qquad w \in W_h.$$

This system can be rewritten in matrix form as for system (11.60) (cf. (11.61)), and the error estimate (11.62) also holds for this system.

A boundary condition of third kind

We now consider a boundary condition of a *third kind*:

$$-\Delta p = f \qquad \text{in } \Omega,$$
$$bp + \frac{\partial p}{\partial \boldsymbol{\nu}} = g \quad \text{on } \Gamma, \tag{11.66}$$

where b is a given (nonzero) function on Γ and g is a given function.

With the linear spaces $\mathbf{V}$ and W defined as in Sec. 11.4.2, the mixed variational formulation of problem (11.66) is

Find $\mathbf{u} \in \mathbf{V}$ and $p \in W$ such that
$$\int_\Omega \mathbf{u} \cdot \mathbf{v}\, d\mathbf{x} + \int_\Gamma b^{-1} \mathbf{u} \cdot \boldsymbol{\nu}\, \mathbf{v} \cdot \boldsymbol{\nu}\, d\ell - \int_\Omega \nabla \cdot \mathbf{v} p\, d\mathbf{x}$$
$$= -\int_\Gamma b^{-1} g \mathbf{v} \cdot \boldsymbol{\nu}\, d\ell, \quad \mathbf{v} \in \mathbf{V}, \tag{11.67}$$
$$\int_\Omega \nabla \cdot \mathbf{u} w\, d\mathbf{x} = \int_\Omega f w\, d\mathbf{x}, \quad w \in W.$$

Similarly, with the mixed finite element spaces $\mathbf{V}_h$ and W_h defined in Sec. 11.4.2, the mixed finite element method for problem (11.66) is

Find $\mathbf{u}_h \in \mathbf{V}_h$ and $p_h \in W_h$ such that
$$\int_\Omega \mathbf{u}_h \cdot \mathbf{v}\, dx + \int_\Gamma b^{-1} \mathbf{u}_h \cdot \boldsymbol{\nu}\, \mathbf{v} \cdot \boldsymbol{\nu}\, d\ell - \int_\Omega \nabla \cdot \mathbf{v} p_h\, dx$$
$$= -\int_\Gamma b^{-1} g \mathbf{v} \cdot \boldsymbol{\nu}\, d\ell, \qquad \mathbf{v} \in \mathbf{V}_h, \tag{11.68}$$
$$\int_\Omega \nabla \cdot \mathbf{u}_h w\, dx = \int_\Omega f w\, dx, \qquad w \in W_h.$$

The matrix form and error estimate for system (11.68) can be obtained in the same fashion as in Sec. 11.4.2 (see Exercise 11.17).

11.4.4. Mixed finite element spaces

Simple mixed finite element spaces on triangles have been used so far, which will be generalized to more general spaces in this section. We consider the model problem for the unknown p

$$-\nabla \cdot (\mathbf{a}\nabla p) = f \quad \text{in } \Omega,$$
$$p = g \quad \text{on } \Gamma, \tag{11.69}$$

where $\Omega \subset \mathbb{R}^d$ ($d = 2$ or 3) is a bounded two- or three-dimensional domain with boundary Γ, the diffusion tensor $\mathbf{a}$ is assumed to be symmetric and positive definite, and f and g are given real-valued piecewise continuous bounded functions in Ω and Γ, respectively. This problem was considered previously (see Sec. 11.3.1). To write it in a mixed variational formulation, the admissible spaces W and $\mathbf{V}$ introduced in Sec. 11.4.2 are exploited. In addition, we introduce the notation

$$\|w\|_W = \left(\int_\Omega w^2\, dx \right)^{1/2}, \qquad w \in W,$$
$$\|\mathbf{v}\|_\mathbf{V} = \left(\int_\Omega \left[|\mathbf{v}|^2 + |\nabla \cdot \mathbf{v}|^2 \right]\, dx \right)^{1/2}, \qquad \mathbf{v} \in \mathbf{V}.$$

They are the respective *norms* of the spaces W and $\mathbf{V}$. In three dimensions, we recall the divergence definition

$$\nabla \cdot \mathbf{v} = \frac{\partial v_1}{\partial x_1} + \frac{\partial v_2}{\partial x_2} + \frac{\partial v_3}{\partial x_3}, \qquad \mathbf{v} = (v_1, v_2, v_3).$$

Introduce the vector
$$\mathbf{u} = -\mathbf{a}\nabla p. \tag{11.70}$$

In the same way as in the derivation of system (11.59), problem (11.69) is written in the mixed variational formulation:

Find $\mathbf{u} \in \mathbf{V}$ and $p \in W$ such that

$$\int_\Omega \mathbf{a}^{-1}\mathbf{u} \cdot \mathbf{v} \, d\mathbf{x} - \int_\Omega \nabla \cdot \mathbf{v}p \, d\mathbf{x} = -\int_\Gamma g\mathbf{v} \cdot \boldsymbol{\nu} \, d\ell, \quad \mathbf{v} \in \mathbf{V}, \tag{11.71}$$

$$\int_\Omega \nabla \cdot \mathbf{u}w \, d\mathbf{x} = \int_\Omega fw \, d\mathbf{x}, \qquad\qquad w \in W,$$

where $\mathbf{a}^{-1}$ indicates the inverse of $\mathbf{a}$. For system (11.71), there is a positive constant C_1 such that the *inf-sup* condition between the spaces $\mathbf{V}$ and W holds (Chen, 2005):

$$\sup_{0 \neq \mathbf{v} \in \mathbf{V}} \frac{\left| \int_\Omega \nabla \cdot \mathbf{v}w \, d\mathbf{x} \right|}{\|\mathbf{v}\|_\mathbf{V}} \geq C_1 \|w\|_W \quad \forall w \in W. \tag{11.72}$$

Because of the condition on the tensor $\mathbf{a}$ (positive definiteness) and condition (11.72), problem (11.71) has a unique solution $\mathbf{u} \in \mathbf{V}$ and $p \in W$ (Brezzi and Fortin, 1991).

Let $\mathbf{V}_h \subset \mathbf{V}$ and $W_h \subset W$ be certain finite dimensional subspaces. The discrete version of system (11.71) is

Find $\mathbf{u}_h \in \mathbf{V}_h$ and $p_h \in W_h$ such that

$$\int_\Omega \mathbf{a}^{-1}\mathbf{u}_h \cdot \mathbf{v} \, d\mathbf{x} - \int_\Omega \nabla \cdot \mathbf{v}p_h \, d\mathbf{x} = -\int_\Gamma g\mathbf{v} \cdot \boldsymbol{\nu} \, d\ell, \quad \mathbf{v} \in \mathbf{V}_h, \tag{11.73}$$

$$\int_\Omega \nabla \cdot \mathbf{u}_h w \, d\mathbf{x} = \int_\Omega fw \, d\mathbf{x}, \qquad\qquad w \in W_h.$$

For this problem to have a unique solution, it is natural to impose a discrete *inf-sup* condition between the finite element spaces $\mathbf{V}_h$ and W_h similar to condition (11.72):

$$\sup_{0 \neq \mathbf{v} \in \mathbf{V}_h} \frac{\left| \int_\Omega \nabla \cdot \mathbf{v}w \, d\mathbf{x} \right|}{\|\mathbf{v}\|_\mathbf{V}} \geq C_2 \|w\|_W \quad \forall w \in W_h, \tag{11.74}$$

where $C_2 > 0$ is a constant independent of h.

Condition (11.74) is also called the *Babuška–Brezzi condition* or sometimes the *Ladyshenskaja–Babuška–Brezzi condition*.

In the previous two sections, we considered the mixed finite element spaces $\mathbf{V}_h$ and W_h over triangles. These spaces are the lowest-order triangular spaces introduced by Raviart and Thomas (1977), and they satisfy condition (11.74). In this section, we describe the more general Raviart–Thomas spaces on triangles and rectangles that satisfy this *stability condition*. In the literature, there exist RTN (Raviart and Thomas, 1977; Nédélec,

1980), BDM (Brezzi *et al.*, 1985), BDDF (Brezzi *et al.*, 1987a), BDFM (Brezzi *et al.*, 1987b), and CD (Chen and Douglas, 1989) spaces that also satisfy this condition.

For simplicity, in this section, let Ω be a polygonal domain in the plane. For a curved domain, the definition of the mixed finite element spaces under consideration is the same, but the degrees of freedom for the vector space $\mathbf{V}_h$ need to be modified (Brezzi and Fortin, 1991).

Mixed finite element spaces on triangles

For a planar domain $\Omega \subset \mathbb{R}^2$, let K_h be a partition of Ω into triangles such that adjacent elements completely share their common edge. For a triangle $K \in K_h$, define the set of polynomials of degree r on K:

$$P_r(K) = \{\text{Functions } v : v \text{ is a polynomial of degree at most } r \text{ on } K\},$$

where $r \geq 0$ is an integer. Mixed finite element spaces $\mathbf{V}_h \times W_h$ are defined locally on each element $K \in K_h$, so let $\mathbf{V}_h(K) = \mathbf{V}_h|_K$ (the restriction of $\mathbf{V}_h$ to K) and $W_h(K) = W_h|_K$.

As an example for mixed spaces on triangles, we consider the mixed finite element spaces introduced by Raviart and Thomas (1977). They are defined for each $r \geq 0$ by

$$\mathbf{V}_h(K) = \big(P_r(K)\big)^2 \oplus \big((x_1, x_2)P_r(K)\big), \quad W_h(K) = P_r(K),$$

where the notation $\oplus$ is a direct sum and $(x_1, x_2)P_r(K) = \big(x_1 P_r(K), x_2 P_r(K)\big)$. The case $r = 0$ was used in the previous two sections. In this case, we observe that $\mathbf{V}_h(K)$ has the form

$$\mathbf{V}_h(K) = \{\text{Functions } \mathbf{v} : \mathbf{v} = (a_K + b_K x_1, c_K + b_K x_2),$$

$$a_K,\ b_K,\ c_K \in \mathbb{R}\},$$

and its dimension is three. As discussed in Sec. 11.4.2, as parameters, or *the degrees of freedom*, to describe the functions in $\mathbf{V}_h$, we use the values of normal components of the functions at the midpoints of edges in K_h (Fig. 11.14). Also, in the case $r = 0$, the degrees of freedom for W_h can be the averages of functions over each element K (see Sec. 11.4.2).

In general, for $r \geq 0$ the dimensions of $\mathbf{V}_h(K)$ and $W_h(K)$ are

$$\dim\big(\mathbf{V}_h(K)\big) = (r+1)(r+3), \quad \dim\big(W_h(K)\big) = \frac{(r+1)(r+2)}{2}.$$

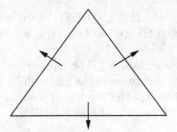

Figure 11.14 The triangular RT.

The degrees of freedom for the space $\mathbf{V}_h(K)$, with $r \geq 0$, are given by (Raviart and Thomas, 1977)

$$\int_e \mathbf{v} \cdot \boldsymbol{\nu} w \, d\ell \quad \forall w \in P_r(e), \quad e \in \partial K,$$

$$\int_K \mathbf{v} \cdot \mathbf{w} \, d\mathbf{x} \quad \forall \mathbf{w} \in (P_{r-1}(K))^2,$$

where $\boldsymbol{\nu}$ is the outward unit normal to the edge $e \in \partial K$. This is a legitimate choice; i.e., a function in the space $\mathbf{V}_h$ is uniquely determined by these degrees of freedom.

Mixed finite element spaces on rectangles

We now consider the case where Ω is a rectangular domain and K_h is a partition of Ω into rectangles such that the horizontal and vertical edges of rectangles are parallel to the x_1- and x_2-coordinate axes, respectively, and adjacent elements completely share their common edge. We define

$$Q_{l,r}(K) = \left\{ \text{Functions } v \colon v(\mathbf{x}) = \sum_{i=0}^{l} \sum_{j=0}^{r} v_{ij} x_1^i x_2^j, \right.$$

$$\left. \mathbf{x} = (x_1, x_2) \in K, v_{ij} \in \mathbb{R} \right\};$$

i.e., $Q_{l,r}(K)$ is the set of polynomials of degree at most l in x_1 and r in x_2, $l, r \geq 0$, defined on the rectangle K.

For each $r \geq 0$, we define

$$\mathbf{V}_h(K) = Q_{r+1,r}(K) \times Q_{r,r+1}(K), \quad W_h(K) = Q_{r,r}(K).$$

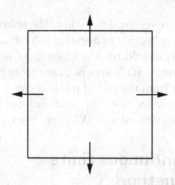

Figure 11.15 The rectangular RT.

These spaces are an extension of the Raviart–Thomas spaces on triangles to rectangles (Raviart and Thomas, 1977). In the case $r = 0$, the vector space $\mathbf{V}_h(K)$ takes the form

$$\mathbf{V}_h(K) = \{\text{Vector functions } \mathbf{v}\colon \mathbf{v} = (a_K^1 + a_K^2 x_1, a_K^3 + a_K^4 x_2),$$
$$a_K^i \in \mathbb{R}, \ i = 1, 2, 3, 4\},$$

and its dimension is four. The degrees of freedom for $\mathbf{V}_h$ are the values of normal components of the functions at the midpoint on each edge in K_h (cf. Fig. 11.15). In this case, $Q_{0,0}(K) = P_0(K)$.

For a general $r \geq 0$, the dimensions of the spaces $\mathbf{V}_h(K)$ and $W_h(K)$ are

$$\dim\big(\mathbf{V}_h(K)\big) = 2(r+1)(r+2), \quad \dim\big(W_h(K)\big) = (r+1)^2.$$

The degrees of freedom for the local space $\mathbf{V}_h(K)$ are

$$\int_e \mathbf{v} \cdot \boldsymbol{\nu} w \, d\ell \quad \forall w \in P_r(e), \quad e \in \partial K,$$

$$\int_K \mathbf{v} \cdot \mathbf{w} \, d\mathbf{x} \quad \forall \mathbf{w} = (w_1, w_2), \ w_1 \in Q_{r-1,r}(K), \quad w_2 \in Q_{r,r-1}(K).$$

For the reference of mixed finite elements in three dimensions and other types of mixed elements, the reader can refer to Brezzi and Fortin (1991) and Chen (2005). We have presented the mixed finite element method only for stationary problems. This method can be extended to transient problems as in Chapter 7. The discretization in time can be carried out using either the backward Euler method or the Crank-Nicolson method and in space using the mixed method. The linear systems of algebraic equations arising from the mixed method are of *saddle type*; i.e., the system matrices have both positive and negative eigenvalues. Thus, the solution of these systems needs special care. For a collection of iterative algorithms suitable for saddle linear systems, the reader should refer to Chen (2005). When the spaces $\mathbf{V}_h$ and W_h are the lowest-order RTN spaces over rectangular parallelepipeds, the linear system arising from the mixed method can be written as a system generated by a *cell-centered* (or *block-centered*) finite difference scheme using certain quadrature rules (Russell and Wheeler, 1983).

11.5. The discontinuous finite element method

In the previous sections, the functions used in the finite element spaces V_h for the discretization of second-order partial differential equations were continuous either across the whole interelement boundaries or at certain points on these boundaries. In this section, we consider the case where the functions in the finite element spaces are totally discontinuous across these boundaries, i.e., *discontinuous finite elements*. *Discontinuous Galerkin (DG) finite element methods* were originally introduced for a linear *advection (hyperbolic) problem* by Reed and Hill (1973). They have established themselves as an important alternative for numerically solving advection (convection) problems for which continuous finite element methods lack robustness. Important features of the DG methods are that they conserve mass locally (on each element) and are of high-order accuracy.

11.5.1. DG methods

We consider the advection (or convetion) problem:

$$\mathbf{b} \cdot \nabla p + Rp = f, \quad \mathbf{x} \in \Omega,$$
$$p = g, \quad \mathbf{x} \in \Gamma_-, \tag{11.75}$$

where the functions $\mathbf{b}$, R, f, and g are given, $\Omega \subset \mathbb{R}^d$ ($d \leq 3$) is a bounded domain with boundary Γ, the *inflow boundary* Γ_- is defined by

$$\Gamma_- = \{\mathbf{x} \in \Gamma : (\mathbf{b} \cdot \boldsymbol{\nu})(\mathbf{x}) < 0\},$$

and $\boldsymbol{\nu}$ is the outward unit normal to Γ. The advection coefficient $\mathbf{b}$ is assumed to be smooth in $(\mathbf{x}, t)$, and the reaction coefficient R is assumed to be bounded and nonnegative.

For $h > 0$, let K_h be a finite element partition of Ω into elements $\{K\}$. The partition K_h is assumed to satisfy the minimum angle condition (2.52). For the DG methods, adjacent elements in the partition K_h are not required to match; a vertex of one element can lie in the interior of the edge or the face of another element. Let $\mathcal{E}_h^o$ denote the set of all interior boundaries e in K_h, $\mathcal{E}_h^b$ the set of the boundaries e on Γ, and $\mathcal{E}_h = \mathcal{E}_h^o \cup \mathcal{E}_h^b$. We tacitly assume that $\mathcal{E}_h^o \neq \emptyset$.

Associated with the partition K_h, we define the finite element space

$$V_h = \{\text{Functions } v\colon v \text{ is a bounded function on } \Omega \text{ and}$$

$$v|_K \in P_r(K), \ K \in K_h\},$$

where we recall that $P_r(K)$ is the set of polynomials on element K of degree at most $r \geq 0$. Note that no continuity across interelement boundaries is required on functions in this space.

For each $K \in K_h$, we split its boundary ∂K into the inflow and outflow parts by

$$\partial K_- = \{\mathbf{x} \in \partial K\colon (\mathbf{b} \cdot \boldsymbol{\nu})(\mathbf{x}) < 0\},$$

$$\partial K_+ = \{\mathbf{x} \in \partial K\colon (\mathbf{b} \cdot \boldsymbol{\nu})(\mathbf{x}) \geq 0\},$$

where $\boldsymbol{\nu}$ is the outward unit normal to ∂K. A triangle K with a boundary made up of ∂K_- and ∂K_+ is shown is Fig. 11.16. For each internal boundary $e \in \mathcal{E}_h^o$, the left- and right-hand limits on e of a function $v \in V_h$ are defined by

$$v_-(\mathbf{x}) = \lim_{\epsilon \to 0^-} v(\mathbf{x} + \epsilon \mathbf{b}), \quad v_+(\mathbf{x}) = \lim_{\epsilon \to 0^+} v(\mathbf{x} + \epsilon \mathbf{b}),$$

for $\mathbf{x} \in e$. The jump of v across e is given by

$$[v] = v_+ - v_-.$$

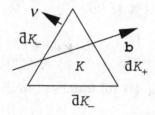

Figure 11.16 An illustration of ∂K_- and ∂K_+.

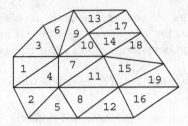

Figure 11.17   An ordering of computation for DG.

For an external boundary $e \in \mathcal{E}_h^b$, we define (from inside Ω)

$$[v] = v.$$

Now, the DG method for problem (11.75) is defined: For $K \in K_h$, given $p_{h,-}$ on ∂K_-, find $p_h = p_h|_K \in P_r(K)$ such that

$$\int_K (\mathbf{b} \cdot \nabla p_h + R p_h) v \, d\mathbf{x} - \int_{\partial K_-} p_{h,+} v_+ \mathbf{b} \cdot \boldsymbol{\nu} \, d\ell$$

$$= \int_K f v \, d\mathbf{x} - \int_{\partial K_-} p_{h,-} v_+ \mathbf{b} \cdot \boldsymbol{\nu} \, d\ell \qquad \forall v \in P_r(K),$$

(11.76)

where $p_{h,-} = g$ on Γ_-. Note that Eq. (11.76) is the standard finite element method for problem (11.75) on each element K, with the boundary condition being *weakly* imposed. Equation (11.76) also holds for the continuous problem (11.75) (Chen, 2005). For a typical triangulation (Fig. 11.17), the numerical solution p_h can be determined first on the triangles K adjacent to Γ_-. Then, this process is continued (working away from known information) until p_h is found in the whole domain Ω. Thus, the computation of Eq. (11.76) is local.

If $\mathbf{b}$ is *divergence free* (or *solenoidal*), i.e., $\nabla \cdot \mathbf{b} = 0$, we can use Green's formula (2.9) to see that (Fig. 11.16)

$$\int_K \mathbf{b} \cdot \nabla p_h \, d\mathbf{x} = \int_{\partial K_-} p_{h,+} \mathbf{b} \cdot \boldsymbol{\nu} \, d\ell + \int_{\partial K_+} p_{h,-} \mathbf{b} \cdot \boldsymbol{\nu} \, d\ell.$$

We substitute this into Eq. (11.76) with $v = 1$ to give

$$\int_K R p_h \, d\mathbf{x} + \int_{\partial K_+} p_{h,-} \mathbf{b} \cdot \boldsymbol{\nu} \, d\ell = \int_K f \, d\mathbf{x} - \int_{\partial K_-} p_{h,-} \mathbf{b} \cdot \boldsymbol{\nu} \, d\ell, \quad (11.77)$$

which expresses a local conservation property (i.e., the difference between inflow and outflow equals the sum of accumulation of mass).

To express method (11.76) in the general form introduced in Chapter 3, we define the local bilinear form

$$a_K(v, w) = \int_K (\mathbf{b} \cdot \nabla v + Rv) \, d\mathbf{x} - \int_{\partial K_-} [v] w_+ \mathbf{b} \cdot \boldsymbol{\nu} \, d\ell, \quad K \in K_h,$$

and the mesh-dependent bilinear form

$$a_h(v, w) = \sum_{K \in K_h} a_K(v, w).$$

Then, Eq. (11.76) is expressed as follows: Find $p_h \in V_h$ such that

$$a_h(p_h, v) = \int_\Omega fv \, d\mathbf{x} \quad \forall v \in V_h, \tag{11.78}$$

where $p_{h,-} = g$ on Γ_-. We consider a couple of examples for method (11.78).

Example 11.1 A one-dimensional example of problem (11.75) is

$$\frac{dp}{dx} + p = f, \quad x \in (0, 1),$$
$$p(0) = g. \tag{11.79}$$

Let $0 = x_0 < x_1 < \cdots < x_M = 1$ be a partition of the unit interval $(0, 1)$ into a set of subintervals $I_i = (x_{i-1}, x_i)$, with length $h_i = x_i - x_{i-1}$, $i = 1, 2, \ldots, M$. In this case, Eq. (11.76) becomes: For $i = 1, 2, \ldots, M$, given $(p_h(x_{i-1}))_-$, find $p_h = p_h|_{I_i} \in P_r(I_i)$ such that

$$\int_{I_i} \left(\frac{dp_h}{dx} + p_h \right) v \, dx + [p_h(x_{i-1})](v(x_{i-1}))_+ = \int_{I_i} fv \, dx \quad \forall v \in P_r(I_i),$$

where $(p_h(x_0))_- = g$. In the case $r = 0$, V_h is the space of piecewise constants, and the DG method reduces to: For $i = 1, 2, \ldots, M$, find $p_i = (p_h(x_i))_-$ such that

$$\frac{p_i - p_{i-1}}{h_i} + p_i = \frac{1}{h_i} \int_{I_i} f \, dx,$$
$$p_0 = g. \tag{11.80}$$

Note that system (11.80) is nothing but a simple *upwind* finite difference method with an averaged right-hand side.

Example 11.2 Set $R = f = 0$ in the advection problem (11.75); the resulting equation simplifies to

$$\mathbf{b} \cdot \nabla p = 0, \quad \mathbf{x} \in \Omega,$$
$$p = g, \qquad \mathbf{x} \in \Gamma_-. \tag{11.81}$$

Also, let $r = 0$. Then, Eq. (11.76) reads: for $K \in K_h$, given $p_{h,-}$ on ∂K_-, find $p_K = p_h|_K$ such that

$$\int_{\partial K_-} p_K \mathbf{b} \cdot \boldsymbol{\nu} \, d\ell = \int_{\partial K_-} p_{h,-} \mathbf{b} \cdot \boldsymbol{\nu} \, d\ell;$$

that is,

$$p_K = \frac{\int_{\partial K_-} p_{h,-} \mathbf{b} \cdot \boldsymbol{\nu} \, d\ell}{\int_{\partial K_-} \mathbf{b} \cdot \boldsymbol{\nu} \, d\ell}. \tag{11.82}$$

Thus, we see that for each element $K \in K_h$, the value p_K is determined by a weighted average of the values $p_{h,-}$ on adjoining elements with edges on ∂K_-. As an example, let Ω be a rectangular domain in $\mathbb{R}^2$, K_h consist of rectangles, and $\mathbf{b} > \mathbf{0}$. In this case, for a configuration shown in Fig. 11.18, we see that

$$p_3 = \frac{b_1}{b_1 + b_2} p_1 + \frac{b_2}{b_1 + b_2} p_2,$$

where $p_i = p_h|_{K_i}$, $i = 1, 2, 3$, and $\mathbf{b} = (b_1, b_2)$. Again, in this case, Eq. (11.82) corresponds to the usual upwind finite difference method for problem (11.81).

If the solution p to problem (11.75) is sufficiently smooth on each $K \in K_h$, an error estimate for method (11.78) is given by

$$\left(\int_\Omega |p - p_h|^2 \, d\mathbf{x} \right)^{1/2} + h^{1/2} \left(\sum_{K \in K_h} \int_\Omega |\mathbf{b} \cdot \nabla(p - p_h)|^2 \, d\mathbf{x} \right)^{1/2} \leq Ch^{r+1/2},$$

$$\tag{11.83}$$

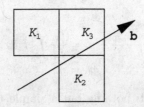

Figure 11.18   Adjoining rectangles.

for $r \geq 0$, where the constant C depends on the $(r+1)$th partial derivatives of p. Note that the $L^2(\Omega)$-estimate (the first term) is half a power of h from being optimal, while the $L^2(\Omega)$-estimate of the derivative in the velocity (or *streamline*) direction (the second term) is in fact optimal. For general triangulations, this $L^2(\Omega)$ estimate is sharp in the sense that the exponent of h cannot be increased (Johnson, 1994).

We end with a remark that a time-dependent advection problem can be written as a system in the same form as for problem (11.75). To see this, consider the problem

$$\phi \frac{\partial p}{\partial t} + \mathbf{b} \cdot \nabla p + Rp = f, \quad \mathbf{x} \in \Omega,\ t > 0,$$

and set $t = x_0$ and $b_0 = \phi$. Then, we see that

$$\bar{\mathbf{b}} \cdot \nabla_{(t,\mathbf{x})} p + Rp = f,$$

where $\bar{\mathbf{b}} = (b_0, \mathbf{b})$ and $\nabla_{(t,\mathbf{x})} = (\frac{\partial}{\partial t}, \nabla_\mathbf{x})$ (treating time as a space-like variable). Thus, the above development of the DG method for problem (11.75) applies.

11.5.2. Stabilized DG methods

We consider a stabilized DG (SDG) method, which modifies method (11.76) as follows: for $K \in K_h$, given $p_{h,-}$ on ∂K_-, find $p_h = p_h|_K \in P_r(K)$ such that

$$\int_K (\mathbf{b} \cdot \nabla p_h + Rp_h)(v + \theta \mathbf{b} \cdot \nabla v)\, d\mathbf{x} - \int_{\partial K_-} p_{h,+} v_+ \mathbf{b} \cdot \boldsymbol{\nu}\, d\ell$$

$$= \int_K f(v + \theta \mathbf{b} \cdot \nabla v)\, d\mathbf{x} - \int_{\partial K_-} p_{h,-} v_+ \mathbf{b} \cdot \boldsymbol{\nu}\, d\ell \quad \forall v \in P_r(K),$$

$$(11.84)$$

where θ is a *stabilization parameter*. The difference between methods (11.76) and (11.84) is that a stabilized term is added in the left- and right-hand sides of the latter. This stabilized method is also called the *streamline diffusion method* due to intuition that the added term

$$\theta \int_K (\mathbf{b} \cdot \nabla p_h)(\mathbf{b} \cdot \nabla v)\, d\mathbf{x}$$

corresponds to the diffusion in the direction of streamlines (or characteristics) (Johnson, 1994). The parameter θ is chosen so that $\theta = \mathcal{O}(h)$, to generate the same convergence rate as for DG. For $r = 0$, DG and SDG are the same.

Now, the bilinear forms $a_K(\cdot,\cdot)$ and $a_h(\cdot,\cdot)$ are defined by

$$a_K(v,w) = \int_K (\mathbf{b} \cdot \nabla v + Rv)(w + \theta \mathbf{b} \cdot \nabla w) \, d\mathbf{x}$$

$$- \int_{\partial K_-} [v] w_+ \mathbf{b} \cdot \boldsymbol{\nu} \, d\ell, \quad K \in K_h,$$

and

$$a_h(v,w) = \sum_{K \in K_h} a_K(v,w).$$

Then, Eq. (11.84) is expressed as follows: find $p_h \in V_h$ such that

$$a(p_h,v) = \sum_{K \in K_h} \int_K f(v + \theta \mathbf{b} \cdot \nabla v) \, d\mathbf{x} \quad \forall v \in V_h, \tag{11.85}$$

where $p_{h,-} = g$ on Γ_-. The convergence result (11.83) holds also for this stabilized DG method (Chen, 2005).

For an appropriate choice of the stabilization parameter θ, the SDG method is much more stable than the DG. For a comparison, see Chen (2005). The DG and SDG methods have been developed here only for the hyperbolic problem (11.75); they can be also used for the solution of diffusion problems (Chen, 2005).

11.6. The characteristic finite element method

In this section, we consider an application of the finite element method to the *reaction–diffusion–advection problem*:

$$\frac{\partial(\phi p)}{\partial t} + \nabla \cdot (\mathbf{b}p - \mathbf{a}\nabla p) + Rp = f, \tag{11.86}$$

for the unknown solution p, where ϕ, $\mathbf{b}$ (vector), $\mathbf{a}$ (tensor), R, and f are given functions. Note that problem (11.86) involves advection (or convection) ($\mathbf{b}$), diffusion ($\mathbf{a}$), and reaction (R). Many equations arise in this form, e.g., saturation and concentration equations for multiphase, multicomponent flows in porous media (see Chapter 10).

When diffusion dominates advection, the finite element method developed in the first seven chapters performs well for problem (11.86). When advection dominates diffusion, it does not perform well. In particular, it exhibits excessive nonphysical oscillations when the solution to Eq. (11.86) is not smooth. Standard upstream weighting approaches have been applied to the finite element method with the purpose of eliminating the nonphysical oscillations, but these approaches smear sharp fronts in the solution. Although extremely fine mesh refinement is possible to overcome this difficulty, it is not feasible due to the excessive computational effort involved.

Many numerical methods have been developed for solving problem (11.86) where advection dominates, such as the *optimal spatial method*. This method employs an *Eulerian approach* that is based on the minimization of the error in the approximation of spatial derivatives and the use of optimal test functions satisfying a local adjoint problem (Brooks and Hughes, 1982; Barrett and Morton, 1984). It yields an upstream bias in the resulting approximation and has the features: (i) time truncation errors dominate the solution; (ii) the solution has significant numerical diffusion and phase errors; and (iii) the *Courant number* (i.e., $|\mathbf{b}|\Delta t/(\phi h)$) is generally restricted to be less than one, where Δt and h are temporal and spatial mesh sizes, respectively.

Other Eulerian methods such as the *Petrov-Galerkin finite element method* have been developed to use nonzero spatial truncation errors to cancel temporal errors and thereby reduce the overall truncation errors (Christie *et al.*, 1976; Westerink and Shea, 1989). While these methods improve accuracy in the approximation of the solution, they still suffer from a strict Courant number limitation.

Another class of numerical methods for the solution of problem (11.86) are the *Eulerian–Lagrangian methods*. Because of the Lagrangian nature of advection, these methods treat the advection by a characteristic tracking approach. They have shown great potential. This class is rich and bears a variety of names, the *method of characteristics* (Garder *et al.*, 1964), the *modified method of characteristics* (Douglas and Russell, 1982), the *transport diffusion method* (Pironneau, 1982), the *Eulerian–Lagrangian method* (Neuman, 1981), the *operator splitting method* (Espedal and Ewing, 1987), the *Eulerian–Lagrangian localized adjoint method* (Celia *et al.*, 1990; Russell, 1990), the *characteristic mixed finite element method* (Yang, 1992; Arbogast and Wheeler, 1995), and the *Eulerian–Lagrangian mixed discontinuous method* (Chen, 2002b). The common features of this class are: (1) the Courant number restriction of the purely Eulerian methods is alleviated because of the Lagrangian nature of the advection step; (2) since the spatial and temporal dimensions are coupled through the characteristic tracking, the effect of time truncation errors present in the optimal spatial method is greatly reduced; and (3) they produce non oscillatory solutions without numerical diffusion, using reasonably large time steps on grids no finer than necessary to resolve the solution on the moving fronts. In this section, we describe the Eulerian–Lagrangian methods.

11.6.1. The modified method of characteristics

The modified method of characteristics (MMOC) was independently developed by Douglas and Russell (1982) and Pironneau (1982) and is based on a non divergence form of Eq. (11.86). It was called the *transport-diffusion method* by Pironneau. In the engineering literature, the name *Eulerian–Lagrangian method* is often used (Neuman, 1981). A characteristic-based splitting scheme was considered for numerical solution of fluid mechanics problems in Chapter 9.

A one-dimensional model problem

For the purpose of introduction, we consider a one-dimensional model problem on the whole real line:

$$\phi(x)\frac{\partial p}{\partial t} + b(x)\frac{\partial p}{\partial x} - \frac{\partial}{\partial x}\left(a(x,t)\frac{\partial p}{\partial x}\right) + R(x,t)p = f(x,t),$$

$$x \in \mathbb{R}, \ t > 0, \qquad (11.87)$$

$$p(x,0) = p_0(x), \quad x \in \mathbb{R}.$$

Set

$$\psi(x) = (\phi^2(x) + b^2(x))^{1/2}.$$

Assume that

$$\phi(x) > 0, \quad x \in \mathbb{R},$$

so $\psi(x) > 0$, $x \in \mathbb{R}$. Let the characteristic direction associated with the hyperbolic part of problem (11.87), $\phi\partial p/\partial t + b\partial p/\partial x$, be denoted by $\tau(x)$, so

$$\frac{\partial}{\partial \tau(x)} = \frac{\phi(x)}{\psi(x)}\frac{\partial}{\partial t} + \frac{b(x)}{\psi(x)}\frac{\partial}{\partial x}.$$

Then, problem (11.87) can be rewritten as

$$\psi(x)\frac{\partial p}{\partial \tau} - \frac{\partial}{\partial x}\left(a(x,t)\frac{\partial p}{\partial x}\right) + R(x,t)p = f(x,t),$$

$$x \in \mathbb{R}, \ t > 0, \qquad (11.88)$$

$$p(x,0) = p_0(x), \quad x \in \mathbb{R}.$$

We assume that the coefficients a, b, R, and ϕ are bounded and satisfy

$$\left| \frac{b(x)}{\phi(x)} \right| + \left| \frac{d}{dx} \left(\frac{b(x)}{\phi(x)} \right) \right| \leq C, \quad x \in \mathbb{R},$$

where C is a positive constant. The admissible function space V is the space of continuous functions on $\mathbb{R}$ (the set of real numbers) that have piecewise continuous and bounded first derivatives in $\mathbb{R}$ and approach zero at $\pm\infty$.

Now, multiplying the first equation of problem (11.88) by any $v \in V$ and applying integration by parts in space, the resulting problem can be written in the equivalent variational formulation:

$$\int_{\mathbb{R}} \psi \frac{\partial p}{\partial \tau} v \, dx + \int_{\mathbb{R}} a \frac{\partial p}{\partial x} \frac{dv}{dx} \, dx + \int_{\mathbb{R}} Rpv \, dx = \int_{\mathbb{R}} fv \, dx, \quad v \in V, \ t > 0,$$

$$p(x,0) = p_0(x), \qquad\qquad\qquad\qquad\qquad\qquad\qquad x \in \mathbb{R}.$$
$$\text{(11.89)}$$

Let $0 = t^0 < t^1 < \cdots < t^n < \cdots$ be a partition in time. The characteristic derivative is approximated in the following way: let

$$\check{x}_n = x - \frac{\Delta t^n}{\phi(x)} b(x), \tag{11.90}$$

and note that, at $t = t^n$,

$$\psi \frac{\partial p}{\partial \tau} \approx \psi(x) \frac{p(x,t^n) - p(\check{x}_n, t^{n-1})}{\left((x - \check{x}_n)^2 + (\Delta t^n)^2 \right)^{1/2}}$$
$$= \phi(x) \frac{p(x,t^n) - p(\check{x}_n, t^{n-1})}{\Delta t^n},$$
$$\text{(11.91)}$$

where $\Delta t^n = t^n - t^{n-1}$ is the time step. That is, a backtracking algorithm is used to approximate the characteristic derivative; $\check{x}_n$ is the foot (at level t^{n-1}) of the characteristic corresponding to x at the head (at level t^n) (Fig. 11.19).

Let V_h be a finite element subspace of the admissible function space V (see Chapters 1–6). When considering the whole line, V_h is necessarily infinite dimensional. In practice, we can assume that the support of p_0 is compact, the portion of the line on which we need to know the solution p is bounded, and p is very small outside that set. Then, V_h can be taken to be finite dimensional.

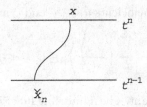

Figure 11.19   An illustration of the definition $\tilde{x}_n$.

The MMOC for problem (11.87) is defined: for $n = 1, 2, \ldots$, find $p_h^n \in V_h$ such that

$$
\int_{\mathbb{R}} \phi \frac{p_h^n - \breve{p}_h^{n-1}}{\Delta t^n} v \, dx + \int_{\mathbb{R}} a^n \frac{dp_h^n}{dx} \frac{dv}{dx} \, dx
$$
$$
+ \int_{\mathbb{R}} R^n p_h^n v \, dx = \int_{\mathbb{R}} f^n v \, dx \qquad \forall v \in V_h, \tag{11.92}
$$

where

$$
\breve{p}_h^{n-1} = p_h(\tilde{x}_n, t^{n-1}) = p_h \left(x - \frac{\Delta t^n}{\phi(x)} b(x), t^{n-1} \right). \tag{11.93}
$$

The initial approximation p_h^0 can be defined as the interpolant of p_0 in the finite element space V_h, e.g.

Note that Eq. (11.92) determines $\{p_h^n\}$ uniquely in terms of the data p_0 and f (at least, for reasonable a and R such that a is uniformly positive with respect to x and t and R is nonnegative). This can be seen as follows: since system (11.92) is a finite-dimensional system, it suffices to show uniqueness of the solution. Let $f = p_0 = 0$, and take $v = p_h^n$ in Eq. (11.92) to see that

$$
\int_{\mathbb{R}} \phi \frac{p_h^n - \breve{p}_h^{n-1}}{\Delta t^n} p_h^n \, dx + \int_{\mathbb{R}} a^n \left(\frac{dp_h^n}{dx} \right)^2 \, dx + \int_{\mathbb{R}} R^n (p_h^n)^2 \, dx = 0;
$$

with an induction assumption that $p_h^{n-1} = 0$, this equation implies $p_h^n = 0$.

It is obvious that the linear system arising from Eq. (11.92) is symmetric positive definite (see Sec. 1.2.2), even in the presence of the advection term. This system has an improved (over that arising from a direct application to problem (11.87) of the finite element method described in Chapters 1–6) condition number of order (see Exercise 11.18):

$$
\mathcal{O} \left(1 + \max_{x \in \mathbb{R}, \, t \geq 0} |a(x,t)| h^{-2} \Delta t \right), \qquad \Delta t = \max_{n=1,2,\ldots} \Delta t^n.
$$

Thus, the system arising from method (11.92) is well suited for the iterative linear solution algorithms.

We end with a remark on a convergence result for method (11.92). Let $V_h \subset V$ be a finite element space with the following approximation property:

$$\inf_{v_h \in V_h} \left(\left(\int_{\mathbb{R}} |v - v_h|^2 \, dx \right)^{1/2} + h \left(\int_{\mathbb{R}} \left| \frac{dv}{dx} - \frac{dv_h}{dx} \right|^2 \, dx \right)^{1/2} \right) \leq C h^{r+1},$$

$$(11.94)$$

where the constant $C > 0$ is independent of h and depends on the $(r+1)$th derivative of v, and $r > 0$ is an integer. Under appropriate assumptions on the smoothness of the solution p and a suitable choice of p_h^0 it can be shown (Douglas and Russell, 1982) that

$$\max_{1 \leq n \leq N} \left(\left(\int_{\mathbb{R}} |p^n - p_h^n|^2 \, dx \right)^{1/2} + h \left(\int_{\mathbb{R}} \left| \frac{dp^n}{dx} - \frac{dp_h^n}{dx} \right|^2 \, dx \right)^{1/2} \right)$$

$$\leq C(p) \left(h^{r+1} + \Delta t \right),$$

$$(11.95)$$

where N is an integer such that $t^N = T < \infty$ and $J = (0, T]$ is the time interval of interest.

This result, by itself, is not different from what we have obtained with the standard finite element methods in Chapters 1–3 and 7. However, the constant C is greatly improved when the MMOC is applied to problem (11.87). In time, C depends on a norm of $\partial^2 p / \partial t^2$ with the standard methods, but on a norm of $\partial^2 p / \partial \tau^2$ with the MMOC (Chen, 2005). The latter norm is much smaller, and thus long time steps with large Courant numbers are possible.

Some matters are raised by method (11.92) and its analog for more complicated differential problems. The first concern is the backtracking scheme that determines $\check{x}_n$ and a numerical quadrature rule that computes the associated integral. For the problem considered in this section, this matter can be resolved; the required computations can be performed exactly. For more complicated problems, there are discussions by Russell and Trujillo (1990). The second matter is the treatment of boundary conditions. In this section, we work on the whole line or on periodic boundary conditions (see the next section). For a bounded domain, if a backtracked characteristic crosses a boundary of the domain, it is not obvious what the meaning of $\check{x}_n$ or of $p_h(\check{x}_n)$ is. The last matter, and perhaps the greatest drawback of the MMOC, is its failure to conserve mass. This issue is discussed in detail below.

Periodic boundary conditions

In the previous section, problem (11.87) was considered on the whole line. For a bounded interval, say, $(0, 1)$, the MMOC has a difficulty in handling general boundary conditions. In this case, it is usually developed for *periodic boundary conditions* (see Exercise 11.19):

$$p(0, t) = p(1, t), \quad \frac{\partial p}{\partial x}(0, t) = \frac{\partial p}{\partial x}(1, t). \tag{11.96}$$

These conditions are also called *cyclic boundary conditions*. In the periodic case, assume that all functions in problem (11.87) are spatially $(0, 1)$-periodic. Accordingly, the linear space V is modified to

$$V = \Big\{ \text{Functions } v \colon \text{Each } v \text{ is continuous on } [0, 1],$$

$$\text{its first derivative } \frac{dv}{dx} \text{ is piecewise continuous}$$

$$\text{and bounded on } (0, 1), \text{ and } v \text{ is } (0, 1)\text{-periodic} \Big\}.$$

With this modification, Eqs. (11.89) and (11.92) remain unchanged.

Extension to multi dimensional problems

We now extend the MMOC to Eq. (11.86) defined on a multi dimensional domain Ω. Let $\Omega \subset \mathbb{R}^d$ $(d \leq 3)$ be a rectangle (respectively, a rectangular parallelepiped), and assume that (11.86) is Ω-periodic; i.e., all functions in problem (11.86) are spatially Ω-periodic. We write (11.86) in non divergence form

$$\phi(\mathbf{x}) \frac{\partial p}{\partial t} + \mathbf{b}(\mathbf{x}, t) \cdot \nabla p - \nabla \cdot (\mathbf{a}(\mathbf{x}, t) \nabla p)$$

$$+ R(\mathbf{x}, t)p = f(\mathbf{x}, t), \quad \mathbf{x} \in \Omega, \ t > 0, \tag{11.97}$$

$$p(\mathbf{x}, 0) = p_0(\mathbf{x}), \quad \mathbf{x} \in \Omega.$$

Set

$$\psi(\mathbf{x}, t) = (\phi^2(\mathbf{x}) + |\mathbf{b}(\mathbf{x}, t)|^2)^{1/2},$$

where $|\mathbf{b}|^2 = b_1^2 + b_2^2 + \cdots + b_d^2$, with $\mathbf{b} = (b_1, b_2, \ldots, b_d)$. Assume that

$$\phi(\mathbf{x}) > 0, \quad \mathbf{x} \in \Omega.$$

Now, the characteristic direction corresponding to the hyperbolic part of problem (11.97), $\phi\partial p/\partial t + \mathbf{b} \cdot \nabla p$, is τ, so

$$\frac{\partial}{\partial \tau} = \frac{\phi(\mathbf{x})}{\psi(\mathbf{x},t)}\frac{\partial}{\partial t} + \frac{1}{\psi(\mathbf{x},t)}\mathbf{b}(\mathbf{x},t) \cdot \nabla.$$

With this definition, system (11.97) becomes

$$\psi(\mathbf{x},t)\frac{\partial p}{\partial \tau} - \nabla \cdot (\mathbf{a}(\mathbf{x},t)\nabla p) + R(\mathbf{x},t)p = f(\mathbf{x},t),$$
$$\mathbf{x} \in \Omega, \ t > 0, \qquad (11.98)$$
$$p(\mathbf{x},0) = p_0(\mathbf{x}), \qquad \mathbf{x} \in \Omega.$$

We define the admissible function space, $d = 2$ or 3,

$$V = \left\{ \text{Functions } v: v \text{ is continuous on } \Omega, \frac{\partial v}{\partial x_1}, \frac{\partial v}{\partial x_2}, \ldots, \frac{\partial v}{\partial x_d} \text{ are} \right.$$
$$\left. \text{piecewise continuous and bounded on } \Omega, \text{ and } v \text{ is } \Omega\text{-periodic} \right\}.$$

Applying Green's formula (2.9) in space and the periodic boundary conditions, problem (11.98) can be written in the equivalent variational form:

$$\int_\Omega \psi\frac{\partial p}{\partial \tau}v \, d\mathbf{x} + \int_\Omega \mathbf{a}\nabla p \cdot \nabla v \, d\mathbf{x} + \int_\Omega Rpv \, d\mathbf{x} = \int_\Omega fv \, d\mathbf{x}, \quad v \in V, \ t > 0,$$
$$p(\mathbf{x},0) = p_0(\mathbf{x}), \qquad\qquad\qquad \mathbf{x} \in \Omega.$$
$$(11.99)$$

The characteristic is approximated by

$$\check{\mathbf{x}}_n = \mathbf{x} - \frac{\Delta t^n}{\phi(\mathbf{x})}\mathbf{b}(\mathbf{x},t^n). \qquad (11.100)$$

Furthermore, we see that, at $t = t^n$

$$\psi\frac{\partial p}{\partial \tau} \approx \psi(\mathbf{x},t^n)\frac{p(\mathbf{x},t^n) - p(\check{\mathbf{x}}_n,t^{n-1})}{(|\mathbf{x} - \check{\mathbf{x}}_n|^2 + (\Delta t^n)^2)^{1/2}}$$
$$(11.101)$$
$$= \phi(\mathbf{x})\frac{p(\mathbf{x},t^n) - p(\check{\mathbf{x}}_n,t^{n-1})}{\Delta t^n}.$$

A backtracking algorithm similar to that employed in one dimension is used to approximate the characteristic derivative (Fig. 11.20).

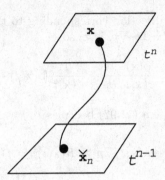

Figure 11.20 An illustration of the definition $\check{x}_n$.

Let $V_h \subset V$ be a finite element space associated with a regular partition K_h of Ω (see Chapters 5 and 6). The MMOC for problem (11.97) is given: for $n = 1, 2, \ldots$, find $p_h^n \in V_h$ such that

$$\int_\Omega \phi \frac{p_h^n - \breve{p}_h^{n-1}}{\Delta t^n} v \, d\mathbf{x} + \int_\Omega \mathbf{a}^n \nabla p_h^n \cdot \nabla v \, d\mathbf{x}$$
$$+ \int_\Omega R^n p_h^n v \, d\mathbf{x} = \int_\Omega f^n v \, d\mathbf{x} \qquad \forall v \in V_h, \tag{11.102}$$

where

$$\breve{p}_h^{n-1} = p_h \left(\check{\mathbf{x}}_n, t^{n-1} \right) = p_h \left(\mathbf{x} - \frac{\Delta t^n}{\phi(\mathbf{x})} \mathbf{b}(\mathbf{x}, t^n), t^{n-1} \right). \tag{11.103}$$

The remarks made earlier for the one-dimensional MMOC (11.92) also apply to system (11.102). In particular, existence and uniqueness of a solution for reasonable choices of $\mathbf{a}$ and R can be shown in an analogous way (see Exercise 11.20), and the error estimate (11.95) under appropriate assumptions on the solution p also holds for system (11.102) (Chen, 2005):

$$\max_{1 \le n \le N} \left(\left(\int_\Omega |p^n - p_h^n|^2 \, d\mathbf{x} \right)^{1/2} + h \left(\int_\Omega |\nabla p^n - \nabla p_h^n|^2 \, d\mathbf{x} \right)^{1/2} \right)$$

$$\le C(p)(h^{r+1} + \Delta t),$$

provided an approximation property similar to estimate (11.94) holds for V_h in the multiple dimensions.

Discussion of a conservation relation

We discuss the MMOC in the simple case where

$$R = f = 0, \quad \nabla \cdot \mathbf{b} = 0 \quad \text{in } \Omega. \tag{11.104}$$

That is, $\mathbf{b}$ is *divergence free* (or *solenoidal*). Application of condition (11.104), the periodicity assumption, and the divergence theorem (2.7) to Eq. (11.97) yields the *conservation relation*

$$\int_\Omega \phi(\mathbf{x}) p(\mathbf{x}, t) \, d\mathbf{x} = \int_\Omega \phi(\mathbf{x}) p_0(\mathbf{x}) \, d\mathbf{x}, \quad t > 0. \tag{11.105}$$

In applications, it is desirable to conserve at least a discrete form of this relation in any numerical approximation of problem (11.97). However, in general, the MMOC does not conserve it. To see this, we take $v = 1$ in Eq. (11.102) and apply relation (11.104) to give

$$\int_\Omega \phi(\mathbf{x}) p(\mathbf{x}, t^n) \, d\mathbf{x} = \int_\Omega \phi(\mathbf{x}) p(\check{\mathbf{x}}_n, t^{n-1}) \, d\mathbf{x}$$
$$\neq \int_\Omega \phi(\mathbf{x}) p(\mathbf{x}, t^{n-1}) \, d\mathbf{x}. \tag{11.106}$$

For each n, define the transformation

$$\mathbf{G}(\mathbf{x}) \equiv \mathbf{G}(\mathbf{x}, t^n) = \mathbf{x} - \frac{\Delta t^n}{\phi(\mathbf{x})} \mathbf{b}(\mathbf{x}, t^n). \tag{11.107}$$

We assume that $\mathbf{b}/\phi$ has bounded first partial derivatives in space. Then, for $d = 3$, the *Jacobian of this transformation*, $\mathbf{J}(\mathbf{G})$, is

$$\begin{pmatrix} 1 - \dfrac{\partial}{\partial x_1}\left(\dfrac{b_1^n}{\phi}\right)\Delta t^n & -\dfrac{\partial}{\partial x_2}\left(\dfrac{b_1^n}{\phi}\right)\Delta t^n & -\dfrac{\partial}{\partial x_3}\left(\dfrac{b_1^n}{\phi}\right)\Delta t^n \\[3mm] -\dfrac{\partial}{\partial x_1}\left(\dfrac{b_2^n}{\phi}\right)\Delta t^n & 1 - \dfrac{\partial}{\partial x_2}\left(\dfrac{b_2^n}{\phi}\right)\Delta t^n & -\dfrac{\partial}{\partial x_3}\left(\dfrac{b_2^n}{\phi}\right)\Delta t^n \\[3mm] -\dfrac{\partial}{\partial x_1}\left(\dfrac{b_3^n}{\phi}\right)\Delta t^n & -\dfrac{\partial}{\partial x_2}\left(\dfrac{b_3^n}{\phi}\right)\Delta t^n & 1 - \dfrac{\partial}{\partial x_1}\left(\dfrac{b_3^n}{\phi}\right)\Delta t^n \end{pmatrix},$$

and its determinant equals (see Exercise 11.21)

$$|\mathbf{J}(\mathbf{G})| = 1 - \nabla \cdot \left(\frac{\mathbf{b}^n}{\phi}\right)\Delta t^n + \mathcal{O}((\Delta t^n)^2). \tag{11.108}$$

Thus, even in the case where ϕ is constant, for the second equality of Eq. (11.106) to hold, it requires that the Jacobian of the transformation (11.107) be identically one. While this is true for constant ϕ and $\mathbf{b}$, it cannot be expected to be true for variable coefficients. In the case where ϕ is constant and $\nabla \cdot \mathbf{b} = 0$, it follows from relation (11.108) that the determinant of this transformation is $1 + \mathcal{O}((\Delta t^n)^2)$, so a systematic error of size $\mathcal{O}((\Delta t^n)^2)$ should be expected. On the other hand, if $\nabla \cdot (\mathbf{b}/\phi) \neq 0$, the determinant is $1 + \mathcal{O}(\Delta t^n)$ and a systematic error of size $\mathcal{O}(\Delta t^n)$ can occur. In particular, in using the MMOC in the solution of a two-phase immiscible flow problem (see Chapter 10), Douglas *et al.* (1997) found that conservation of mass failed by as much as 10%, in simulations with stochastic rock properties and about half that much with uniform rock properties. Errors of this magnitude obscure the relevance of numerical approximations to an unacceptable level and have motivated the search for a modification of the MMOC that both conserves relation (11.105) and is a little more expensive computationally than the MMOC. Another method, the *modified method of characteristics with adjusted advection*, was defined by Douglas *et al.* (1997) that satisfies these criteria. This method is derived from the MMOC by perturbing the foot of characteristics in an ad hoc fashion. We do not introduce this method in this chapter. Instead, we describe the *Eulerian–Lagrangian localized adjoint method* (ELLAM) (Celia *et al.*, 1990; Russell, 1990).

11.6.2. The Eulerian–Lagrangian localized adjoint method

We consider the ELLAM for problem (11.86) in divergence form:

$$
\begin{aligned}
\frac{\partial(\phi p)}{\partial t} + \nabla \cdot (\mathbf{b}p - \mathbf{a}\nabla p) + Rp &= f, & \mathbf{x} \in \Omega,\ t > 0, \\
(\mathbf{b}p - \mathbf{a}\nabla p) \cdot \boldsymbol{\nu} &= g, & \mathbf{x} \in \Gamma,\ t > 0, \\
p(\mathbf{x}, 0) &= p_0(\mathbf{x}), & \mathbf{x} \in \Omega,
\end{aligned}
\tag{11.109}
$$

where $\Omega \subset \mathbb{R}^d$ ($d \leq 3$) is a bounded domain and $\phi = \phi(\mathbf{x}, t)$ and $\mathbf{b} = \mathbf{b}(\mathbf{x}, t)$ are now variable. We consider a flux boundary condition in this problem; an extension to Dirichlet conditions is possible (Chen, 2005).

For any $\mathbf{x} \in \Omega$ and two times $0 \leq t^{n-1} < t^n$, the hyperbolic part of problem (11.109), $\phi \partial p/\partial t + \mathbf{b} \cdot \nabla p$, defines the characteristic $\check{\mathbf{x}}_n(\mathbf{x}, t)$ along the interstitial velocity $\varphi = \mathbf{b}/\phi$ (Fig. 11.20):

$$\frac{\partial}{\partial t}\check{\mathbf{x}}_n = \varphi(\check{\mathbf{x}}_n, t), \quad t \in J^n,$$

$$\check{\mathbf{x}}_n(\mathbf{x}, t^n) = \mathbf{x}. \tag{11.110}$$

In general, the characteristics in system (11.110) can be determined only approximately. There are many methods to solve this first-order ordinary differential equation for the approximate characteristics. We consider only the Euler method.

The Euler method to solve system (11.110) for the approximate characteristics is: For any $\mathbf{x} \in \Omega$,

$$\check{\mathbf{x}}_n(\mathbf{x}, t) = \mathbf{x} - \varphi(\mathbf{x}, t^n)(t^n - t), \quad t \in [\check{t}(\mathbf{x}), t^n], \tag{11.111}$$

where $\check{t}(\mathbf{x}) = t^{n-1}$ if $\check{\mathbf{x}}_n(\mathbf{x}, t)$ does not backtrack to the boundary Γ for $t \in [t^{n-1}, t^n]$; $\check{t}(\mathbf{x}) \in J^n = (t^{n-1}, t^n]$ is the time instant when $\check{\mathbf{x}}_n(\mathbf{x}, t)$ intersects Γ, i.e., $\check{\mathbf{x}}_n(\mathbf{x}, \check{t}(\mathbf{x})) \in \Gamma$, otherwise. Let

$$\Gamma_+ = \{\mathbf{x} \in \Gamma : (\mathbf{b} \cdot \boldsymbol{\nu})(\mathbf{x}) \geq 0\}.$$

For $(\mathbf{x}, t) \in \Gamma_+ \times J^n$, the approximate characteristic emanating backward from $(\mathbf{x}, t)$ is

$$\check{\mathbf{x}}_n(\mathbf{x}, \theta) = \mathbf{x} - \varphi(\mathbf{x}, t)(t - \theta), \quad \theta \in [\check{t}(\mathbf{x}, t), t], \tag{11.112}$$

where $\check{t}(\mathbf{x}, t) = t^{n-1}$ if $\check{\mathbf{x}}_n(\mathbf{x}, \theta)$ does not backtrack to the boundary Γ for $\theta \in [t^{n-1}, t]$; $\check{t}(\mathbf{x}, t) \in (t^{n-1}, t]$ is the time instant when $\check{\mathbf{x}}_n(\mathbf{x}, \theta)$ intersects Γ, otherwise.

If Δt^n is sufficiently small (depending upon the smoothness of φ), the approximate characteristics do not cross each other, which is assumed. Then $\check{\mathbf{x}}_n(\cdot, t)$ is a one-to-one mapping of $\mathbb{R}^d$ to $\mathbb{R}^d$ ($d \leq 3$); we indicate its inverse by $\hat{\mathbf{x}}_n(\cdot, t)$.

For any $t \in J^n$, we define

$$\tilde{\varphi}(\mathbf{x}, t) = \varphi(\hat{\mathbf{x}}_n(\mathbf{x}, t), t^n), \quad \tilde{\mathbf{b}} = \tilde{\varphi}\phi. \tag{11.113}$$

We assume that $\tilde{\mathbf{b}} \cdot \boldsymbol{\nu} \geq 0$ on Γ_+.

Let K_h be a partition of Ω into elements $\{K\}$. For each $K \in K_h$, let $\check{K}(t)$ represent the trace-back of K to time t, $t \in J^n$:

$$\check{K}(t) = \{\mathbf{x} \in \Omega : \mathbf{x} = \check{\mathbf{x}}_n(\mathbf{y}, t) \text{ for some } \mathbf{y} \in K\},$$

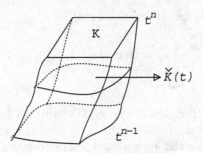

Figure 11.21 An illustration of $\mathcal{K}^n$.

and $\mathcal{K}^n$ be the space-time region that follows the characteristics (Fig. 11.21):

$$\mathcal{K}^n = \{(\mathbf{x}, t) \in \Omega \times J : t \in J^n \text{ and } \mathbf{x} \in \check{K}(t)\}.$$

Also, define $\mathcal{B}^n = \{(\mathbf{x}, t) \in \partial \mathcal{K}^n : \mathbf{x} \in \partial \Omega\}$.

We write the hyperbolic part of problem (11.109) as

$$\frac{\partial(\phi p)}{\partial t} + \nabla \cdot (\mathbf{b}p) = \frac{\partial(\phi p)}{\partial t} + \nabla \cdot (\tilde{\mathbf{b}}p) + \nabla \cdot ([\mathbf{b} - \tilde{\mathbf{b}}]p). \qquad (11.114)$$

With $\boldsymbol{\tau}(x, t) = (\tilde{\mathbf{b}}, \phi)$ and a smooth test function $v(\mathbf{x}, t)$, application of Green's formula in space and time gives (see Exercise 11.22)

$$\int_{\mathcal{K}^n} \left(\frac{\partial(\phi p)}{\partial t} + \nabla \cdot (\tilde{\mathbf{b}}p) \right) v \, d\mathbf{x} \, dt$$

$$= \int_K \phi^n p^n v^n \, d\mathbf{x} - \int_{\check{K}(t^{n-1})} \phi^{n-1} p^{n-1} v^{n-1,+} \, d\mathbf{x} \qquad (11.115)$$

$$+ \int_{\mathcal{B}^n} p\tilde{\mathbf{b}} \cdot \boldsymbol{\nu} v \, d\ell - \int_{\mathcal{K}^n} p\boldsymbol{\tau} \cdot \left(\nabla v, \frac{\partial v}{\partial t} \right) \, d\mathbf{x} \, dt,$$

where we used the fact that $\boldsymbol{\tau} \cdot \boldsymbol{\nu}_{\mathcal{K}^n} = 0$ on the space–time edges $(\partial \mathcal{K}^n \cap (\check{K} \times J^n)) \setminus \mathcal{B}^n$ and $v^{n-1,+} = v(x, t^{n-1,+}) = \lim_{\epsilon \to 0^+} v(x, t^{n-1} + \epsilon)$ to take into account the fact that $v(x, t)$ can be discontinuous at the time levels.

Similarly, the diffusion part of problem (11.109) gives

$$\int_{\mathcal{K}^n} \nabla \cdot (\mathbf{a}\nabla p) v \, d\mathbf{x} \, dt$$

$$= \int_{J^n} \left\{ \int_{\partial \check{K}(t)} \mathbf{a}\nabla p \cdot \boldsymbol{\nu}_{\check{K}(t)} v \, d\ell - \int_{\check{K}(t)} (\mathbf{a}\nabla p) \cdot \nabla v \, d\mathbf{x} \right\} dt. \qquad (11.116)$$

We assume that the test function $v(\mathbf{x}, t)$ is constant along the approximate characteristics. Then, combining Eqs. (11.114)–(11.116) yields the space–time variational formulation of problem (11.109)

$$
\int_\Omega \phi^n p^n v^n \, d\mathbf{x} - \int_\Omega \phi^{n-1} p^{n-1} v^{n-1,+} \, d\mathbf{x}
$$
$$
+ \int_{J^n} \left\{ \int_\Omega \mathbf{a} \nabla p \cdot \nabla v \, d\mathbf{x} + \int_\Omega R p v \, d\mathbf{x} \right\} dt
$$
$$
= \int_{J^n} \left\{ \int_\Omega f v \, d\mathbf{x} - \int_\Gamma g v \, d\ell \right\} dt
$$
$$
+ \int_{J^n} \left\{ \int_\Omega \nabla \cdot \left[(\tilde{\mathbf{b}} - \mathbf{b}) p \right] \hat{v} \, d\mathbf{x} - \int_\Gamma p \left[\tilde{\mathbf{b}} - \mathbf{b} \right] \cdot \boldsymbol{\nu} v \, d\ell \right\} dt.
$$
$$(11.117)$$

If we apply backward Euler time integration along characteristics to the diffusion, reaction, and source term in Eq. (11.117), we see that

$$
\int_\Omega \left(\phi^n p^n v^n + \Delta t^n \mathbf{a}^n \nabla p^n \cdot \nabla v^n + \Delta t^n R^n p^n v^n \right) \, d\mathbf{x}
$$
$$
= \int_\Omega \left(\phi^{n-1} p^{n-1} v^{n-1,+} + \Delta t^n f^n v^n \right) \, d\mathbf{x} - \int_{J^n} \int_\Gamma g v \, d\ell dt
$$
$$
+ \int_{J^n} \left\{ \int_\Omega \nabla \cdot \left[(\tilde{\mathbf{b}} - \mathbf{b}) p \right] \hat{v} \, d\mathbf{x} - \int_\Gamma p \left[\tilde{\mathbf{b}} - \mathbf{b} \right] \cdot \boldsymbol{\nu} v \, d\ell \right\} dt,
$$
$$(11.118)$$

where $\Delta t^n(\mathbf{x}) = t^n - \check{t}(\mathbf{x})$. The $\mathbf{x}$-dependent Δt^n seems quite appropriate, since the diffusion at each point is weighted by the length of time over which it acts.

Let $V_h \subset V$ be a finite element space (see Chapters 1–6). For any $w \in V_h$, we define a test function $v(\mathbf{x}, t)$ to be a constant extension of $w(\mathbf{x})$ into the space–time region $\Omega \times J^n$ along the approximate characteristics (Eqs. (11.111) and (11.112)):

$$
v(\check{\mathbf{x}}_n(\mathbf{x}, t), t) = w(\mathbf{x}), \quad t \in [\check{t}(\mathbf{x}), t^n], \ \mathbf{x} \in \Omega,
$$
$$
v(\check{\mathbf{x}}_n(\mathbf{x}, \theta), \theta) = w(\mathbf{x}), \quad \theta \in [\check{t}(\mathbf{x}, t), t], \ (\mathbf{x}, t) \in \Gamma_+ \times J^n.
$$
$$(11.119)$$

Now, based on Eq. (11.118), an ELLAM procedure is defined: for $n = 1, 2, \ldots$, find $p_h^n \in V_h$ such that

$$
\int_\Omega \left(\phi^n p_h^n v^n + \Delta t^n \mathbf{a}^n \nabla p_h^n \cdot \nabla v^n + \Delta t^n R^n p_h^n v^n \right) \, d\mathbf{x}
$$
$$(11.120)$$
$$
= \int_\Omega \left(\phi^{n-1} p_h^{n-1} v^{n-1,+} + \Delta t^n f^n v^n \right) \, d\mathbf{x} - \int_{J^n} \int_\Gamma g v \, d\ell dt.
$$

The test function $v = 1$ in Eq. (11.120) yields the statement of global mass conservation. The remarks made earlier on accuracy and efficiency of the MMOC also apply to relation (11.120) (see Exercise 11.23). In particular, when V_h is the space of piecewise linear functions defined on a regular triangulation K_h, the following convergence result holds (Wang, 2000).

Assume that Ω is a convex polygonal domain or has a smooth boundary Γ, and the coefficients $\mathbf{a}$, $\mathbf{b}$, ϕ, f, and R are sufficiently smooth. If the solution p to problem (11.109) and its temporal derivative $\partial p/\partial t$ are also smooth, the initialization error satisfies

$$\left(\int_\Omega |p_0 - p_h^0|^2 \, d\mathbf{x} \right)^{1/2} \leq C(p_0)h^2,$$

and Δt is sufficiently small, then

$$\max_{1 \leq n \leq N} \left(\int_\Omega |p^n - p_h^n|^2 \, d\mathbf{x} \right)^{1/2} \leq C(p, f, p_0)(\Delta t + h^2),$$

where p_h is the solution of Eq. (11.120).

With advection on the right-hand side of Eq. (11.120) only, the linear system arising from this equation is well suited for iterative linear solution algorithms in multiple space dimensions. The characteristic idea can be combined with other finite element methods presented in Secs. 11.1–11.5; see Yang (1992) and Arbogast and Wheeler (1995) for characteristic mixed methods and Chen (2002b) for Eulerian–Lagrangian discontinuous methods.

11.7. The adaptive finite element method

In engineering problems, many important physical and chemical phenomena are sufficiently localized and transient that *adaptive numerical methods* are necessary to resolve them. Adaptive numerical methods have become increasingly important because researchers have realized the great potential of the concepts underlying these methods. They are numerical schemes that automatically adjust themselves to improve approximate solutions. These methods are not exactly new in the computational area even in the finite element literature. The adaptive adjustment of time steps in the numerical solution of ordinary differential equations has been the subject of research for many decades. Furthermore, the search for optimal finite element grids dates back to the early 1970s (Oliveira, 1971). But modern interest in this subject began in the late 1970s, mainly thanks to important contributions by Babuška and Rheinboldt (1978a,b) and many others.

The overall accuracy of numerical approximations often deteriorates due to local singularities like those arising from reentrant corners of domains, interior or boundary layers, and sharp moving fronts. An obvious strategy is to refine the grids near these critical regions, i.e., to insert more grid points near where the singularities occur. The question is then how we identify those regions, refine them, and obtain a good balance between the refined and unrefined regions such that the overall accuracy is optimal. To answer this question, we need to utilize *adaptivity*. That is, we need somehow to *restructure a numerical scheme* to improve the quality of its approximate solutions. This puts a great demand on the choice of numerical methods. Restructuring a numerical scheme includes changing the number of elements, refining local grids, increasing the local order of approximation, moving nodal points, and modifying algorithm structures.

Another closely related question is how to obtain reliable estimates of the accuracy of computed approximate solutions. *A priori* error estimates, as obtained so far, are often insufficient because they produce information only on the asymptotic behavior of errors and they require a solution regularity that is not satisfied in the presence of the above mentioned singularities. To answer this question, we need to assess the quality of approximate solutions *a posteriori*, i.e., after an initial approximation is obtained. This requires that we compute *a posteriori* error estimates. Of course, the computation of the *a posteriori* estimates should be far less expensive than that of the approximate solutions. Moreover, it must be possible to compute *dynamically* local error indicators that lead to some estimate of the local quality of the solution.

The aim of this section is to present a brief introduction of some of basic topics on the two components of the adaptive finite element methods: the *adaptive strategy* and *a posteriori error estimation*. We focus on these two components for the standard finite element methods considered in Chapters 1–7.

11.7.1. Local grid refinement in space

There are three basic types of adaptive strategies: (1) local refinement of a fixed grid, (2) addition of more degrees of freedom locally by utilizing higher-order basis functions in certain elements, and (3) adaptively moving a computational grid to achieve better local resolution.

Local grid refinement of a fixed grid is called an *h-scheme*. In this scheme, the mesh is automatically refined or unrefined depending upon a local error indicator. Such a scheme leads to a very complex data management problem because it involves the dynamic regeneration of a grid, renumbering of nodal points and elements, and element connectivity.

However, the h-scheme can be very effective in generating near-optimal grids for a given error tolerance. Efficient h-schemes with fast data management procedures have been developed for complex problems (Diaz *et al.*, 1984; Ewing, 1986; Bank, 1990). Moreover, the h-scheme can also be employed to *unrefine* a grid (or *coarsen* a grid) when a local error indicator becomes smaller than a preassigned tolerance.

The addition of more degrees of freedom locally by utilizing higher-order basis functions in certain elements is referred to as a *p-scheme* (Babuška *et al.*, 1983; Szabo, 1986). As discussed in the previous chapters, finite element methods for a given problem attempt to approximate a solution by functions in a finite-dimensional space of polynomials. The p-scheme generally utilizes a fixed grid and a fixed number of grid elements. If the error indicator in any element exceeds a given tolerance, the local order of the polynomial degree is increased to reduce the error. This scheme can be very effective in modeling thin boundary layers around bodies moving in a flow field, where the use of very fine grids is impractical and costly. However, the data management problem associated with the p-scheme, especially for regions of complex geometry, can be very difficult.

Adaptively moving a computational grid to get better local resolution is usually termed an *r-scheme* (Miller and Miller, 1981). It employs a fixed number of grid points and attempts to move them dynamically to areas where the error indicator exceeds a preassigned tolerance. The r-scheme can be easily implemented, and does not have the difficult data management problem associated with the h- and p-schemes. On the other hand, it suffers from several deficiencies. Without special care in its implementation, it can be unstable and result in grid tangling and local degradation of approximate solutions. It can never reduce the error below a fixed limit since it is not capable of handling the migration of regions where the solution is singular. However, by an appropriate combination with other adaptive strategies, the r-scheme can lead to a useful scheme for controlling solution errors.

Combinations of these three basic strategies such as the *hr-, hp-,* and *hpr-schemes* are also possible (Babuška and Dorr, 1981; Oden *et al.*, 1989). In this chapter, as an example, we study the widely applied h-scheme.

Regular h-schemes

We focus on a two-dimensional domain. An extension of the concept in this section to three dimensions is simple to visualize. However, the modification of the supporting algorithms in the next section is not straightforward.

In the two-dimensional case, a grid can be triangular, quadrilateral, or of mixed-type (i.e., consisting of both triangles and quadrilaterals; see Chapter 5). A vertex is *regular* if it is a vertex of each of its neighboring

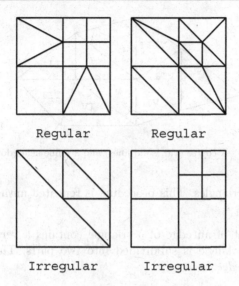

Regular   Regular

Irregular   Irregular

Figure 11.22   Examples of regular and irregular vertices.

elements, and a grid is *regular* if its every vertex is regular. All other vertices are said to be *irregular* (*slave nodes* or *hanging nodes*; see Fig. 11.22). The *irregularity index* of a grid is the maximum number of irregular vertices belonging to the same edge of an element.

If all elements in a grid are subdivided into an equal number (usually four) of smaller elements simultaneously, the refinement is referred to as *global*. For example, a refinement is global by connecting the opposite midpoints of the edges of each triangle or quadrilateral in the grid. Global refinement does not introduce irregular vertices. In all the previous sections except in Sec. 11.5, all the refinements were global and regular. In contrast, in the case of a *local refinement* where only some of the elements in a grid are subdivided into smaller elements, irregular vertices may appear (Fig. 11.22).

In this section, we study a regular local refinement. The following *refinement rule* can be used to convert irregular vertices to regular ones (Bank, 1990; Braess, 1997). This rule is designed for a triangular grid and guarantees that each of the angles in the original grid is bisected at most once. We may think of starting with a triangulation as in Fig. 11.23. It contains six irregular vertices, which need to be converted to regular vertices.

A refinement rule for a triangulation is defined as follows:

1. If an edge of a triangle contains two or more vertices of other triangles (not counting its own vertices), then this triangle is subdivided into four

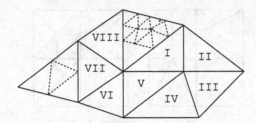

Figure 11.23   A coarse grid (solid lines) and a refinement (dotted lines).

equal smaller triangles. This procedure is repeated until such triangles no longer exist.

2. If the midpoint of an edge of a triangle contains a vertex of another triangle, this triangle is subdivided into two parts. The new edge is called a *green edge*.

3. If a further refinement is needed, the green edges are first eliminated before the next iteration.

For the triangulation in Fig. 11.23, we apply the first step to triangles I and VIII. This requires the use of the refinement rule twice on triangle VII. Next, we construct green edges on triangles II, V, and VI and on three subtriangles (see Exercise 11.24).

Despite its recursive nature, this procedure stops after a finite number of iterations. Let k be the maximum number of levels in the underlying refinement, where the maximum is taken over all elements ($k = 2$ in Fig. 11.23). Then, every element is subdivided at most k times, which presents an upper bound on the number of times step 1 is used. We emphasize that this procedure is purely two-dimensional. A generalization to three dimensions is not straightforward. For a triangulation of the domain Ω into tetrahedra, see a technique due to Rivara (1984a).

Irregular h-schemes

Irregular grids leave more freedom for local refinement. In the general case of arbitrary irregular grids, an element may be refined locally without any interference from its neighbors. As for regular local refinements, some desirable properties should be preserved for irregular refinements.

First, in the process of consecutive refinements no distorted elements should be generated. That is, the minimal angle of every element should be bounded away from zero by a common bound that probably depends only on the initial grid (see inequality (2.52)).

Second, a new grid resulting from a local refinement should contain all the nodes of the old grid. In particular, if continuous finite element spaces $\{V_{h_k}\}$ are exploited for a second-order partial differential problem in all levels, consecutive refinements should lead to a *nested* sequence of these spaces:

$$V_{h_1} \subset V_{h_2} \subset \cdots \subset V_{h_k} \subset V_{h_{k+1}} \subset \cdots,$$

where $h_{k+1} < h_k$ and h_k is the mesh size at the kth grid level. In the case of irregular local refinements, to preserve continuity of functions in these spaces the function values at the irregular nodes of a new grid are obtained by *polynomial interpolation* of the values at the old grid nodes.

Third, as defined before, the *irregularity index* of a grid is the maximum number of irregular vertices belonging to an edge of an element. There are reasons to restrict ourselves to 1-irregular grids. In practice, it seems to be very unlikely that grids with a higher irregularity index can be useful for a local h-scheme. In general, the stiffness matrix arising from the finite element discretization of a problem should be sparse. It turns out that the sparsity cannot be guaranteed for a general irregular grid (Bank *et al.*, 1983). To produce 1-irregular grids, we can employ the *1-irregular rule*: Refine any unrefined element for which any of the edges contains more than one irregular node.

Unrefinements

As noted, an h-scheme can be also employed to *unrefine* a grid. There are two factors that decide if an element needs to be unrefined: (1) a local error indicator and (2) a structural condition imposed on the grid resulting from the regularity or 1-irregularity requirement. Both of these factors must be examined before an element is unrefined.

When an element is refined, it produces a number of new smaller elements; the old element is called a *father* and the smaller ones are termed its *sons*. A *tree structure* (or *family structure*) consists of remembering for each element its father (if there is one) and its sons. Figure 11.24 shows a typical tree structure, together with a corresponding current grid generated by consecutive refinements of a single square. The *root* of the tree originates at the initial element and the *leaves* are those elements being not refined.

The tree structure provides for easy and fast unrefinements. When the tree information is stored, a local unrefinement can be done by simply "cutting the corresponding branch" of the tree, i.e., unrefining previously refined elements and restoring locally the previous grid.

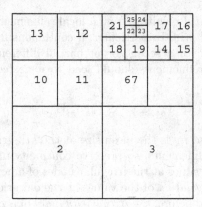

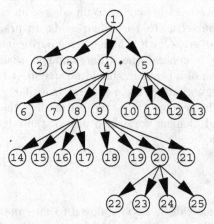

Figure 11.24   A local refinement and the corresponding tree structure.

11.7.2. Data structures

In the finite element methods developed in the previous chapters and the previous sections in this chapter, all elements and nodes are usually numbered in a consecutive fashion so that a minimal band in the stiffness matrix of a finite element system can be produced. When a computational code identifies an element to evaluate its contribution to this matrix, the minimal information required is the set of node numbers corresponding to this element (see Sec. 2.5).

Adaptive local refinements and unrefinements require much more complex data structures than the classical global ones in Sec. 2.5. When elements and nodes are added and deleted adaptively, it is often impossible to number them in a consecutive fashion. Hence, we need to establish some kind of *natural order of elements*. In particular, all elements must be placed in an order and a code must recognize, for a given element, the next element (or the previous element if necessary) in the sequence. Therefore, for an element, the following information should be stored:

- nodes,

- neighbors,

- father,

- sons, and

- level of refinement.

For a given node, its coordinates are also needed. The logic of a data structure corresponding to a particular local refinement may need additional information. The above listed information seems to be the minimal requirement for all existing data structures. Several data structures are available for adaptive local grid refinements and unrefinements (Rheinboldt and Mesztenyi, 1980; Bank *et al.*, 1983; Rivara, 1984b).

11.7.3. *A posteriori* error estimates

We now study the second component of the adaptive finite element method: *a posteriori* error estimation. *A posteriori error estimators* and *indicators* can be utilized to give a specific assessment of errors and to form a solid basis for local refinements and unrefinements. *A posteriori* error estimators can be roughly classified as follows (Verfürth, 1996):

1. *Residual estimators*. These estimators bound the error of the computed approximate solution by a suitable norm of its residual with respect to the strong form of a differential equation (Babuška and Rheinboldt, 1978a).

2. *Local problem-based estimators*. This approach solves locally discrete problems, which are similar to, but simpler than the original problem, and uses appropriate norms of the local solutions for error estimation (Babuška and Rheinboldt, 1978b; Bank and Weiser, 1985).

3. *Averaging-based estimators*. This approach utilizes some local extrapolation or averaging technique to define error estimation (Zienkiewicz and Zhu, 1987).

4. *Hierarchical basis estimators.* This approach calculates the residual of the computed approximate solution with respect to another finite element space of higher-order polynomials or with respect to a refined grid (Deuflhard *et al.*, 1989).

As an example, we briefly study the residual estimators for the model problem in two dimensions for the solution p:

$$-\Delta p = f \quad \text{in } \Omega,$$
$$p = 0 \quad \text{on } \Gamma_D, \qquad\qquad (11.121)$$
$$\frac{\partial p}{\partial \boldsymbol{\nu}} = g \quad \text{on } \Gamma_N,$$

where Ω is a bounded domain in the plane with boundary $\bar{\Gamma} = \bar{\Gamma}_D \cup \bar{\Gamma}_N$, $\Gamma_D \cap \Gamma_N = \emptyset$, f and g are given functions, and the *Laplacian operator* Δ is defined as in Sec. 2.1.

Assume that the Dirichlet boundary part Γ_D is closed relative to the entire boundary Γ and has a positive length. We recall the admissible function space

$$V = \left\{ \text{Functions } v\colon v \text{ is continuous on } \Omega, \frac{\partial v}{\partial x_1} \text{ and } \frac{\partial v}{\partial x_2} \text{ are} \right.$$

$$\left. \text{piecewise continuous and bounded on } \Omega, \text{ and } v = 0 \text{ on } \Gamma_D \right\}.$$

Also, introduce the notation (the bilinear and linear forms; see Chapter 3)

$$a(p,v) = \int_\Omega \nabla p \cdot \nabla v \, d\mathbf{x}, \quad L(v) = \int_\Omega fv \, d\mathbf{x} + \int_{\Gamma_N} gv \, d\ell, \quad v \in V.$$

As in Eq. (2.10), problem (11.121) can be recast in the variational form:

Find $p \in V$ such that $a(p,v) = L(v) \quad \forall v \in V.$ $\qquad (11.122)$

Let Ω be a convex polygonal domain (or its boundary Γ be smooth), and let K_h be a triangulation of Ω into triangles K of diameter h_K, as in Sec. 2.2. To the triangulation K_h, we associate a grid function $h(\mathbf{x})$ such that, for some positive constant C_1,

$$C_1 h_K \leq h(\mathbf{x}) \leq h_K \quad \forall \mathbf{x} \in K, \quad K \in K_h. \qquad (11.123)$$

Moreover, assume that there exists a positive constant C_2 such that

$$C_2 h_K^2 \leq |K| \quad \forall K \in K_h, \qquad (11.124)$$

where $|K|$ is the area of the triangle K. Recall that inequality (11.124) is the *minimum angle* condition stating that the angles of triangles in K_h are bounded below by C_2 (see inequality (2.52)).

To keep the notation to a minimum, let the finite element space $V_h \subset V$ be defined by

$$V_h = \{v \in V : v|_K \in P_1(K), \ K \in K_h\}.$$

An extension to finite element spaces of higher-order polynomials will be noted at the end of this section. The finite element method for problem (11.121) is formulated:

$$\text{Find } p_h \in V_h \text{ such that } a(p_h, v) = L(v) \quad \forall v \in V_h. \tag{11.125}$$

It follows from Eqs. (11.122) and (11.125) that

$$a(p - p_h, v) = L(v) - a(p_h, v) \quad \forall v \in V. \tag{11.126}$$

The right-hand side of Eq. (11.126) implicitly defines the *residual* of the approximate solution p_h as an element in the dual space of the admissible space V. Because the Dirichlet boundary part Γ_D has a positive length, *Poincaré's inequality* holds (Chen, 2005):

$$\left(\int_\Omega |v|^2 \, d\mathbf{x} \right)^{1/2} \leq C(\Omega) \left(\int_\Omega |\nabla v|^2 \, d\mathbf{x} \right)^{1/2} \quad \forall v \in V, \tag{11.127}$$

where C depends on the domain Ω and the length of Γ_D. Set the notation (the norm of the admissible space)

$$\|v\|_V = \left(\int_\Omega (|v|^2 + |\nabla v|^2) \, d\mathbf{x} \right)^{1/2}, \quad \forall v \in V.$$

Using inequality (11.127) and Cauchy's inequality (1.30), we have

$$\frac{1}{1 + C^2(\Omega)} \|v\|_V \leq \sup\{a(v, w) : w \in V, \ \|w\|_V = 1\} \leq \|v\|_V. \tag{11.128}$$

Consequently, it follows from Eq. (11.126) and inequality (11.128) that

$$\sup\{L(v) - a(p_h, v) : v \in V, \ \|v\|_V = 1\}$$

$$\leq \|p - p_h\|_V \tag{11.129}$$

$$\leq \left(1 + C^2(\Omega)\right) \sup\{L(v) - a(p_h, v) : v \in V, \ \|v\|_V = 1\}.$$

Since the supremum term in inequality (11.129) is equivalent to the norm of the residual in the dual space of V, this inequality implies that the norm in the space V of the error is, up to multiplicative constants, bounded from above and below by the norm of the residual in the dual space of V. Most a *posteriori* error estimators attempt to bound this dual norm of the residual by quantities that can be more easily evaluated from f, g, and p_h.

Let $\mathcal{E}_h^o$ denote the set of all interior edges e in the triangulation K_h, $\mathcal{E}_h^b$ the set of the edges e on the boundary Γ, and $\mathcal{E}_h = \mathcal{E}_h^o \cup \mathcal{E}_h^b$. Furthermore, let $\mathcal{E}_h^D$ and $\mathcal{E}_h^N$ be the sets of edges e on Γ_D and Γ_N, respectively.

With each edge $e \in \mathcal{E}_h$, we associate a unit normal vector $\boldsymbol{\nu}$. For $e \in \mathcal{E}_h^b$, $\boldsymbol{\nu}$ is just the outward unit normal to Γ. For $e \in \mathcal{E}_h^o$, with $e = \bar{K}_1 \cap \bar{K}_2$, $K_1, K_2 \in K_h$, the direction of $\boldsymbol{\nu}$ is associated with the definition of jumps across e; if the jump of any function v across e is defined by

$$[v] = (v|_{K_2})|_e - (v|_{K_1})|_e, \tag{11.130}$$

then $\boldsymbol{\nu}$ is defined as the unit normal exterior to the element K_2 (Fig. 11.25).

Note that, by Green's formula (2.9), the definition of the linear form $L(\cdot)$, and the fact that $\Delta p_h = 0$ on all $K \in K_h$,

$$L(v) - a(p_h, v)$$

$$= L(v) - \sum_{K \in K_h} \int_K \nabla p_h \cdot \nabla v \, d\mathbf{x}$$

$$= L(v) - \sum_{K \in K_h} \left[\int_{\partial K} \nabla p_h \cdot \boldsymbol{\nu}_K v \, d\ell - \int_K \Delta p_h v \, d\mathbf{x} \right] \tag{11.131}$$

$$= \int_\Omega fv \, d\mathbf{x} + \sum_{e \in \mathcal{E}_h^N} \int_e (g - \nabla p_h \cdot \boldsymbol{\nu}) v \, d\ell - \sum_{e \in \mathcal{E}_h^o} \int_e [\nabla p_h \cdot \boldsymbol{\nu}] v \, d\ell.$$

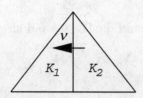

Figure 11.25   An illustration of $\boldsymbol{\nu}$.

Applying inequality (11.129) and Eq. (11.131), one can show that (see Exercise 11.25)

$$\|p - p_h\|_V \leq C \Bigg\{ \sum_{K \in K_h} h_K^2 \int_K f^2 \, d\mathbf{x} + \sum_{e \in \mathcal{E}_h^N} h_e \int_e |g - \nabla p_h \cdot \boldsymbol{\nu}|^2 \, d\ell$$

$$+ \sum_{e \in \mathcal{E}_h^o} h_e \int_e \|[\nabla p_h \cdot \boldsymbol{\nu}]\|^2 \, d\ell \Bigg\}^{1/2},$$

(11.132)

where C depends on C_2 in inequality (11.124) and $C(\Omega)$ in inequality (11.127), and h_K and h_e represent the diameter and length, respectively, of the triangle K and the edge e.

The right-hand side in inequality (11.132) can be utilized as an *a posteriori* error estimator because it involves only the known data f and g, the approximate solution p_h, and the geometrical data of the triangulation K_h. For general functions f and g, the exact computation of the integrals in the first and second terms of the right-hand side of (11.132) is often impossible. These integrals must be approximated by appropriate quadrature formulas (see Sec. 6.7). On the other hand, it is also possible to approximate f and g by polynomials in suitable finite element spaces. Both approaches, numerical quadrature and approximation by simpler functions combined with exact integration of the latter functions, are often equivalent and generate analogous *a posteriori* estimators. We restrict ourselves to the simpler function approximation approach. In particular, let f_h and g_h be the L^2-projections of f and g into the spaces of piecewise constants with respect to K_h and $\mathcal{E}_h^N$, respectively; i.e., on each $K \in K_h$ and $e \in \mathcal{E}_h^N$, $f_K = f_h|_K$ and $g_e = g_h|_e$ are given by the local mean values

$$f_K = \frac{1}{|K|} \int_K f \, d\mathbf{x}, \quad g_e = \frac{1}{h_e} \int_e g \, d\ell. \tag{11.133}$$

Then, we define a *residual a posteriori error estimator*:

$$\mathcal{R}_K = \Bigg\{ h_K^2 \int_K |f_K|^2 \, d\mathbf{x} + \sum_{e \in \partial K \cap \mathcal{E}_h^N} h_e \int_e |g_e - \nabla p_h \cdot \boldsymbol{\nu}|^2 \, d\ell$$

$$+ \frac{1}{2} \sum_{e \in \partial K \cap \mathcal{E}_h^o} h_e \int_e \|[\nabla p_h \cdot \boldsymbol{\nu}]\|^2 \, d\ell \Bigg\}^{1/2}. \tag{11.134}$$

The first term in $\mathcal{R}_K$ is related to the residual of the approximate solution p_h with respect to the strong form of the differential equation. The second and third terms reflect the facts that p_h does not exactly satisfy the Neumann boundary condition and that $\nabla p_h \cdot \boldsymbol{\nu}$ is not continuous across internal edges. Since the interior edges are counted twice, combining inequality (11.132), relation (11.134), and the triangle inequality, we obtain (see Exercise 11.27)

$$
\begin{aligned}
\|p - p_h\|_V \le C \bigg\{ &\sum_{K \in K_h} \left(\mathcal{R}_K^2 + h_K^2 \int_K |f - f_K|^2 \, d\mathbf{x} \right) \\
&+ \sum_{e \in \mathcal{E}_h^N} h_e \int_e |g - g_e|^2 \, d\ell \bigg\}^{1/2}.
\end{aligned}
\tag{11.135}
$$

Based on inequality (11.135), with a given tolerance $\epsilon > 0$, an *adaptive algorithm* (termed algorithm I) can be defined (below RHS denotes the right-hand side of Eq. (11.135)):

- Choose an initial grid K_{h_0} with grid size h_0, and find a finite element solution p_{h_0} using equality (11.125) with $V_h = V_{h_0}$;

- Given a solution p_{h_k} in V_{h_k} with grid size h_k, stop if the following stopping criterion is satisfied:

$$
\text{RHS} \le \epsilon; \tag{11.136}
$$

- If inequality (11.136) is violated, find a new grid K_{h_k} with grid size h_k such that the following equation is satisfied:

$$
\text{RHS} = \epsilon, \tag{11.137}
$$

and continue.

Inequality (11.136) is the stopping criterion and Eq. (11.137) defines the adaptive strategy. It follows from inequality (11.135) that the estimate $\|p - p_h\|_{H^1(\Omega)}$ is bounded by ϵ if Eq. (11.136) is reached with $p_h = p_{h_k}$. Equation (11.137) determines a new grid size h_k by maximality. Namely, we seek a grid size h_k as large as possible (to maintain efficiency) such that Eq. (11.137) is satisfied. The maximality is generally determined by *equidistribution* of an error such that the error contributions from the individual elements K are approximately equal. Let M_{h_k} be the number of elements in K_{h_k}; equidistribution means that

$$
(\text{RHS}|_K)^2 = \frac{\epsilon^2}{M_{h_k}}, \quad K \in K_{h_k}.
$$

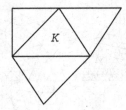

Figure 11.26   An illustration of Ω_K.

Since the solution p_{h_k} depends on K_{h_k}, this is a nonlinear problem. The nonlinearity can be simplified by replacing M_{h_k} by $M_{h_{k-1}}$ (the number at the previous level), e.g.

The following inequality implies, in a sense, that the converse of inequality (11.135) also holds (Verfürth, 1996; Chen, 2005): for $K \in K_h$

$$
\begin{aligned}
\mathcal{R}_K \leq C \Big\{ &\sum_{K' \in \Omega_K} \Big(\int_{K'} |p - p_h|^2 \, d\mathbf{x} + h_{K'}^2 \int_{K'} |f - f_{K'}|^2 \, d\mathbf{x} \Big) \\
&+ \sum_{e \in \partial K \cap \mathcal{E}_h^N} h_e \int_e |g - g_e|^2 \, d\ell \Big\}^{1/2},
\end{aligned}
\tag{11.138}
$$

where (Fig. 11.26)

$$
\Omega_K = \cup \{ K' \in K_h : \partial K' \cap \partial K \neq \emptyset \}.
$$

Estimate (11.138) indicates that algorithm I is *efficient* in the sense that the computational grid produced by this algorithm is not overly refined for a given accuracy, while estimate (11.136) implies that this algorithm is *reliable* in the sense that the V-error is guaranteed to be within a given tolerance.

We end this section with three remarks. First, it is possible to control the error in norms other than the above V-norm; we can control the gradient error in the maximum norm, for example. Second, the results in this section carry over to finite element spaces of polynomials of degree $r \geq 2$. In this case, f_h and g_h are the L^2 projections of f and g into the spaces of piecewise polynomials of degree $r - 1$ with respect to K_h and $\mathcal{E}_h^N$, respectively, and f_K in the first term of $\mathcal{R}_K$ is replaced by $\Delta p_h|_K + f_K$ (see Exercise 11.28). Finally, the adaptive finite element methods presented in this section can be extended to transient problems (Chen, 2005).

Figure 11.27 Uniform (left) and adaptive (right) triangulations.

Example 11.3 Consider problem (11.121) on a circular segment centered at the origin, with radius one and angle $3\pi/2$ (Fig. 11.27; Verfürth, 1996). The function f is zero, and the solution p vanishes on the straight parts of the boundary Γ and has a normal derivative $\frac{2}{3}\cos(\frac{2}{3}\theta)$ on the curved part of Γ. In terms of polar coordinates, the exact solution p to problem (11.121) is $p = r^{2/3}\sin(\frac{2}{3}\theta)$. We calculate the finite element solution p_h using method (11.125) with the space of piecewise linear functions V_h associated with the two triangulations shown in Fig. 11.27. The left triangulation is constructed by five uniform refinements of an initial triangulation K_{h_0}, which is composed of three right-angled isosceles triangles with short edges of unit length. In each refinement step, every triangle is divided into four smaller triangles by connecting the midpoints of its edges. The midpoint of an edge having its two endpoints on Γ is projected onto Γ. The right triangulation in Fig. 11.27 is obtained from K_{h_0} by using Algorithm I based on the error estimator in the definition (11.134). A triangle $K \in K_h$ is divided into four smaller triangles if $\mathcal{R}_K \geq 0.5\max_{K'\in K_h} \mathcal{R}_{K'}$. The midpoint of an edge having its two endpoints on the boundary Γ is projected onto Γ. For both triangulations, Table 11.1 lists the number of triangles (NT), the number of unknowns (NN), the relative error $e_r = \|p - p_h\|_V/\|p\|_V$, and the measurement $m_q = (\sum_{K\in K_h} \mathcal{R}_K^2)^{1/2}/\|p - p_h\|_V$ of the quality of the error estimator. From this table, we clearly see the advantage of the adaptive method and the reliability of the error estimator.

Table 11.1 A comparison of uniform and adaptive refinements.

Refinement	NT	NN	e_r	m_q
Uniform	3072	1552	3.8%	0.7
Adaptive	298	143	2.8%	0.6

11.8. The multiscale finite element method

Complex chemical and physical problems of practical interest often involve a wide range of active scales. Typical examples include heterogeneous porous media (see Chapter 10) and composite materials with fine micro structures. A complete analysis of these problems is really difficult. The difficulty in analyzing porous media flow is attributed to the heterogeneity of subsurface formations that span over many scales. This heterogeneity is often represented by the multiscale fluctuation in the permeability of the media. For composite materials, the dispersed phases (particles or fibers), which can be randomly distributed in the matrix, cause fluctuations in the thermal or electrical conductivity. In addition, the conductivity is often discontinuous across the phase boundaries.

A direct numerical solution of multiscale problems is hard even with today's computing power. The major difficulty of the direct solution stems from the scale of computation. For porous media flow simulation, millions of grid blocks are needed to resolve the required solution, with each block typically having a dimension of tens of meters, whereas the media permeability measured from cores is at a scale of several centimeters. This gives rise to more than 10^5 degrees of freedom for each spatial dimension in computation. Thus, a considerable amount of computer memory and CPU time is required. This problem can be lessened to some extent using parallel computing but the size of discrete problems cannot be reduced. Whenever one can afford to resolve all the small scale features of a physical problem, direct solution methods provide quantitative information of the physical processes at all scales. On the other hand, from the engineering point of view, it usually suffices to predict the macroscopic material properties of the multiscale problems, such as permeability and conductivity. Therefore, it is desirable to develop a numerical method that can capture the effect of small scales on large scales, without resolving all the small-scale features.

Many multiscale methods have been introduced for years, such as averaging methods in classical mechanics, statistical mechanics, homogenization techniques, kinetic theory, and various methods based on Taylor, Fourier, and wavelet expansion techniques. Most of these classical methods obtain analytically or empirically explicit equations for the scale of interest by eliminating other scales. These equations are then used for computer simulations. Despite their success, these classical methods often need to find analytical or empirical closures for some systems that are not always understood or justified. Consequently, their applications to complex multiscale chemical and physical problems are limited.

Here, we briefly consider a multiscale finite element method for solving partial differential equations with a multiscale solution. The central idea of this method is to incorporate the microscale information of the differential problem into finite element basis functions (Hou and Wu, 1997). It is through these modified bases and finite element formulations that the effect of microscales on macroscales can be correctly captured. A similar multiscale finite element method, called the heterogeneous multiscale method was proposed by E and Engquist (2003).

11.8.1. The multiscale finite element method

We consider the second-order stationary problem

$$
\begin{aligned}
-\nabla \cdot (\mathbf{a}_\epsilon \nabla p^\epsilon) &= f \quad \text{in } \Omega, \\
p^\epsilon &= 0 \qquad \text{on } \Gamma,
\end{aligned} \tag{11.139}
$$

where the function f is given and $\mathbf{a}_\epsilon = (a_{ij}(\mathbf{x}, \mathbf{x}/\epsilon))$ is a permeability or conductivity tensor. In practice, this tensor may be random or highly oscillatory. Hence, the solution of Eq. (11.139) can display a multiscale feature.

For $0 < h \le 1$, let K_h be a partition of the domain Ω into elements $\{K\}$ with their diameter bounded by h. The mesh size h resolves the variations of Ω, f, and the slow variable of $\mathbf{a}_\epsilon$. On each element $K \in K_h$, we define a set of nodal basis $\{\varphi_i^K\}$ $(i = 1, 2, \ldots, l)$. In the multiscale finite element method under consideration, each basis (shape) function φ_i^K satisfies

$$
-\nabla \cdot \left(\mathbf{a}_\epsilon \nabla \varphi_i^K \right) = 0 \quad \text{in } K, \ i = 1, 2, \ldots, l. \tag{11.140}
$$

Let $\{\mathbf{x}_i\}$ be the nodes on element K, $i = 1, 2, \ldots, l$. As usual, we require that $\varphi_i^K(\mathbf{x}_j) = \delta_{ij}$ (see Chapter 2), where δ_{ij} represents the Kronecker symbol, $i, j = 1, 2, \ldots, l$. One needs to specify the boundary condition for φ_i^K so that problem (11.140) possesses a unique solution φ_i^K for each i, which is addressed in the next section. For the time being, we assume that the basis functions φ_i^K are continuous on the entire domain Ω and satisfy the homogeneous boundary condition on the external boundary Γ of Ω. Then, we define the finite element space

$$
V_h = \left\{ \text{Functions } v \colon v(\mathbf{x}) = \sum_{K \in K_h} \sum_{i=1}^{l} v_i^K \varphi_i^K(\mathbf{x}), \ \mathbf{x} \in \Omega \right\},
$$

where the coefficients v_i^K are real constants. Now, the finite element solution $p_h \in V_h$ is required to satisfy

$$
\int_\Omega \mathbf{a}_\epsilon \nabla p_h \cdot \nabla v \, d\mathbf{x} = \int_\Omega f v \, d\mathbf{x} \quad \forall v \in V_h. \tag{11.141}
$$

Note that the major difference between the method (11.141) and the standard finite element method lies in the modification of the basis functions for the former, which are required to satisfy Eq. (11.140), along with certain boundary conditions. It is through these modified basis functions and the finite element formulation that the effect of microscales on macroscales can be correctly captured. When the element K is a triangle or tetrahedron, the simplest basis functions are linear; in this case, the number of basis functions is $l = d + 1$, where $d = 2$ or 3 (the dimension number).

The multiscale finite element method (11.141) is developed to capture the large-scale solution. It can be shown that this method provides the same rate of convergence as the standard finite element method when the small scales are well resolved, i.e., $h << \epsilon$. If the mesh size h does not resolve the small scales, these two methods behave very differently. The standard method does not converge to the correct solution, while, in contrast, the multiscale method captures the correct large-scale solution. This method may fail to converge when the mesh scale h is close to the physical small scale ϵ due to the presence of a *resonance* between these two scales. For a two-scale problem, the error due to this resonance manifests as a ratio between the wave length of the small scale oscillation and the mesh size. An *over sampling technique* was proposed to reduce the resonance error (Hou and Wu, 1997), but resonance persists, which is typical for any multiscale method.

11.8.2. Boundary conditions of basis functions

The importance of imposing proper boundary conditions for the basis functions $\{\varphi_i^K\}$ is obvious because they are required to satisfy the homogeneous equation (11.140). A good selection of these boundary conditions can greatly improve the accuracy of the finite element solution. In fact, the boundary conditions determine how well the local property of the multiscale system is sampled into the basis functions. Here we consider a method, which is easy to implement and analyze. Other choices can be seen in the paper by Hou and Wu (1997).

For each element $K \in K_h$, let $\{\phi_i^K\}$ be a standard set of basis functions as defined in Chapters 4–6, $i = 1, 2, \ldots, l$. Then, on the boundary of K, we set

$$\varphi_i^K = \phi_i^K \quad \text{on } \partial K, \ i = 1, 2, \ldots, l. \tag{11.142}$$

For example, for a triangle or tetrahedron K, if $\lambda_1, \lambda_2, \ldots, \lambda_{d+1}$ are the basis functions associated with the vertices of K, then Eq. (11.142) becomes

$$\varphi_i^K = \lambda_i \quad \text{on } \partial K, \ i = 1, 2, \ldots, d + 1. \tag{11.143}$$

That is, each φ_i^K varies linearly along the boundary ∂K. For an analysis of the multiscale finite element method considered, the reader can refer to Hou *et al.* (1999).

11.9. Exercises

11.1 It follows from Sec. 5.2 that the linear approximation p_h to the solution p on the triangle K (Fig. 11.2) is given by

$$p_h = p_i\lambda_i + p_j\lambda_j + p_k\lambda_k, \quad p_l = p_h(\mathbf{m}_l), \quad l = i, j, k, \qquad (11.144)$$

where we recall that the local basis functions λ_i (natural coordinates) are defined by

$$\lambda_i(\mathbf{m}_j) = \begin{cases} 1 & \text{if } i = j, \\ 0 & \text{if } i \neq j, \end{cases}$$

with

$$\lambda_i + \lambda_j + \lambda_k = 1. \qquad (11.145)$$

Recall that they are also the area or *barycentric coordinates* of the triangle K. If the coefficient $\mathbf{a}$ is a constant tensor on the triangle K, show that the flux defined in Eq. (11.4) can be written as

$$f_i = |K| \sum_{l=i}^{k} \mathbf{a}\nabla\lambda_l \cdot \nabla\lambda_i \, p_l. \qquad (11.146)$$

11.2 For $v \in P_1(K)$, where K is a triangle, show that v is uniquely determined by its values at the midpoints of the three edges of K (refer to Sec. 11.3).

11.3 Use Fig. 11.28 to construct the linear basis functions φ_i at the three nodes $\mathbf{x}_i$ according to the definition given in Sec. 11.3. Then, use

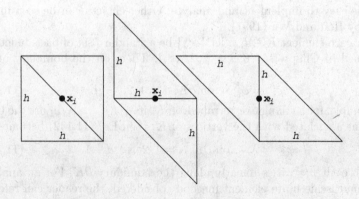

Figure 11.28   The support of a basis function at node $\mathbf{x}_i$.

this result to determine the stiffness matrix $\mathbf{A}$ in Eq. (11.41) for problem (11.37), with $\mathbf{a} = \mathbf{I}$ (the identity tensor), $g = 0$, and a uniform partition of the unit square $(0,1) \times (0,1)$ as given in Fig. 2.6 of Chapter 2.

11.4 Write a computer program to solve problem (11.37) approximately using the nonconforming finite element method developed in Sec. 11.3. Use $\mathbf{a} = \mathbf{I}$ (the identity tensor), $f(x_1, x_2) = 8\pi^2 \sin(2\pi x_1) \sin(2\pi x_2)$, $g = 0$, and a uniform partition of $\Omega = (0,1) \times (0,1)$ as given in Fig. 2.6. Also, compute the errors

$$\|p - p_h\| = \left(\int_\Omega (p - p_h)^2 \, d\mathbf{x} \right)^{1/2},$$

with $h = 0.1$, 0.01, and 0.001, and compare them. Here, p and p_h are the exact and approximate solutions, respectively, and h is the mesh size in the x_1- and x_2 directions (see Sec. 2.5.3).

11.5 Consider the problem

$$-\Delta p = f \quad \text{in } \Omega,$$
$$p = g_D \quad \text{on } \Gamma_D,$$
$$\frac{\partial p}{\partial \nu} = g_N \quad \text{on } \Gamma_N,$$

where Ω is a bounded domain in the plane with boundary Γ, $\bar{\Gamma} = \bar{\Gamma}_D \cup \bar{\Gamma}_N$, $\Gamma_D \cap \Gamma_N = \emptyset$, and f, g_D, and g_N are given functions. Write down a variational formulation for this problem and formulate a nonconforming finite element method using the P_1-nonconforming element discussed in Sec. 11.3.

11.6 With the pair of spaces V and W defined in Sec. 11.4.1, show that if $u \in V$ and $p \in W$ satisfy system (11.46) and p is twice continuously differentiable, then p satisfies system (11.43).

11.7 Write a computer program to solve problem (11.43) approximately using the mixed finite element method introduced in Sec. 11.4.1. Use $f(x) = 4\pi^2 \sin(2\pi x)$ and a uniform partition of $(0,1)$ with $h = 0.1$. Also, compute the errors

$$\|p - p_h\| = \left(\int_0^1 (p - p_h)^2 dx \right)^{1/2},$$

$$\|u - u_h\| = \left(\int_0^1 (u - u_h)^2 dx \right)^{1/2},$$

with $h = 0.1$, 0.01, and 0.001, and compare them. Here p, u and p_h, u_h are the solutions to systems (11.46) and (11.48), respectively (see Sec. 11.4.1). (If necessary, refer to Chen (2005) for a linear solver.)

11.8 Consider the problem with an inhomogeneous boundary condition:

$$-\frac{d^2p}{dx^2} = f(x), \quad 0 < x < 1,$$

$$p(0) = p_{D0}, \qquad p(1) = p_{D1},$$

where f is a given real-valued piecewise continuous bounded function in $(0, 1)$, and p_{D0} and p_{D1} are real numbers. Write this problem in a mixed variational formulation, and construct a mixed finite element method using the finite element spaces described in Sec. 11.4.1. Determine the corresponding linear system of algebraic equations for a uniform partition.

11.9 Consider the problem with a Neumann boundary condition at $x = 1$:

$$-\frac{d^2p}{dx^2} = f(x), \quad 0 < x < 1,$$

$$p(0) = \frac{dp}{dx}(1) = 0.$$

Express this problem in a mixed variational formulation, formulate a mixed finite element method using the finite element spaces introduced in Sec. 11.4.1, and determine the corresponding linear system of algebraic equations for a uniform partition.

11.10 With the pair of spaces V and W defined in Sec. 11.4.1, construct the finite element subspaces $V_h \subset V$ of $W_h \subset W$ that consist, respectively, of piecewise quadratic and linear functions on a partition of the unity interval $(0, 1)$. How can the parameters (degrees of freedom) be chosen to describe such functions in V_h and W_h? Find the corresponding basis functions. Then, define a mixed finite element method for Eq. (11.43) using these spaces $V_h \times W_h$, and express the corresponding linear system of algebraic equations for a uniform partition of the interval $(0, 1)$.

11.11 Show that the matrix $\mathbf{M}$ defined in Eq. (11.53) has both positive and negative eigenvalues.

11.12 Define the space

$$\mathbf{V} = \Big\{ \text{Vector functions } \mathbf{v} : \mathbf{v} \text{ and } \nabla \cdot \mathbf{v} \text{ are defined and square}$$

$$\text{integrable on } \Omega; \text{ i.e., } \int_\Omega \left(|\mathbf{v}|^2 + |\nabla \cdot \mathbf{v}|^2 \right) \, d\mathbf{x} < \infty \Big\}.$$

Show that for any decomposition of $\Omega \subset \mathbb{R}^2$ (the plane) into sub-domains such that the interiors of these subdomains are pairwise disjoint, $\mathbf{v} \in \mathbf{V}$ if and only if its normal components are continuous across the interior edges in this decomposition.

11.13 With the pair of spaces $\mathbf{V}$ and W defined in Sec. 11.4.2, prove that if $\mathbf{u} \in \mathbf{V}$ and $p \in W$ satisfy system (11.59) and if the second partial derivatives of p are bounded in the domain Ω, then the solution p satisfies problem (11.56).

11.14 Let the basis functions $\{\varphi_i\}$ and $\{\psi_i\}$ of the spaces $\mathbf{V}_h$ and W_h be defined as in Sec. 11.4.2. For a uniform partition of $\Omega = (0,1) \times (0,1)$ given as in Fig. 2.6, determine the matrices $\mathbf{A}$ and $\mathbf{B}$ in system (11.61).

11.15 Consider problem (11.56) with an inhomogeneous boundary condition, i.e.

$$-\Delta p = f \quad \text{in } \Omega,$$
$$p = g \quad \text{on } \Gamma,$$

where Ω is a bounded domain in the plane with boundary Γ, and f and g are given functions. Express this problem in a mixed variational formulation, formulate a mixed finite element method using the finite element spaces given in Sec. 11.4.2, and determine the corresponding linear system of algebraic equations for a uniform partition of the unit domain $\Omega = (0,1) \times (0,1)$ as displayed in Fig. 2.6.

11.16 Consider the problem

$$-\Delta p = f \quad \text{in } \Omega,$$
$$p = g_D \quad \text{on } \Gamma_D,$$
$$\frac{\partial p}{\partial \nu} = g_N \quad \text{on } \Gamma_N,$$

where Ω is a bounded domain in the plane with boundary Γ, $\bar{\Gamma} = \bar{\Gamma}_D \cup \bar{\Gamma}_N$, $\Gamma_D \cap \Gamma_N = \emptyset$, and f, g_D, and g_N are given functions. Write down a mixed variational formulation for this problem and formulate a mixed finite element method using the finite element spaces $\mathbf{V}_h$ and W_h given in Sec. 11.4.2 (the vector space $\mathbf{V}_h$ needs to be modified).

11.17 Let $\{\varphi_i\}$ and $\{\psi_i\}$ be the basis functions of the spaces $\mathbf{V}_h$ and W_h in system (11.68), respectively. Write system (11.68) in matrix form.

11.18 Show that after multiplying both sides of Eq. (11.92) by Δt^n (time step), the condition number of the stiffness matrix corresponding to the left-hand side of Eq. (11.92) is of order

$$\mathcal{O}\left(1 + \max_{x \in \mathbb{R},\ t \geq 0} |a(x,t)| h^{-2} \Delta t\right), \quad \Delta t = \max_{n=1,2,\ldots} \Delta t^n.$$

11.19 Let $v \in C^1(\mathbb{R})$ (the set of continuously differentiable functions on the set of real numbers) be a $(0,1)$-periodic function. Show that the condition $v(0) = v(1)$ implies

$$\frac{\partial v(0)}{\partial x} = \frac{\partial v(1)}{\partial x}.$$

11.20 Let $\mathbf{a}$ be positive semi definite, ϕ be uniformly positive with respect to x and t, and R be nonnegative. Show that Eq. (11.102) has a unique solution $p_h^n \in V_h$ for every $n \geq 1$.

11.21 Derive relation (11.108) in detail.

11.22 Derive Eq. (11.115) in detail.

11.23 Let $\mathbf{a}$ be positive semi definite, ϕ be uniformly positive with respect to x and t, and R be nonnegative. Show that Eq. (11.120) has a unique solution $p_h^n \in V_h$ for each n.

11.24 For the example in Fig. 11.23, use the **refinement rule** defined in Sec. 11.7.1 to convert irregular vertices to regular vertices.

11.25 Show inequality (11.132) using inequality (11.129) and Eq. (11.131).

11.26 For the problem

$$-\nabla \cdot (\mathbf{a}\nabla p) = f \quad \text{in } \Omega,$$

$$p = 0 \quad \text{on } \Gamma_D,$$

$$\mathbf{a}\nabla p \cdot \boldsymbol{\nu} = g_N \quad \text{on } \Gamma_N,$$

derive an inequality similar to inequality (11.132).

11.27 Apply definitions (11.133) and (11.134) to derive estimate (11.135) from inequality (11.132).

11.28 For the problem

$$-\nabla \cdot (\mathbf{a}\nabla p) = f \quad \text{in } \Omega,$$

$$p = 0 \quad \text{on } \Gamma_D,$$

$$\mathbf{a}\nabla p \cdot \boldsymbol{\nu} = g_N \quad \text{on } \Gamma_N,$$

define an error estimator similar to estimator (11.134).

Nomenclature

A	cross-sectional area
$\mathbf{A}$	coefficient matrix of a system (stiffness matrix)
$\mathbf{A}(\mathbf{p})$	nonlinear matrix
$\mathbf{A}^{(k)}$	restriction of $\mathbf{A}$ to kth element
$\mathbf{a}$	diffusion coefficient
a	end point or constant
$a(\cdot,\cdot)$	bilinear form
$a_h(\cdot,\cdot)$	mesh-dependent bilinear form
$a_K(\cdot,\cdot)$	restriction of $a(\cdot,\cdot)$ on K
$a_-(\cdot,\cdot)$	bilinear form for symmetric DG
$a_+(\cdot,\cdot)$	bilinear form for nonsymmetric DG
a_{ij}	entries of matrix $\mathbf{A}$
$a_{ij}^{(k)}$	restriction of a_{ij} to kth element
$a(p)$	nonlinear coefficient
$\mathbf{B}$	mass matrix
$\mathbf{B}$	differential operator tensor
$\mathbf{B}_1$	matrix in an affine mapping
b	end point or constant
$b(\cdot,\cdot)$	bilinear form
$\mathbf{b}$	convection or advection coefficient
b_e	edge bubble function
b_{ij}	entries of matrix $\mathbf{B}$
b_K	triangle bubble function
c	reaction coefficient
c	speed of sound
c_p	specific heat at constant pressure
c_v	specific heat at constant volume
$c(p)$	nonlinear coefficient

$\mathbf{C}$	coefficient matrix associated with time
C	generic constant
$\mathbf{C}(\mathbf{p})$	nonlinear matrix
d	dimension number ($d = 1$, 2, or 3)
$\mathbf{D}$	elasticity matrix
d_{ij}	entries of matrix $\mathbf{D}$
D_α	partial derivative operator
E	specific energy
$\mathbf{E}$	strain tensor
$\mathcal{E}_h^D$	set of edges on Γ_D
$\mathcal{E}_h^N$	set of edges on Γ_N
$\mathcal{E}_h^o$	set of internal edges in K_h
$\mathcal{E}_h^b$	set of edges on Γ
$\mathcal{E}$	set of edges of the partition K_h
E_i	elasticity constant
E_r	number of nodes
F	functional or total potential energy
$F_k(\cdot)$	local coordinate transformation
$\mathbf{F}(\cdot)$	local coordinate transformation
$\mathbf{F}$	mapping
$\mathbf{F}(\mathbf{p})$	nonlinear vector
f	right-hand function or load
$\mathbf{f}$	right-hand vector of a system
$\mathbf{f}$	load vactor
f_K	local mean value on K
f_i	ith entry of $\mathbf{f}$
$f_j^{(k)}$	restriction of f_j to kth element
f_α	fractional flow function
$f(p)$	nonlinear coefficient
$\mathbf{f}(\mathbf{p})$	nonlinear vector
G	function
G	planar domain
G_i	elasticity constant
$\mathbf{G}$	Jacobian matrix or a mapping
g	boundary datum
g_D	boundary datum
g_N	boundary datum
g_i	boundary datum
g_e	local mean value on e
h	mesh or grid size
h_i	mesh or grid size of ith interval
h_e	length of edge e

h_k	mesh size at the kth level
h_K	diameter of K (element)
I	interval in $\mathbb{R}$
I_i	ith subinterval
I_r	number of internal nodes
$\mathbf{I}$	identity matrix or operator
$\check{I}_i(t)$	trace-back of I_i to time t
$\mathcal{I}_i^n$	space–time region following characteristics
$\mathbf{ID}$	identity function
$\mathbf{IZ}(i,j)$	jth node of ith element
J	time interval of interest $(J = (0,T])$
J^n	nth subinterval of time (t^{n-1}, t^n)
$\mathbf{J}_n$	electron current density
$\mathbf{J}_p$	hole current density
k	permeability of porous media
k	elastic bulk modulus
K	element (triangle, rectangle, etc.)
$|K|$	area or volume of element
$\hat{K}$	reference element
K_h	triangulation (partition)
K_i	ith element
$\check{K}(t)$	trace-back of K to time t
$\mathcal{K}^n$	space–time region following characteristics
$L(\cdot)$	linear functional
$L_-(\cdot)$	linear functional for symmetric DG
$L_+(\cdot)$	linear functional for nonsymmetric DG
L_h	space of Lagrange multipliers
L_i	band width of ith row
L_m	maximum band width
$\mathbf{L}$	lower triangular matrix
$\mathcal{L}$	linear operator
l_i	real numbers
M	number of grid points (nodes)
$\tilde{M}$	entire number of grid points (nodes)
$\mathbf{M}$	coefficient matrix arising from mixed method
$\mathbf{m}_i$	vertices of elements
$\mathcal{N}_h$	set of vertices in K_h
p	unknown variable (pressure, displacement)
p_a	boundary value
p_b	boundary value
p_c	capillary pressure
p_{D0}	boundary value

p_{D1}	boundary value
$\mathbf{p}$	unknown vector of a system
p_h	approximate solution
p_0	initial datum
$\mathbf{p}_0$	initial datum vector
$P(K)$	finite dimensional linear space on element K
$\breve{p}_h^{n-1}$	value of p_h at $\left(\breve{x}_n, t^{n-1}\right)$: $p_h\left(\breve{x}_n, t^{n-1}\right)$
$\tilde{p}_h$	interpolant of p
P_r	set of polynomials of total degree $\leq r$
$P_{l,r}$	set of polynomials defined on prisms
p_α	pressure of α-phase
$\mathbf{P}$	pressure tensor
R	reaction coefficient
Re	Reynolds number
R_h	projection operator
$\mathcal{R}_{D,\mathbf{m}}$	local Dirichlet estimator I
$\mathcal{R}_{D,K}$	local Dirichlet estimator II
$\mathcal{R}_H$	hierarchical basis estimator
$\mathcal{R}_K$	residual *a posteriori* estimator
$\mathcal{R}_{N,K}$	local Neumann estimator
$\mathcal{R}_Z$	averaging-based estimator
$\mathbf{R}$	mapping
r	polynomial degree
$\mathbf{q}$	volumetric flow rate
$\mathbf{q}$	heat flux
q_α	source/sink term of α-phase
Q_r	set of polynomials of degree $\leq r$ in each variable
$\mathbf{Q}$	upper triangular matrix
s	number of nodes
s_α	saturation of α-phase
t	time variable
t^n	nth time step
T	temperature or final time
T_0	surrounding medium temperature
T_1	temperature at one end
T_2	temperature at the other end
T_r	number of coefficients
u	velocity variable in $\mathbb{R}$
$\mathbf{u}$	velocity variable in $\mathbb{R}^d$
$\mathbf{u}$	displacement
$\mathbf{u}$	approximate solution
$\mathbf{u}_\alpha$	velocity of α-phase

$\mathbf{U}$	unknown vector for u or $\mathbf{u}$
$\mathbf{U}$	mass flow flux
U_T	thermal voltage
v	generic function
v_{ij}	coefficient (real number)
v_{ijm}	coefficient (real number)
v_-	left-hand limit notation
v_+	right-hand limit notation
V	linear vector space
V'	dual space to V
V_h	finite element space
$\mathbf{V}$	vector space
$\mathbf{V}$	vector space in a pair of mixed spaces
$\mathbf{V}_h$	finite element space
$\mathbf{V}_h$	vector space in a pair of mixed finite element spaces
$\mathbf{V}_h(K)$	restriction of $\mathbf{V}_h$ on K
w_i	integration weight
W	scalar space in a pair of mixed spaces
W_h	scalar space in a pair of mixed finite element spaces
$W_h(K)$	restriction of W_h on K
x	independent variable in $\mathbb{R}$
x	ith node
$\mathbf{x}$	independent variable in $\mathbb{R}^d$: $\mathbf{x} = (x_1, x_2, \ldots, x_d)$
$\check{x}_n$	foot of a characteristic corresponding to x at t^n
Z	subspace of V induced by $b(\cdot, \cdot)$
$Z^\perp$	orthogonal complement of Z
Z^0	polar set of Z
Z_h	discrete counterpart of Z
$\mathbf{Z}(i,j)$	jth coordinate of ith node
Z_+^2	2-tuple set
Z_+^3	3-tuple set
$\mathrm{cond}(\mathbf{A})$	condition number of $\mathbf{A}$
$\mathbb{R}$	set of real numbers
$\mathbb{R}^M$	M-dimensional Euclidean space
Ω	open set in $\mathbb{R}^d$ ($d = 2$ or 3)
$\bar{\Omega}$	closure of Ω
Ω_e	union of elements with common edge e
Ω_K	union of elements adjacent to K
$\Omega_\mathbf{m}$	union of elements with common vertex $\mathbf{m}$
Γ	boundary of Ω ($\partial\Omega$)
Γ_-	inflow boundary of Γ
Γ_+	outflow boundary of Γ

Γ_D	Dirichlet boundary of Γ
Γ_N	Neumann boundary of Γ
Γ_i	ith boundary
∂K	boundary of K
∇	gradient operator
$\nabla\cdot$	divergence operator (div)
Δ	Laplacian operator
Δ	volume of element
Δ_K	area or volume of element K
∇^2	Laplacian operator
Δ^2	biharmonic operator ($\Delta\Delta$)
Δt	time step size
Δt_c	stability constant
Δt_d	stability constant
Δt^n	time step size at nth step
$\dfrac{\partial}{\partial x}$	derivative with respect to x
$\dfrac{\partial^2}{\partial x^n}$	second derivative with respect to x
$\dfrac{\partial}{\partial x_i}$	partial derivative with respect to x_i
$\dfrac{\partial}{\partial t}$	partial derivative with respect to t (time)
$\dfrac{\partial}{\partial \boldsymbol{\nu}}$	normal derivative
$\dfrac{\partial}{\partial \mathbf{t}}$	tangential derivative
$\dfrac{\partial}{\partial \boldsymbol{\tau}}$	directional derivative along characteristics
$\dfrac{D}{Dt}$	material derivative
D^α	partial derivative notation
D^α_w	weak derivative notation
D^r	multilinear form of rth-order derivative
$C^\infty(\Omega)$	space of functions infinitely differentiable
$\mathcal{D}(\Omega)$	subset of $C^\infty(\Omega)$ having compact support in Ω
$C_0^\infty(\Omega)$	same as $\mathcal{D}(\Omega)$
$\mathrm{diam}(K)$	diameter of K
$L_{loc}^1(\Omega)$	integrable functions on any compact set inside Ω
$L^q(\Omega)$	Lebesgue space
$W^{r,q}(\Omega)$	Sobolev spaces
$W_0^{r,q}(\Omega)$	completion of $\mathcal{D}(\Omega)$ with respect to $\|\cdot\|_{W^{r,q}(\Omega)}$

$\lVert \cdot \rVert$	norm
$\lVert \cdot \rVert_h$	norm on a nonconforming space
$\lVert \cdot \rVert_{L^q(\Omega)}$	norm of $L^q(\Omega)$
$\lVert \cdot \rVert_{W^{r,q}(\Omega)}$	norm of $W^{r,q}(\Omega)$
$\lvert \cdot \rvert_{W^{r,q}(\Omega)}$	seminorm of $W^{r,q}(\Omega)$
$H^r(\Omega)$	same as $W^{r,2}(\Omega)$
$H_0^r(\Omega)$	same as $W_0^{r,2}(\Omega)$
$H^l(K_h)$	piecewise smooth space
$\mathbf{H}(\mathrm{div}, \Omega)$	divergence space
α	multi index (a d-tuple): $\alpha = (\alpha_1, \alpha_2, \ldots, \alpha_d)$
$\boldsymbol{\beta}$	convection or advection coefficient
β	dissipation coefficient
β_1	measure of smallest angle over $K \in K_h$
β_2	quasi-uniform triangulation constant
γ	a given constant
γ_{ij}	shear strain
ϵ	real number
ϵ	integration error
ϵ_i	normal strain
$\boldsymbol{\epsilon}$	strain tensor
π_h	interpolation operator
π_K	restriction of π_h on element K
Π_h	projection operator
δ	distance traveled by particle
$\delta(\mathbf{x} - \mathbf{x}^{(l)})$	Dirac delta function at $\mathbf{x}^{(l)}$
δ_{ij}	Kronecker symbol
κ	thermal diffusion coefficient
$\boldsymbol{\kappa}$	reservoir permeability
$\kappa_{r\alpha}$	relative permeability of α-phase
μ_α	viscosity of α-phase
μ	fluid viscosity
ρ	fluid density
ρ_α	density of α-phase
ρ_K	diameter of largest circle inscribed in K
Σ_K	set of degrees of freedom on element K
$\boldsymbol{\nu}$	outward unit normal
ν_i	elasticity constant
$\varphi, \boldsymbol{\varphi}$	interstitial velocity
φ_i	ith basis function of V_h
$\varphi_{i,j}$	basis function

ψ_i	ith basis function on reference element
$\psi_{i,j}$	basis function on reference element
φ_i	basis function of $\mathbf{V}_h$
λ	viscosity constant
λ_d	Lagrange multipliers
λ_i	barycentric coordinates $(i = 1, 2, 3)$
λ_α	phase mobility
$\boldsymbol{\lambda}$	degrees of freedom of λ_h
$B_{r+1}(K)$	same as $\lambda_1\lambda_2\lambda_3 P_{r-2}(K)$
σ	Poisson's ratio
σ_i	normal stress
$\boldsymbol{\sigma}$	stress tensor or vector
τ_{ij}	shear stress
$\tau, \boldsymbol{\tau}$	characteristic direction
$\tau_\mathbf{p}$	momentum relaxation time
τ_w	energy relaxation time
θ	implicity (explicitly) constant
θ_i	implicity (explicitly) constant
ξ	variable on reference element
$\boldsymbol{\xi}$	variable on reference element
$\hbar$	quantum expansion parameter
$[\,\cdot\,]$	jump operator notation
$\{\!\!\{\,\cdot\,\}\!\!\}$	averaging operator notation
$\det(\cdot)$	determinant of a matrix

Bibliography

[1] I. Aavatsmark (2002), An introduction to multipoint flux approximations for quadrilateral grids, *Comput. Geosci.* **6**, 405–432.

[2] A. Adini and R. Clough (1961), Analysis of plate bending by the finite element method, NSF Report G. 7337, University of California, Berkeley, CA.

[3] S. N. Antontsev (1972), On the solvability of boundary value problems for degenerate two-phase porous flow equations, *Dinamika Splošnoĭ Sredy Vyp.* **10**, 28–53 (in Russian).

[4] T. Arbogast and Z. Chen (1995), On the implementation of mixed methods as nonconforming methods for second order elliptic problems, *Math. Comp.* **64**, 943–972.

[5] T. Arbogast and M. F. Wheeler (1995), A characteristics-mixed finite element for advection-dominated transport problems, *SIAM J. Numer. Anal.* **32**, 404–424.

[6] D. N. Arnold, F. Brezzi, and M. Fortin (1984b), A stable finite element for the Stokes equations, *Calcolo* **21**, 337–344.

[7] O. Axelsson (1994), *Iterative Solution Methods*, Cambridge University Press, Cambridge.

[8] K. Aziz and A. Settari (1979), *Petroleum Reservoir Simulation*, Applied Science Publishers Ltd, London.

[9] I. Babuška and M. R. Dorr (1981), Error estimates for the combined h and p versions of the finite element method, *Numer. Math.* **37**, 257–277.

[10] I. Babuška and W. C. Rheinboldt (1978a), Error estimates for adaptive finite element computations, *SIAM J. Numer. Anal.* **15**, 736–754.

[11] I. Babuška and W. C. Rheinboldt (1978b), *A-posteriori* error estimates for the finite element method, *Int. J. Num. Meth. Eng.* **12**, 1597–1615.

[12] I. Babuška, A. Miller, and M. Vogelius (1983), Adaptive methods and error estimation for elliptic problems of structural mechanics, in *Adaptive Computational Methods for Partial Differential Equations*, I. Babuška, *et al.*, eds., SIAM, PA, pp. 35–56.

[13] R. E. Bank (1990), *PLTMG: A Software Package for Solving Elliptic Partial Differential Equations, User's Guide 6.0*, SIAM, PA.

[14] R. E. Bank and A. Weiser (1985), Some a posteriori error estimators for elliptic partial differential equations, *Math. Comp.* **44**, 283–301.

[15] R. E. Bank, A. H. Sherman, and A. Weiser (1983), Refinement algorithms and data structures for regular local mesh refinement, in *Scientific Computing*, R. Stepleman, *et al.*, eds., North Holland, Amsterdam, New York, Oxford, pp. 3–17.

[16] P. Bochev, C. R. Dohrmann, and M. D. Gunzburger (2006), Stabilization of low-order mixed finite elements for the Stokes equations, *SIAM J. Numer. Anal.* **44**, 82–101.

[17] J. W. Barrett and K. W. Morton (1984), Approximate symmetrization and Petrov-Galerkin methods for diffusion-convection problems, *Comp. Mech. Appl. Mech. Eng.* **45**, 97–122.

[18] D. Braess (1997), *Finite Elements, Theory, Fast Solvers, and Applications in Solid Mechanics*, Cambridge University Press, Cambridge.

[19] S. C. Brenner and L. R. Scott (1994), *The Mathematical Theory of Finite Element Methods*, Springer-Verlag, New York.

[20] F. Brezzi, J. Douglas, Jr., R. Durán, and M. Fortin (1987a), Mixed finite elements for second order elliptic problems in three variables, *Numer. Math.* **51**, 237–250.

[21] F. Brezzi, J. Douglas Jr., and M. Fortin (1987b), Efficient rectangular mixed finite elements in two and three space variables, *RAIRO Modèl. Math. Anal. Numér* **21**, 581–604.

[22] F. Brezzi, J. Douglas Jr., and L. D. Marini (1985), Two families of mixed finite elements for second order elliptic problems, *Numer. Math.* **47**, 217–235.

[23] F. Brezzi and M. Fortin (1991), *Mixed and Hybrid Finite Element Methods*, Springer-Verlag, New York.

[24] A. Brooks and T. J. Hughes (1982), Streamline upwind Petrov-Galerkin formulations for convection dominated flows with particular emphasis on the incompressible Navier–Stokes equations, *Comp. Mech. Appl. Mech. Eng.* **32**, 199–259.

[25] M. A. Celia, T. F. Russell, and I. Herrera (1990), An Eulerian–Lagrangian localized adjoint method for the advection–diffusion equation, *Adv. Water Resour.* **13**, 187–206.

[26] G. Chavent and J. Jaffré (1978), *Mathematical Models and Finite Elements for Reservoir Simulation*, North-Holland, Amsterdam.

[27] Z. Chen (2001b), Degenerate two-phase incompressible flow I: Existence, uniqueness and regularity of a weak solution, *J. Differ. Equ.* **171**, 203–232.

[28] Z. Chen (2002a), Degenerate two-phase incompressible flow II: Regularity, stability and stabilization, *J. Differ. Equ.* **186**, 345–376.

[29] Z. Chen (2002b), Characteristic mixed discontinuous finite element methods for advection-dominated diffusion problems, *Comp. Meth. Appl. Mech. Eng.* **191**, 2509–2538.

[30] Z. Chen (2005), *Finite Element Methods and their Applications*, Springer-Verlag, Heidelberg and New York.

[31] Z. Chen (2007), *Reservoir Simulation: Mathematical Techniques in Oil Recovery*, in the CBMS-NSF Regional Conference Series in Applied Mathematics, Vol. 77, SIAM, Philadelphia, PA.

[32] Z. Chen and J. Douglas Jr. (1989), Prismatic mixed finite elements for second order elliptic problems, *Calcolo* **26**, 135–148.

[33] Z. Chen and J. Douglas Jr. (1991), Approximation of coefficients in hybrid and mixed methods for nonliner parabolic problems, *Mat. Aplic. Comp.* **10**, 137–160.

[34] Z. Chen and R. E. Ewing (1997a), Comparison of various formulations of three-phase flow in porous media, *J. Comp. Phys.* **132**, 362–373.

[35] Z. Chen and R. E. Ewing (1997b), Fully-discrete finite element analysis of multiphase flow in groundwater hydrology, *SIAM J. Numer. Anal.* **34**, 2228–2253.

[36] Z. Chen, G. Huan, and Y. Ma (2006), *Computational Methods for Multiphase Flows in Porous Media*, Computational Science and Engineering Series, Vol. 2, SIAM, Philadephia, PA.

[37] Z. Chen, G. Qin, and R. E. Ewing (2000), Analysis of a compositional model for fluid flow in porous media, *SIAM J. Appl. Math.* **60**, 747–777.

[38] A. J. Chorin (1968), Numerical solution of Navier–Stokes equations, *Math. Comput.* **22**, 745–762.

[39] I. Christie, D. F. Griffiths, and A. R. Mitchell (1976), Finite element methods for second order differential equations with significant first derivatives, *Int. J. Num. Eng.* **10**, 1389–1396.

[40] P. G. Ciarlet (1978), *The Finite Element Method for Elliptic Problems*, North-Holland, Amsterdam.

[41] P. G. Ciarlet and P.-A. Raviart (1972), The combined effect of curved boundaries and numerical integration in isoparametric finite element methods, in *The Mathematical Foundations of the Finite Element Method with Applications to Partial Differential Equations*, A. K. Aziz, ed., Academic Press, New York, pp. 409–474.

[42] R. Clough and C. Felippa (1968), A refined quadrilateral element for the analysis of plate bending, *Proc. Conf. Matrix Methods Struct. Mech., 2nd*, Wringht-Patterson AFB, Ohio, October 15–17, 1968, AFFDL-TR-68-150, 399–440.

[43] G. Comini and S. Del Guidice (1972), Finite element solution of incompressible Navier–Stokes equations, *Numer. Heat Transfer A* **5**, 463–478.

[44] M. Crouzeix and P. Raviart (1973), Conforming and nonconforming finite element methods for solving the stationary Stokes equations, *RAIRO* **3**, 33–75.

[45] R. Courant (1943), Variational methods for the solution of problems of equilibrium and vibrations, *Bull. Am. Math. Soc.* **49**, 1–23.

[46] E. F. D'Azevedo and R. B. Simpson (1989), On optimal interpolation triangle incidences, *SIAM J. Sci. Stat. Comput.* **10**, 1063–1075.

[47] H. Darcy (1856), *Les Fontaines Publiques de la Ville de Dijon*, Victor Dalmond.

[48] P. Deuflhard, P. Leinen, and H. Yserentant (1989), Concepts of an adaptive hierarchical finite element code, *IMPACT Comput. Sci. Eng.* **1**, 3–35.

[49] J. C. Diaz, R. E. Ewing, and R. W. Jones (1984), Self-adaptive local grid-refinement for time-dependent, two-dimensional simulation, in *Finite Elements in Fluids*, Vol. VI, Wiley, New York, pp. 479–484.

[50] J. Douglas Jr. (1961), A survey of numerical methods for parabolic differential equations, in *Advances in Computers*, F. L. Alt, ed., Vol. 2, Academic Press, New York, pp. 1–54.

[51] J. Douglas Jr., F. Furtado, and F. Pereira (1997), On the numerical simulation of water flooding of heterogeneous petroleum reservoirs, *Comput. Geosci.* **1**, 155–190.

[52] J. Douglas Jr., D. W. Peaceman, and H. H. Rachford Jr. (1959), A method for calculating multi-dimensional immiscible displacement, *Trans. SPE AIME* **216**, 297–306.

[53] J. Douglas Jr. and T. F. Russell (1982), Numerical methods for convection dominated diffusion problems based on combining the method of characteristics with finite element or finite difference procedures, *SIAM J. Numer. Anal.* **19**, 871–885.

[54] W. E and B. Engquist (2003), The heterogeneous multiscale methods, *Comm. Math. Sci.* **1**, 87–132.

[55] N. S. Espedal and R. E. Ewing (1987), Characteristic Petrov-Galerkin subdomain methods for two phase immiscible flow, *Comput. Methods Appl. Mech. Eng.* **64**, 113–135.

[56] R. E. Ewing (1986), Efficient adaptive procedures for fluid flow applications, *Comp. Meth. Appl. Mech. Eng.* **55**, 89–103.

[57] P. A. Forsyth (1991), A control volume finite element approach to NAPL groundwater contamination, *SIAM J. Sci. Stat. Comput.* **12**, 1029–1057.

[58] M. Fortin and Soulie (1983), A nonconforming piecewise quadratic finite element on triangles, *Int. J. Numer. Methods Eng.* **19**, 505–520.

[59] B. Fraeijs de Veubeke (1965), Displacement and equilibrium models in the finite element method, in *Stress Analysis*, O. C. Zienkiewicz and G. Holister, eds., John Wiley and Sons, New York.

[60] B. Fraeijs de Veubeke (1968), A conforming finite element for plate bending, *Int. J. Mech. Sci.* **10**, 563–570.

[61] A. O. Garder, D. W. Peaceman, and A. L. Pozzi (1964), Numerical calculations of multidimensional miscible displacement by the method of characteristics, *Soc. Pet. Eng. J.* **4**, 26–36.

[62] V. Girault and P.-A. Raviart (1981), *Finite Element Approximation of the Navier–Stokes Equations*, Springer-Verlag, Berlin, Heidelberg, New York.

[63] R. Glowinski (2003), *Handbook of Numerical Analysis: Numerical Methods for Fluids*, Elsevier Science Publishing Company.

[64] R. Codina, M. Vázquez, and O. C. Zienkiewicz (1995), A fractional step method for compressible flows: Boundary conditions and incompressible limit, *Proc. of 9th Int. Conf. Finite Elements Fluids-New Trends and Applications*, Venezia, pp. 409–418.

[65] K. Hellan (1967), Analysis of elastic plates in flexure by a simplified finite element method, in *Acta Polytechnica Scandinavia*, Civil Engineering Series, Trondheim **46**.

[66] B. Heinrich (1987), *Finite Difference Methods on Irregular Networks*, Birkhauser, Basel, Boston, Stuttgart.

[67] L. R. Herrmann (1967), Finite element bending analysis for plates, *J. Eng. Mech. Div. ASCE* **93**, 13–26.

[68] J. D. Hoffman (1992), *Numerical Methods for Engineers and Scientists*, 2nd edition, CRC Press, Taylor and Francis Group, Boca Raton.

[69] T. Hou and X. Wu (1997), A multiscale finite element method for elliptic problems in composite materials and porous media, *J. Comp. Phys.* **134**, 169–189.

[70] T. Hou, X. Wu, and Z. Cai (1999), Convergence of a multiscale finite element method for elliptic problems with rapidly oscillating coefficients, *Math. Comp.* **68**, 913–943.

[71] M. K. Hubbert (1956), Darcy's law and the field equations of the flow of underground fluids, *Trans. SPE AIME* **207**, 222–239.

[72] B. Joe (1986), Delaunay triangular meshes in convex polygons, *SIAM J. Sci. Stat. Comput.* **7**, 514–539.

[73] C. Johnson (1994), *Numerical Solutions of Partial Differential Equations by the Finite Element Method*, Cambridge University Press, Cambridge.

[74] W. Kaplan (1991), *Advanced Calculus*, 4th edition, Addison Wesley, Publishing Company Inc.

[75] S. G. Lekhnitskii (1963), *Theory of Elascicity of an Anisotropic Elastic Body*, Translation from Russian by P. Fern, Holden Day, San Francisco.

[76] P. A. Lemonnier (1979), Improvement of reservoir simulation by a triangular discontinuous finite element method, SPE paper 8249 presented at the *1979 Annual Fall Technical Conference and Exhibition of SPE of AIME*, Las Vegas, Sept. 23–26.

[77] F. W. Letniowski (1992), Three dimensional Delaunay triangulation for finite element approximations to a second order diffusion operator, *SIAM J. Sci. Stat. Comput.* **13**, 765–770.

[78] J. Li and Y. T. Chen (2008), *Computational Partial Differential Equations Using MATLAB*, Chapman & Hall/CRC Applied Mathematics & Nonlinear Science, Chapman and Hall/CRC.

[79] J. Li and Y. He (2008), A stabilized finite element method based on two local Gauss integrations for the Stokes equations, *J. Comp. Appl. Math.* **214**, 58–65.

[80] J. Li, Y. He, and Z. Chen (2007), A new stabilized finite element method for the transient Navier–Stokes equations, *Comput. Methods Appl. Mech. Eng.* **197**, 22–35.

[81] K. Miller and R. N. Miller (1981), Moving finite elements, *SIAM J. Numer. Anal.* **18**, 79–95.

[82] J. C. Nédélec (1980), Mixed finite elements in $\mathbb{R}^3$, *Numer. Math.* **35**, 315–341.

[83] S. P. Neuman (1981), An Eulerian–Lagrangian numerical scheme for the dispersion–convection equation using conjugate-time grids, *J. Comp. Phys.* **41**, 270–294.

[84] J. T. Oden, L. Demkowicz, and W. Rachowicz (1989), Toward a universal $h-p$ adaptive finite element strategy, Part 2. A posteriori error estimation, *Comp. Meth. Appl. Mech. Eng.* **77**, 113–180.

[85] E. R. Oliveira (1971), Optimization of finite element solutions, *Proceedings of the Third Conference on Matrix Methods in Structural Mechanics*, Wright-Patterson Air Force Base, Ohio, October.

[86] A. M. Ostrowski (1973), *Solution of Equations in Euclidean and Banach Spaces*, 3rd edition, Academic Press, New York.

[87] O. Pironneau (1982), On the transport-diffusion algorithm and its application to the Navier–Stokes equations, *Numer. Math.* **38**, 309–332.

[88] R. Raviart and J.-M. Thomas (1977), *A Mixed Finite Element Method for Second Order Elliptic Problems*, Lecture Notes in Mathematics, Vol. 606, Springer, Berlin, pp. 292–315.

[89] W. H. Reed and T. R. Hill (1973), Triangular mesh methods for the neutron transport equation, *Technical Report*, LA-UR-73-479, Los Alamos Scientific Laboratory.

[90] W. C. Rheinboldt (1998), *Methods for Solving Systems of Nonlinear Equations*, 2nd edition, Society for Industrial and Applied Mathematics, Philadelphia.

[91] W. C. Rheinboldt and C. Mesztenyi (1980), On a data structure for adaptive finite element mesh refinement, *ACM Trans Math. Softw.* **6**, 166–187.

[92] M. C. Rivara (1984a), Algorithms for refining triangular grids suitable for adaptive and multigrid techniques, *Int. J. Num. Meth. Eng.* **20**, 745–756.

[93] M. C. Rivara (1984b), Design and data structure of fully adaptive, multigrid, finite element software, *ACM Trans. Math. Softw.* **10**, 242–264.

[94] T. F. Russell (1990), Eulerian–Lagrangian localized adjoint methods for advection-dominated problems, in *Numerical Analysis*, Pitman Research Notes in Mathematical Series, Vol. 228, D. F. Griffiths and G. A. Watson, eds., Longman Scientific and Technical, Harlow, England, pp. 206–228.

[95] T. F. Russell and R. V. Trujillo (1990), *Eulerian–Lagrangian Localized Adjoint Methods with Variable Coefficients in Multiple Dimensions*, Gambolati, *et al.*, eds., Computational Methods in Surface Hydrology, Springer-Verlag, Berlin, pp. 357–363.

[96] T. F. Russell and M. F. Wheeler (1983), Finite element and finite difference methods for continuous flows in porous media, in *The Mathematics of Reservoir Simulation*, R. E. Ewing, ed., SIAM, Philadelphia, pp. 35–106.

[97] Y. Saad (2004), *Iterative Methods for Sparse Linear Systems*, 2nd edition, SIAM Publications, Philadelphia.

[98] B. A. Szabo (1986), Mesh design for the p-version of the finite element method, *Comp. Meth. Appl. Mech. Eng.* **55**, 86–104.

[99] J. W. Thomas, *Numerical Partial Differential Equations, Finite Difference Methods*, Texts in Applied Mathematics, Vol. 22, Springer-Verlag, New York, 1995.

[100] V. Thomée (1984), *Galerkin Finite Element Methods for Parabolic Problems*, Lecture Notes in Mathematics, Vol. 1054, Springer-Verlag, Berlin.

[101] J. F. Thompson, B. K. Soni, and N. P. Weatherill (1998), *Handbook of Grid Generation*, CRC Press, Boca Raton.

[102] R. Verfürth (1996), *A Review of a Posteriori Error Estimation and Adaptive Mesh-Refinement Techniques*, Wiley-Teubner, Chichester-Stuttgart.

[103] S. Verma and K. Aziz (1996), Two- and three-dimensional flexible grids for reservoir simulation, *Proc. of the 5th European Conf. on the Mathematics of Oil Recovery*, Z. E. Heinemann and M. Kriebernegg, eds., Leoben, Austria, pp. 143–156.

[104] H. Wang (2000), An optimal-order error estimate for an ELLAM scheme for two-dimensional linear advection-diffusion equations, *SIAM J. Numer. Anal.* **37**, 1338–1368.

[105] J. J. Westerink and D. Shea (1989), Consistent higher degree Petrov-Galerkin methods for the solution of the transient convection-diffusion equation, *Int. J. Num. Meth. Eng.* **13**, 839–941.

[106] M. F. Wheeler (1973), A priori L_2 error estimates for Galerkin approximation to parabolic partial differential equations, *SIAM J. Numer. Anal.* **10**, 723–759.

[107] D. Yang (1992), A characteristic mixed method with dynamic finite element space for convection-dominated diffusion problems, *J. Comput. Appl. Math.* **43**, 343–353.

[108] O. C. Zienkiewicz, R. L. Taylor, and P. Nithiarasu (2005), *The Finite Element Method for Fluid Dynamics*, 6th edition, Elsevier, Amsterdam.

[109] O. C. Zienkiewicz, R. Löhner, and K. Morgan (1986), High speed compressible flow and other advection dominated problems of fluid mechanics, in *Finite Elements in Fluids*, R. H. Gallagher *et al.*, eds., Vol. 6, John Wiley & Sons, Chichester, pp. 41–88.

[110] O. C. Zienkiewicz and J. Zhu (1987), A simple error estimator and adaptive procedure for practical engineering analysis, *Int. J. Num. Meth. Eng.* **24**, 337–357.

Index

Absolute permeability tensor, 198

Adaptive finite element method, 278

Adaptive methods, 44

Admissible function space, 30, 191

Admissible functions, 91

admissible real function space, 69

Admissible space, 4

Advection coefficient, 258

Advection problem, 258

Anisotropic material, 156, 167

A-posteriori error estimates, 279, 285

Approximation theory, 59

A-priori error estimates, 279

Area coordinates, 98

Argyris triangle, 78, 108

Argyris triangular element, 78

Artificial compressibility, 184, 185

Assemble the global matrix, 45

Assembly by elements, 15

Automatic grid generation, 44

Averaging-based estimators, 285

Axisymmetric solids, 165

Babuška-Brezzi condition, 254

Backward Euler method, 140

Bandwidth, 49

Basis, 7, 70

Basis functions, 14, 34, 46, 244

Bell's triangle, 108

Biharmonic operator, 77

Biharmonic problem, 77

Bilinear functions, 62

Bilinear polynomials, 90

Block-centered finite difference, 258

Block transmissibility, 220

Boundary condition of third kind, 252

Boundary conditions, 157, 181, 203, 210

Buckley-Leverett equation, 209, 210

Capillary pressure, 204

Cauchy inequality, 20

Cauchy-Green strain tensor, 152

Cauchy-Schwartz inequality, 20

Cell-centered finite difference, 258

Center of gravity, 99

Central difference scheme, 9

Chain rule, 47

Characteristic finite element method, 264

Characteristic functions, 244

Characteristic mixed finite element method, 265

Characteristic-based finite element methods, 174

Choice of an element, 124

Classification of differential
  equations, 208
Clough and Felippa quadrilateral
  element, 96
Compact support, 9
Compatibility condition, 40, 159,
  251
Compressibility, 179
Compressible flow, 202
Computer program, 15, 48, 51
Condition number, 50
Condition numbers, 139
Conditionally stable, 144
Conjugate gradient method, 15,
  49
Conservation of energy, 172
Conservation of mass, 172
Conservation of momentum, 172
Conservation relation, 273
Contraction ratio, 155
Control volume finite element
  method, 221
Control volume finite elements,
  222
Control volumes, 222, 230
Convection, 2
Convection-diffusion problem,
  175
Convection-diffusion-reaction
  problem, 72
Convergence, 21
Convetion problem, 258
Corrector, 179
Crank-Nicholson method, 136,
  142, 146
Cubic bubble functions, 193
Curvature, 204
CVFE grid construction, 226

Dankwerts boundary condition,
  203

Darcy velocity, 197
Darcy's Law, 205
Darcy's law, 3, 29, 198
Data structures, 284
Deformation, 152
Degrees of freedom, 75
Delaunay empty sphere criterion,
  228
Delaunay triangulation, 226
Density, 197
DG methods, 258
Different forms of flow equations,
  199
Differential operator, 160
Differential operator tensor, 158
Dirichlet boundary condition,
  11, 203
Dirichlet type, 3
Discontinuous finite element
  method, 258
Discontinuous finite elements,
  258
Discrete inf-sup condition, 192,
  254
Discrete minimization problem,
  33
Displacement, 2, 28, 152
Dissipation coefficient, 2, 28
Divergence form, 172
Divergence operator, 31, 198
Divergence theorem, 30
Divergence-free, 260, 273
Dot product, 31
Dynamic viscosity, 171
Dynamically local error
  indicators, 279

Edge swap, 228
Effective permeability, 205
Elastic bar, 2
Elastic bulk modulus, 179

Elastic membrane, 27
Element classification, 82
Element matrix, 13, 42
Element-oriented, 45
Element-oriented approach, 12
ELLAM, 274
Elliptic, 208
Elliptic equation, 200
Empty circle criterion, 226
Entropy, 179
Enumeration, 36
Equation of state, 199
Equations of fluid dynamics, 172
Equidistribution of an error, 290
Equilibrium, 153
Equivalence, 18
Error estimate, 19, 58, 140, 159
Essential condition, 12, 41
Euler method, 136
Eulerian approach, 172, 265
Eulerian-Lagrangian localized
    adjoint method, 265, 274
Eulerian-Lagrangian method,
    265
Eulerian-Lagrangian mixed
    discontinuous method, 265
Explicit scheme, 143, 178
Explicit time approximation,
    149, 216
Extrapolation technique, 215
Extrapolation techniques, 146
Extrapolation time approach,
    146

Family structure, 283
Finite difference method, 4
Finite element, 124
Finite element method, 30, 158,
    187, 213
Finite element space, 5, 32, 70
First kind, 203

Five-point difference stencil
    scheme, 37
Fluid compressibility, 199
Fluid density, 3
Fluid viscosity, 3
Flux type, 3
Flux-based upstream weighting,
    230
Formation volume factor, 198
Formulation in hyperbolic form,
    209
Formulation in phase pressure
    and saturation, 206
Formulation in phase pressures,
    205
Forward Euler method, 143
Fourier coefficients, 136
Fourth-order problem, 74
Fractional flow, 206
Fully discrete approximation
    schemes, 136
Fully discrete schemes, 140
Fully explicit scheme, 184
Fully implicit scheme, 175
Functional, 4, 31, 242
Fundamental principle of
    minimum potential energy, 4

Galerkin finite element method,
    6
Galerkin variational, 5
Galerkin variational formulation,
    32
Gas compressibility factor, 202
Gas law, 202
Gauss quadrature rule, 128
Gaussian elimination, 15, 49
General boundary condition, 11
General domains, 124
Geometric conformity, 91
Global coordinate approach, 82

Global coordinate system, 13
Global coordinates, 82
Global formulation, 208
Global matrix, 43
Global pressure, 208
Global transmissibility matrix,
    224
Globally Lipschitz continuous,
    145
Gradient operator, 31, 198
Gravity equilibrium condition,
    204, 213
Green's formula, 30, 73

H^1-conforming finite element
    method, 235
H^2-conforming finite element
    method, 235
Hanging nodes, 281
Harmonic average, 228, 229
Heat conduction, 2, 28
Heat conduction equation, 209
Hermitian elements, 82, 86
Hermitian rectangular elements,
    94
Hermitian tetrahedral elements,
    119
Hermitian triangular elements,
    104
Hexahedral elements, 111
Hierarchical basis estimators,
    286
Homogeneous anisotropic
    medium, 226
Homogeneous boundary
    condition, 2
Homogeneous deformation, 154
Homogeneous Dirichlet
    boundary condition, 28, 38
Homogeneous Neumann type, 29
h-scheme, 279

Hyperbolic, 208
Hyperbolic equation, 209

Immiscible flow, 204
Impervious boundary, 203, 211
Implicit time approximation,
    148, 215
Incompressible flow, 200
Indicators, 285
Inf-sup condition, 192, 254
Inflow boundary, 258
Initial condition, 204
Initial conditions, 203, 211, 213
Integration points, 128
Internal elastic energy, 4
Interpolant, 20
Interpolation error, 21
Interpolation function approach,
    82, 85
Irregular h-schemes, 282
Irregularity index, 281
Isoparametric elements, 122
Isoparametrical equivalence, 122
Isotropic material, 154, 155, 167
Iterative algorithms, 146
Iterative methods, 51

Jump, 259

Kinematic viscosity, 185
Kinematics, 152

Ladyshenskaja-Babuška-Brezzi
    condition, 254
Lagrange polynomial, 86
Lagrange polynomials, 90
Lagrangian approach, 172
Lagrangian elements, 82, 85
Lagrangian hexahedral elements,
    112

Lagrangian rectangular elements, 90
Lagrangian tetrahedral elements, 118
Lagrangian triangular elements, 100
Lagrangian-Eulerian approach, 173
Laminar flow, 174
Laplace equation, 200
Laplacian operator, 28, 200, 286
Linear elasticity, 152
Linear form, 69
Linear function, 5
Linear functional, 69
Linear system, 8
Linearization approach, 145, 214
Load potential, 4
Load vector, 8
Local conservation, 261
Local coordinate system transformation, 42
Local coordinate transformation, 13, 76
Local coordinate transformation approach, 84
Local coordinates, 82
Local grid refinement in space, 279
Local problem-based estimators, 285
Local refinement, 43, 281
Locking, 160

Mass conservation, 3, 29, 178, 204
Mass conservation equation, 197
Mass matrices, 139
Material derivative, 172
Material laws, 154
Material property matrix, 163

Matrix form, 36
Matrix storage, 51
Mesh parameters, 32, 72
Mesh-dependent bilinear form, 237
Metallic rod, 2
Metallic rod problem, 28
MINI element, 193
Minimization problem, 4, 31, 70
Minimum angle condition, 287
Mixed finite element method, 191, 240
Mixed finite element spaces, 243, 253
Mixed finite element spaces on rectangles, 256
Mixed finite element spaces on triangles, 255
Mixed kind, 203
Mixed variational form, 242
Mixed variational formulation, 248
MMOC, 266
Modified method of characteristics, 265
Modified method of characteristics with adjusted advection, 274
Modulus of elasticity, 155
Momentum conservation, 178
Multipoint flux approximations, 230
Multiscale feature, 294
Multiscale finite element method, 293

Natural condition, 12, 41
Natural coordinates, 98, 115, 117
Natural order of elements, 285
Navier-Stokes equations, 194
Navier-Stokes law, 173

Nested sequence of grids, 283
Neumann boundary condition,
    11, 203, 250
Neumann type, 3
Newton law, 173
Newton-Raphson's method, 148
No flow boundary, 211
No-flow boundary condition, 203
No-slip condition, 181
Node, 7
Node-oriented, 45
Node-oriented approach, 12
Nodes, 35
Nonconforming finite element
    method, 189, 235
Nonconforming finite element
    space, 237
Nonconforming finite elements,
    236
Nondivergence form, 172
Nonhomogeneous boundary
    condition, 3
Nonhomogeneous Dirichlet
    boundary condition, 9, 38
nonhomogeneous Dirichlet type,
    29
Nonhomogeneous problem, 61
Nonlinear transient problems,
    144
Nonphysical oscillations, 265
Nonwetting phase, 204
Normal derivative, 29, 31
Numerical integration formula,
    48

Operator splitting method, 265
Optimal spatial method, 265
Orthogonal grids, 233
Orthonormal system, 136
Over-sampling technique, 295

P_1-nonconforming element, 240
Parabolic, 208
Parabolic equation, 201
Parabolic problems, 135
Partition, 5
PEBI, 222
Pentahedral elements, 120
Periodic boundary condition, 11
Periodic boundary conditions,
    270
Permeability, 3, 30
Perpendicular bisection, 222
Petrov-Galerkin finite element
    method, 265
Phase formulation, 206
Phase mobilities, 206
Piecewise continuous, 4
Piezometric head, 200
Plan stress, 151
Plane strain, 151
Plane stress state, 169
Poisson equation, 28, 71, 200,
    208
Poisson problem, 38
Poisson ratio, 155
Porosity, 197
Positive definite, 8
Positive definite matrix, 71
Positive transmissibilities, 225
Potential, 200
Potential-based upstream
    weighting, 230
Practical Issues, 217
preconditioned, 51
Predictor, 179
Prescribed mass flux boundary
    condition, 30
Prescribed traction boundary,
    181
Pressure, 3, 173
Pressure equation, 207

Pressure-specified boundary
  condition, 30
Pressure-volume-temperature,
  202
Principle of virtual work, 5
Property matrix, 156
p-scheme, 280
Pseudo-potential, 199
Pseudo-pressure, 203

Quadratic functions, 22
Quadrature rule, 48
Quadrature rules, 128
Quadrilateral elements, 96
Quasi-implicit scheme, 185
Quasi-uniform grids, 15, 43
Quasi-uniform triangulation, 140

r-scheme, 280
Reaction-diffusion-advection
  problem, 264
Rectangular elements, 89
Reduced Argyris triangle, 108
Reference element, 13
Refinement rule, 281
Regular h-schemes, 280
Relative permeabilities, 205
Residual a-posteriori error
  estimator, 289
Residual estimators, 285
Resonance, 295
Reynolds number, 174
Ritz finite element method, 6
Ritz variational form, 4
Robin boundary condition, 203
Rock compressibility, 201

Saddle point problem, 242
Saturation, 204
Saturation equation, 207
Second kind, 203

Second-order problem, 72
Semi-discrete approximation
  scheme, 136
Semi-discrete scheme, 138
Semi-implicit methods, 150
Semi-implicit scheme, 175, 185
Semipervious boundary, 203
Separation of variables, 136
Serendipity elements, 93, 113
Shape functions, 7, 14, 34, 46,
  244
Shape-regular, 59
Shape-regular grids, 240
Shape-regular triangulation, 72
Simultaneous flow, 204
Simultaneous solution, 205
Single phase flow, 29, 212
Single phase flow in porous
  media, 3
Slightly compressible flow, 200
Slightly compressible fluid, 199
Slip condition, 181
Solenoidal, 260, 273
Sources and sinks, 197
Sparse matrix, 9
Specific energy, 179
Speed of sound, 179
Split I approach, 179
Split II approach, 179
Splitting method, 174
Stability condition, 143, 149,
  178, 186, 192, 254
Stability estimate, 141
Stability result, 139
Stabilization terms, 177
Stabilized DG methods, 263
Step size, 5
Stiff system, 139
Stiffness matrix, 8, 139
Stokes equation, 174
Stokes equations, 77

Stream-function vorticity formulation, 174
Streamline diffusion method, 263
Subparametic elements, 123
Superparametic elements, 123
Surface tension, 204
Symmetric bilinear form, 69
Symmetric matrix, 8
Symmetry, 39
System matrix, 217

Taylor-Hood element, 194
Tensor products, 90
Test function, 5
Tetrahedral elements, 114
Thermal diffusion, 3, 29
Thin elastic plate, 77
Third boundary condition, 11, 40
Third kind, 203
Three-dimensional solids, 161
Total compressibility, 201
Total mobility, 206
Total velocity, 207
Transient problems, 135
Transmissibilities, 233
Transmissibility coefficients, 224, 225
Transport diffusion method, 265
Tree structure, 283
Trial function, 6, 7, 33, 70
Triangular elements, 89, 97

Triangulation, 32
Tridiagonal matrix, 8
Turbulent flow, 174
Two-phase flow, 204
Two-point boundary value problem, 2

Unconditionally stable, 142
Units, 198
Universal gas constant, 202
Unrefinements, 283
Upstream weighting, 229
Upwind finite difference, 261

Variable thermal diffusion coefficient, 29
Variational form, 145
Variational formulation, 4
Variational problem, 70
Velocity, 3
Viscid flow, 181
Viscosity, 198
Viscous stress tensor, 173
Voronoi grids, 222

Wave equation, 209
Weak formulation, 5
Weighted formulation, 207
Wetting phase, 204

Young modulus, 155